ISBN 978-0-282-97881-5
PIBN 10875428

MANUAL

OF

GEOMETRY

AND

CONIC SECTIONS,

WITH APPLICATIONS TO

TRIGONOMETRY AND MENSURATION.

BY

WILLIAM G. PECK, PH.D., LL.D.,

PROFESSOR OF MATHEMATICS AND ASTRONOMY IN COLUMBIA COLLEGE, AND OF MECHANICS IN THE SCHOOL OF MINES.

A. S. BARNES & COMPANY,

NEW YORK AND CHICAGO.

PUBLISHERS' NOTICE.

PECK'S MATHEMATICAL SERIES.

CONCISE, CONSECUTIVE, AND COMPLETE.

I. FIRST LESSONS IN NUMBERS.
II. MANUAL OF PRACTICAL ARITHMETIC.
III. COMPLETE ARITHMETIC.
IV. MANUAL OF ALGEBRA.
V. MANUAL OF GEOMETRY.
VI. TREATISE ON ANALYTICAL GEOMETRY.
VII. DIFFERENTIAL AND INTEGRAL CALCULUS.
VIII. ELEMENTARY MECHANICS (without the Calculus).
XI. ELEMENTS OF MECHANICS (with the Calculus).

NOTE.—Teachers and others, discovering errors in any of the above works, will confer a favor by communicating them to us.

PREFACE.

IN completing the last volume of the condensed course of Mathematics, the author deems it proper to say a few words in explanation of the general plan of the series.

There is a growing belief among educators that our text-books contain a great deal of matter which is not essential, either to continuity or to completeness, and that they are to this extent overcrowded and unnecessarily voluminous. The desire for condensed, but not emasculated, books is well expressed by an eminent authority in educational matters who says: "Small text-books, containing only the essentials of the subjects treated of—only those parts that have life in them, parts that cannot be eliminated without leaving the subjects imperfect—are rare. * * * Such books we must have, if we use text-books at all." The demand for small and complete books comes from every department of instruction, but from none more emphatically, than from that of Mathematics. The present series is designed to meet this demand by furnishing a set of books, embracing within moderate limits every mathematical principle necessary to the fullest academic, or technical education. Commencing with the simplest ideas of number, the successive volumes conduct the student by easy gradation through all the essential principles of Arithmetic, Algebra, Geometry, Trigonometry, Analytics, and the Calculus.

The entire course is treated as a unit, each part being adapted to all the others, and all the parts being so arranged as to form a symmetrical whole. No effort has been spared to make the definitions clear, and the illustrations full, and the treatment of the subject throughout has been intended to afford the best results with the least expenditure of time and labor—in a word, to make the series *concise* in demonstration, *consecutive* in arrangement, *complete* in detail.

GREENWICH, CONN., *July* 4, 1876.

CONTENTS.

GEOMETRY.

PAGE

Introduction... 7

BOOK I.

Lines, Angles, and Polygons................................... 12

BOOK II.

The Circle, and the Measure of Angles............. 41

BOOK III.

Ratios, Proportions, and Comparison of Lines............ 68

BOOK IV.

Measure and Comparison of Areas........................ 91

BOOK V.

Regular Polygons and Circles............................ 108

BOOK VI.

Planes, and Angles formed by Planes.................... 122

BOOK VII.

Comparison and Measure of Polyedrons.................. 139

BOOK VIII.

PAGE

MEASURE OF THE CYLINDER, CONE, AND SPHERE 159

BOOK IX.

SPHERICAL GEOMETRY 173

CONIC SECTIONS.

INTRODUCTION 189

I.—THE PARABOLA 190

II.—THE ELLIPSE 198

III.—THE HYPERBOLA 213

TRIGONOMETRY.

I.—LOGARITHMS 223

II.—CIRCULAR FUNCTIONS 230

III.—PLANE TRIGONOMETRY 236

IV.—ANALYTICAL TRIGONOMETRY 252

V.—SPHERICAL TRIGONOMETRY 261

MENSURATION.

INTRODUCTION 281

I.—MENSURATION OF LINES 282

II.—MENSURATION OF SURFACES 234

III.—MENSURATION OF VOLUMES 297

MANUAL OF GEOMETRY.

INTRODUCTION.

FUNDAMENTAL CONCEPTIONS.

1. A portion of space, limited in all directions, is called a **volume**; that which separates a volume from the rest of space is called a **surface**. If two surfaces intersect, or cut each other, that which is common to both is called a **line**; if two lines intersect, that which is common to both is called a **point**.

General Definitions.

2. A **magnitude** is anything that can be measured, that is, whose value can be expressed in terms of some thing of the same kind, taken as a unit.

3. A **point** is that which has position, but no magnitude.

4. A **line** is that which has length, without breadth or thickness.

5. A **surface** is that which has length and breadth without thickness.

6. A **volume** is that which has length, breadth, and thickness.

Straight Lines and Curved Lines.

7. A **straight line** is a line whose direction does not change at any point; thus, AC is a straight line.

A C

It is assumed that one straight line, and only one, can be drawn through any two points.

8. A **broken line** is a line made up of straight lines lying in different directions; thus, ACDE is a broken line.

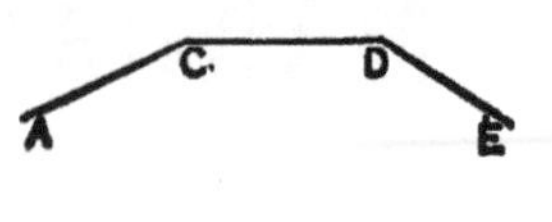

9. A **curved line**, or a **curve**, is a line whose direction changes at every point; thus, ACD is a curved line.

When the term *line* is used by itself, a *straight line* is always meant.

Planes and Curved Surfaces.

10. A **plane** is a surface such that a straight line through any two of its points lies wholly in the surface.

11. A **curved surface** is a surface of which no part is plane.

Generation of Lines and Surfaces.

12. If a point moves through space in accordance with a fixed law it is said to **generate**, or trace out, a line; the moving point is called the **generatrix**, and the law that governs its motion is called the **law of generation**.

The law of generation determines the kind of lines generated.

13. If a line moves through space in accordance with a fixed law it is said to **generate** a surface; the moving line is the **generatrix**, and the law which governs its motion is the **law of generation**.

Angles.

14. An **angle** is the inclination of two lines that meet at a common point; the two lines are called **sides** and their common point is the **vertex**: thus, the inclination of OA and OC is an angle, whose sides are OA and OC, and whose vertex is O.

If an angle stands by itself it may be designated by a single letter placed at the vertex; if two angles have a common vertex each is designated by three letters, of which the middle one denotes the vertex and the others the two sides: thus, the angle between QP and QR is called the angle Q; that between OA and OC is called the angle AOC; and that between OC and OD is called the angle COD.

Definition and Division of the Subject.

15. Lines, surfaces, volumes, and angles are called geometrical **magnitudes.**

16. **Geometry** is that branch of mathematics which treats of the properties and relations of geometrical magnitudes.

17. Geometry is divided into two parts: **geometry of two dimensions,** or **plane geometry,** in which the magnitudes considered lie in the same plane; and **geometry of three dimensions** in which the magnitudes considered do not lie in the same plane.

All the magnitudes considered in the first five books are supposed to lie in the same plane.

Definitions of Terms.

18. A **theorem** is an assertion whose truth is to be proved; the reasoning employed in making the proof is called a **demonstration.**

A **problem** is a question proposed, requiring an answer; the operation of finding the answer is called a **solution.**

Both theorems and problems are called **propositions.**

19. An **axiom** is an assertion whose truth is universally admitted.

20. A **postulate** is a problem whose solution is taken for granted.

21. A **lemma** is a proposition introduced out of the

regular course to aid in the demonstration of a theorem, or in the solution of a problem.

22. A **corollary** is an obvious consequence of one or more propositions.

A **scholium** is an explanatory remark made with reference to one or more preceding propositions.

23. An **hypothesis** is a supposition made either in the enunciation of a proposition or in the course of a demonstration.

24. The following are some of the most important axioms and postulates.

Axioms.

1°. Things equal to the same thing are equal to each other.

2°. If equals are added to equals, the sums are equal.

3°. If equals are subtracted from equals, the remainders are equal.

4°. If equals are multiplied by equals, the products are equal.

5°. If equals are divided by equals, the quotients are equal.

6°. The whole is greater than any of its parts.

7°. The whole is equal to the sum of all its parts.

8°. A straight line is the shortest path between two points.

Postulates.

1°. A straight line can be drawn through any two points.

2°. A straight line can be prolonged to any extent.

3°. A straight line can be constructed equal to a given straight line.

4°. A straight line can be bisected, that is, divided into two equal parts.

5°. An angle can be constructed equal to a given angle.

6°. An angle can be bisected.

Notation.

25. The signs and methods of notation used in Algebra are also employed in Geometry. The principal signs are the ordinary signs of *addition*, +; of *subtraction*, −; of *multiplication*, ×; of *division*, ÷; the *radical sign*, $\sqrt{\ }$; and the *signs of proportion*, :, : :, :. The *vinculum* and the *parenthesis* are used to connect quantities that are to be operated on as a single quantity; *coefficients* are used to denote multiples; and *exponents* to denote powers. The symbols 1°, 2°, 3°, &c. are read *first, second, third,* &c. In addition to these methods of denotation, geometrical magnitudes are represented by pictorial symbols, called *geometrical figures*, in which points and lines are designated by letters placed so as to be convenient for reference.

NOTE.—In the references throughout this work, *P.* stands for proposition, *B.* for book, *Cor.* for corollary, *Ax.* for axiom, *Prob.* for problem, *Post.* for postulate, and *Scho.* for scholium. In referring to subjects in the same book the number of the book is not given.

BOOK I.

LINES, ANGLES, AND POLYGONS.

Generation and Classification of Angles.

26. An angle may be generated by a line revolving about one of its points in the same manner that one leg of a pair of compasses turns about the hinge. Thus, if we suppose OC to start from the position OA and to revolve about O in the direction indicated by the arrow, it will generate an angle AOC, whose magnitude will obviously depend on the amount of turning, and not at all on the length of OC.

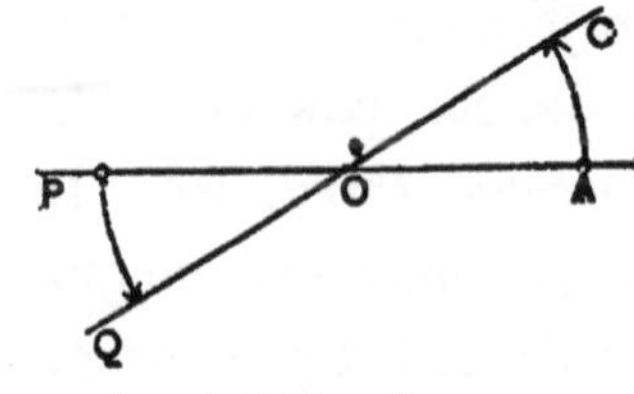

In this explanation we only consider that part of the revolving line which lies on one side of the vertex; if we suppose the line to be prolonged through the vertex, the prolongation may be regarded as a second line lying in an opposite direction. Thus, if we suppose QC to start from the position PA and to revolve about O in the direction indicated by the arrows, the part OC will generate an angle AOC and the opposite part, OQ, will generate another angle POQ.

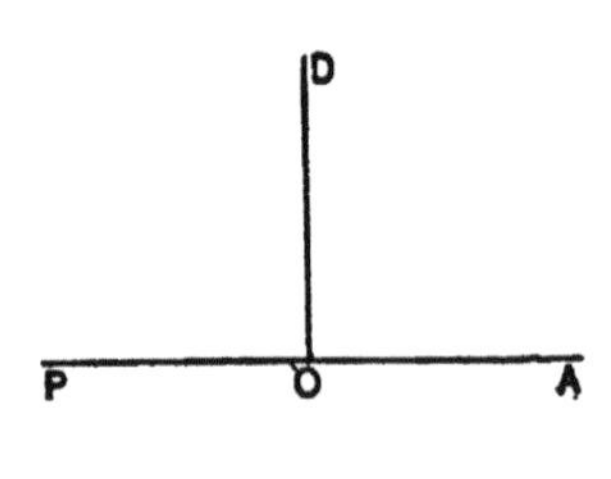

27. If a line DO meets a second line PA so as to form two angles, these angles are said to be **adjacent.** If the adjacent angles are equal, each is called a **right angle**, and the first line is said to be **perpendicular** to the second. Thus, if DOA and DOP are equal, each is a right angle, and OD is perpendicular to PA, at O.

The point O is called the **foot** of the perpendicular OD.

28. An **acute angle** is one that is less than a right angle; an **obtuse angle** is one that is greater than a right angle: thus, AOC is *acute*, and COP is *obtuse*.

The angles AOC and COP are both called **oblique**, and the line OC is said to be *oblique* to PA.

29. If two lines **intersect**, that is, cross each other, they form four angles about the point of intersection. Those that lie on the same side of one line and on opposite sides of the other are called **adjacent**, and those that lie on opposite sides of both are called **opposite**, or **vertical angles**: thus, AOC, COP, are *adjacent*, and AOC, POQ, are *opposite* angles.

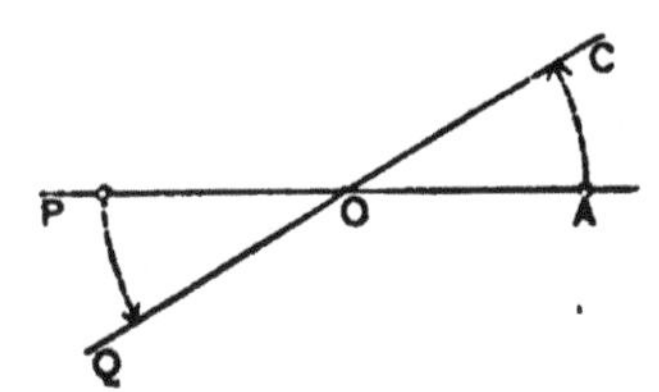

Any one of the four angles about O has two adjacent angles and one opposite angle.

Proposition I. Theorem.

If two lines intersect, the opposite angles which they form are equal.

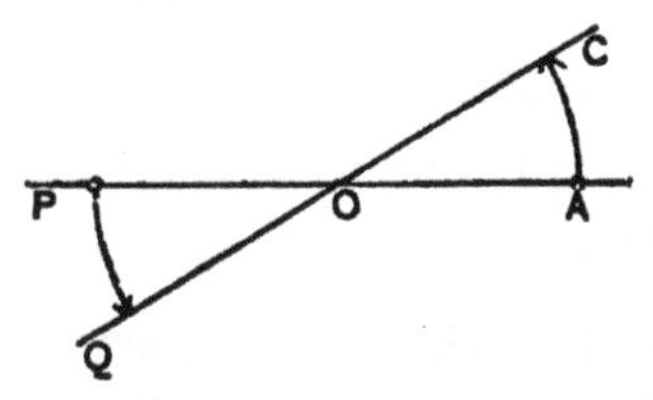

Let PA and QC be two lines that intersect, and suppose PA to remain fixed whilst QC revolves about their common point, O. If we suppose the moving line to start from the position PA and to revolve in the direction indicated by the arrows till it comes into some other position, QC, the part OC will generate the angle AOC and the opposite part OQ will generate the opposite angle POQ; but the amount

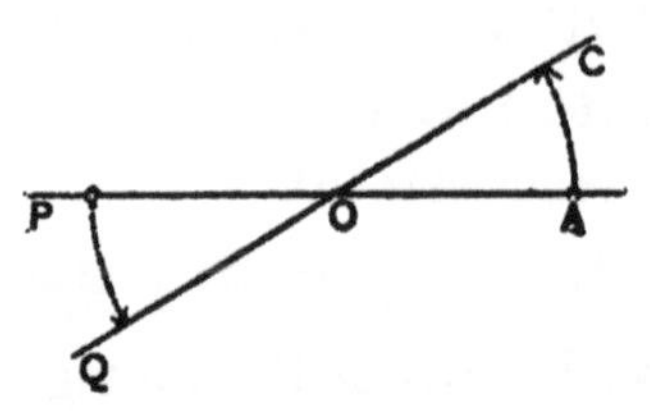

of turning of OC is obviously the same as that of OQ; hence, AOC and POQ are equal. If we now suppose the motion to continue until OC falls on OP and OQ on OA, the former will generate the angle COP and the latter will generate the angle QOA, and for the same reason as before these angles are equal. Hence, in every position of QC, we have, AOC equal to POQ and COP to QOA, *which was to be proved.*

Proposition II. Theorem.

One perpendicular, and one only, can be drawn to a given line at a given point.

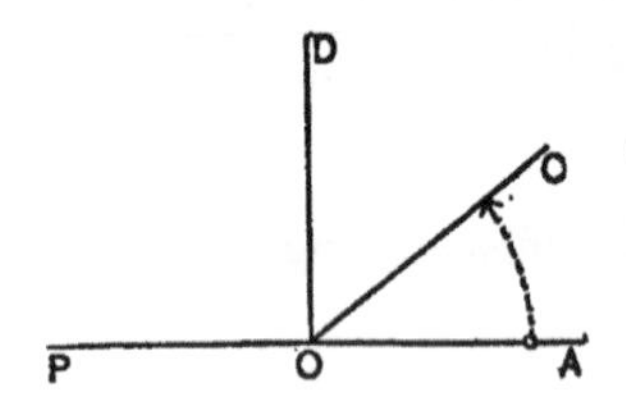

Let PA be the given line and O the given point. Suppose a line OC to start from the position OA and to revolve about O in the indicated direction. At the beginning of the motion, AOC is less than COP; as the motion progresses AOC increases, COP decreases, and their difference continually diminishes, until the revolving line comes to some position, OD, in which the two angles are equal; after passing this position the first angle becomes greater than the second, and their difference continually increases till the revolving line comes to the position OP: there is therefore one position of the revolving line, and but one, in which the two angles are equal. Hence, there is one perpendicular to PA at O, and only one, *which was to be proved.*

Cor. 1. If OD is perpendicular to PA, its prolongation OQ is also perpendicular to PA. For, from P. 1, AOD is equal to

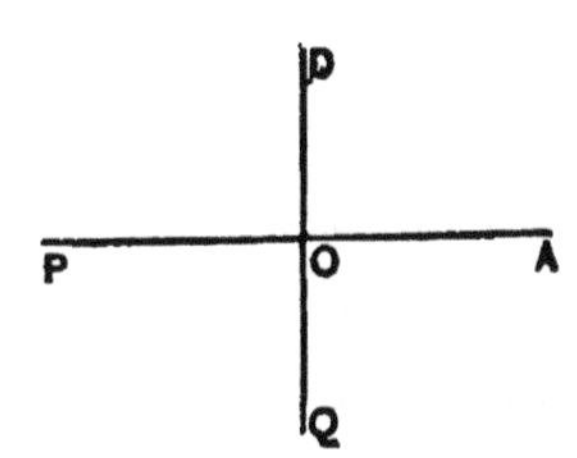

POQ, and DOP to QOA; consequently POQ is equal to QOA, (Ax. 1°): hence, OQ is perpendicular to PA.

Cor. 2. If one of the four angles formed by two lines that intersect is a right angle, the other three are right angles and the two lines are mutually perpendicular to each other.

Scho. A right angle is an angle that may be generated by a line OC in revolving through a quarter of a complete revolution ; hence, all right angles are equal.

PROPOSITION III. THEOREM.

The sum of the adjacent angles formed by one line meeting another is equal to two right angles.

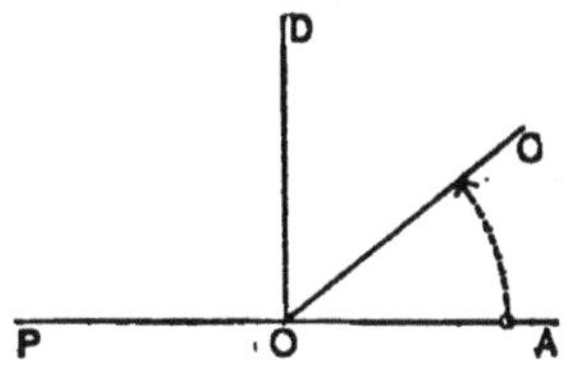

Let PA be any line and let OC be another line meeting it at O. Suppose OC to start from the position OA, and to revolve about O in the indicated direction. As the angle AOC increases, its adjacent angle COP diminishes by an equal amount; hence, the sum of the two is always the same whatever may be the position of OC. But, when OC is perpendicular to PA, the sum of the two angles is equal to two right angles; it is therefore equal to two right angles for every position of OC, *which was to be proved.*

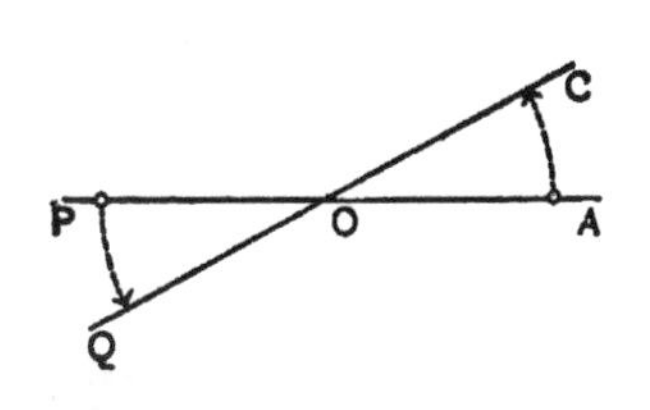

Cor. 1. If PA and QC intersect, the sum of any two of the adjacent angles which they form is equal to two right angles. For, from P. 1, AOC is equal to POQ, and COP to QOA; hence, (Ax. 2°), the sum of POQ and QOA is equal to the sum of AOC and COP, that is, to two right angles. In like manner it may be shown that the sum of QOA and AOC, and that the sum of COP and POQ is equal to two right angles.

Cor. 2. The sum of all the angles formed by any number of lines meeting at any point of a given line, and lying on the same side of that line, is equal to two right angles. For, the sum of the angles AOC, COE, EOF, and FOP, is obviously equal to the sum of AOC and COP, which is equal to two right angles.

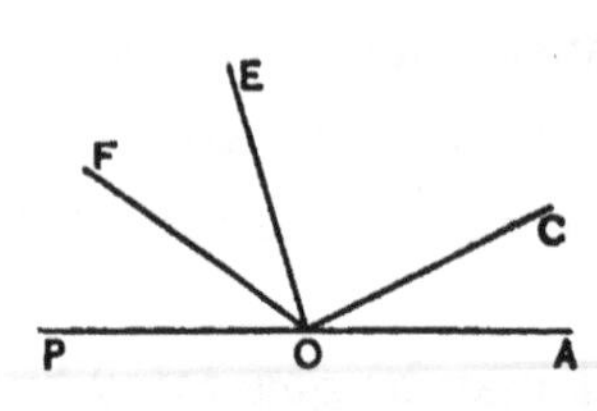

Cor. 3. The sum of all the angles formed by any number of lines meeting at a common point is equal to four right angles. For the sum of the angles AOC, COF, FOQ, QOS, and SOA is equal to the angular space generated by OA whilst turning through a complete revolution, that is, it is equal to four right angles, (P. 2, *Scho.*).

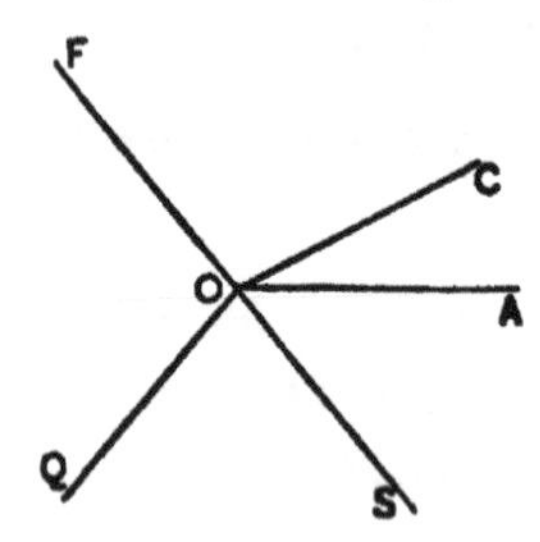

Definitions and Remarks.

30. Two angles are said to be **complementary** when their sum is equal to a right angle; they are said to be **supplementary** when their sum is equal to two right angles.

Complementary angles are always acute. Of two supplementary angles both may be right angles, or one may be acute and the other obtuse.

It follows from the preceding definitions, that complements of equal angles are equal, and that supplements of equal angles are equal.

Proposition IV. Theorem.

If the sum of two adjacent angles is equal to two right angles, their exterior sides are in the same straight line.

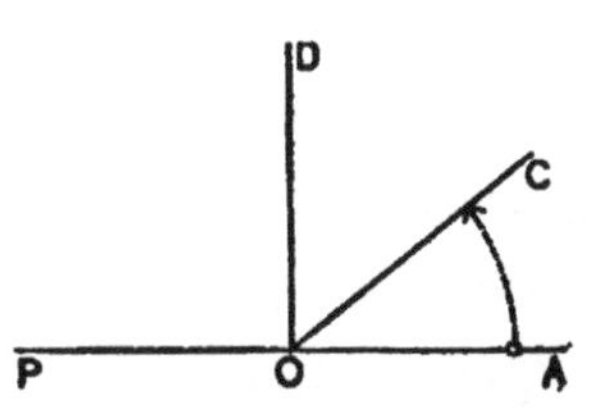

Let the sum of AOC and COP be equal to two right angles; then is COP the supplement of AOC, by the definition just given; but the angle that OC makes with the prolongation of AO is the supplement of AOC, by P. 3. Hence, the angle COP is identical with the angle that CO makes with the prolongation of AO; consequently, the side OP must be identical with the prolongation of AO, *which was to be proved.*

Scho. Proposition IV is the converse of Proposition III, that is, the *hypothesis* and *conclusion* of the one are respectively the *conclusion* and *hypothesis* of the other.

PROPOSITION V. THEOREM.

Through a point without a line one perpendicular can be drawn to that line, and but one.

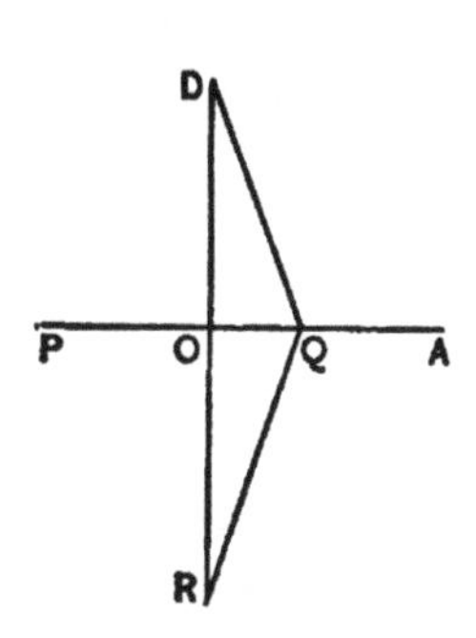

1°. Let PA be any line and D a point without the line. Suppose that part of the plane of the figure which lies above PA to revolve about PA (as a door turns on its hinges), till it falls on the part below PA, and let R be the revolved position of D; then suppose the plane to revolve back to its primitive position and draw DR cutting PA at O. The line DR will be perpendicular to PA; for, if we again suppose the upper part of the plane to revolve about PA till it falls on the lower part, the point D will fall at R, and because O remains fixed, the angle DOP will fall upon, and coincide with, POR; consequently DOP is equal to POR; but POR is equal to AOD, (P. 1.); POD is therefore equal to AOD, that is, DO is perpendicular to PA, *which was to be proved.*

2°. Furthermore, no other line can be drawn through D perpendicular to PA; for, if possible, let DQ be such a line, and draw QR. It may be shown in the same manner as before that the angles DQP and PQR are equal, and since DQP is a right angle, by hypothesis, the sum of DQP and PQR is equal to two right angles, and consequently DQR is a straight line (P. 4.). But this is impossible, because only one straight line can be drawn through two points; hence, the supposition that a second line can be drawn through D perpendicular to PA is false; DR is therefore the only line that can be drawn through D perpendicular to PA, *which was to be proved.*

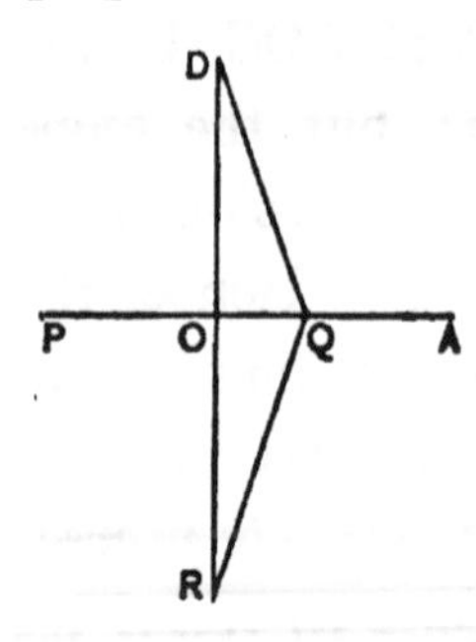

Scho. The second part of the preceding demonstration is an example of the method of reasoning called the reductio ad absurdum. This method consists in assuming a proposition such that either it or the one to be demonstrated *must* be true, the other one being false; the assumed proposition is then shown to be false, from which it follows that the one to be demonstrated *must* be true.

Definitions.

31. A **polygon** is a portion of a plane bounded by straight lines; the bounding lines are called **sides**; the points in which the sides meet are called **vertices**; and the entire broken line that bounds the polygon is called the **perimeter.** Thus, ACDE is a polygon whose sides are AC, CD, DE, and EA; whose vertices are A, C, D, and E; and whose perimeter is the broken line ACDEA.

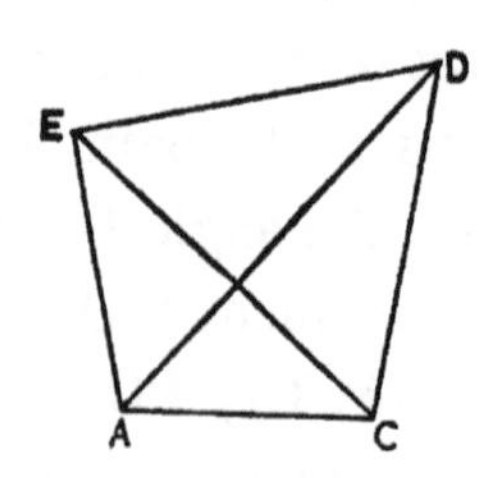

32. The side on which a polygon is supposed to stand is called its **base**; a line joining any two vertices which are not consecutive is called a **diagonal**; thus, AC is the base of the polygon, ACDE and AD, CE, are diagonals.

33. If all the sides of a polygon are equal, it is said to be **equilateral**; if all the angles are equal, it is said to be **equiangular**.

34. Polygons are named either from the number of their sides, or from the number of their angles: a polygon of three sides, is called a **triangle**; one of four sides, a **quadrilateral**; one of five sides, a **pentagon**; one of six sides, a **hexagon**; one of seven sides, a **heptagon**; one of eight sides, an **octagon**; and so on.

Classification of Triangles.

35. Triangles may be classified either with respect to the relations among their sides, or with respect to the relations among their angles.

36. When classified with respect to sides there are two classes; *scalene*, and *isosceles triangles*.

A **scalene triangle** is one that has no two sides equal; as ACD.

An **isosceles triangle** is one that has two equal sides; as EFG.

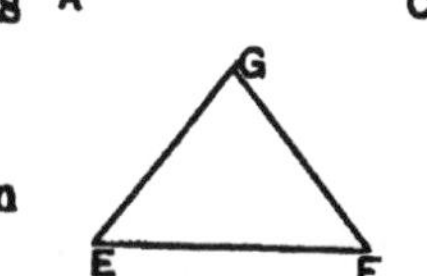

An *equilateral triangle* is a particular case of an isosceles triangle; its three sides are all equal.

37. When classified with respect to angles, there are two classes; *right-angled* and *oblique triangles*.

A **right-angled triangle** is one that has one right angle; as ACD.

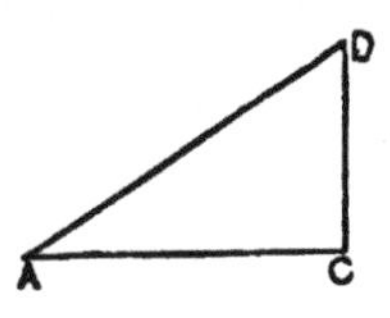

The side AD, opposite the right angle, is called the **hypothenuse**.

An **oblique triangle** is one whose angles are all oblique; as, EFG, HKL.

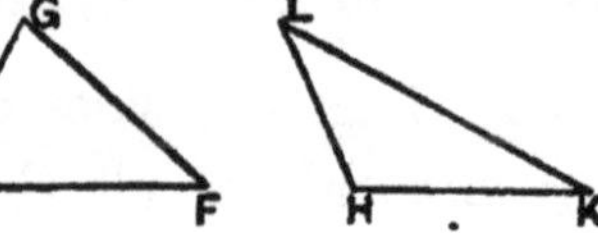

If one angle is obtuse, the triangle is **obtuse-angled**; if all the angles are acute, the triangle is **acute-angled.**

PROPOSITION VI. THEOREM.

The sum of any two sides of a triangle is greater than the third side.

Let ACD be any triangle.

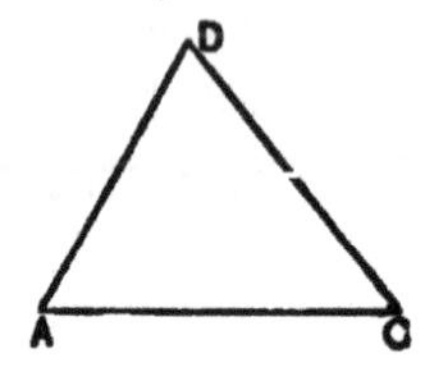

The distance from A to D, measured on the broken line ACD, is greater than the distance measured on the straight line AD; hence,

$$AC + CD > DA; \ldots\ldots\ldots (1)$$

which was to be proved.

Cor. The difference of any two sides of a triangle is less than the third side. For, if from both members of (1), we subtract either AC, or CD, say AC, we have

$$CD > DA - AC, \text{ or } DA - AC < CD.$$

Scho. In order that three lines may be sides of a triangle the sum of any two must be greater than the third, and the difference of any two must be less than the third.

PROPOSITION VII. THEOREM.

The sum of the lines drawn from a point within a triangle to the extremities of any side is less than the sum of the other two sides.

Let ACD be a triangle, and P a point within it. Draw PA and PC, and prolong AP to meet DC at Q. From P. 6, we have,

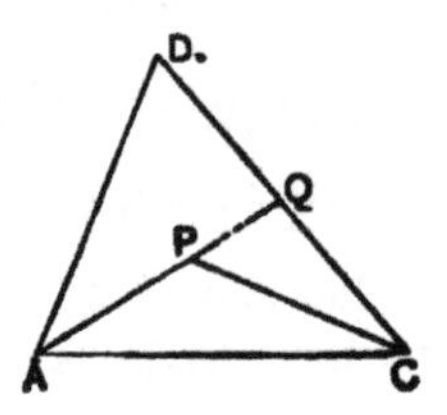

$$CP < CQ + QP; \ldots\ldots\ldots (1)$$

Adding PA to both members of (1), remembering that QP + PA is equal to QA, we have,

$$CP + PA < CQ + QA. \quad \ldots\ldots\ldots (2)$$

Again, we have,

$$QA < QD + DA; \quad \ldots\ldots\ldots (3)$$

Adding CQ to both members of (3), remembering that CQ + QD is equal to CD, we have,

$$CQ + QA < CD + DA. \quad \ldots\ldots\ldots (4)$$

Comparing (2) and (4), we have,

$$CP + PA < CD + DA,$$

which was to be proved.

PROPOSITION VIII. THEOREM.

If two sides and their included angle, in one triangle, are equal to two sides and their included angle, in another triangle, each to each, the triangles are equal in all their parts.

Let ACD and PQR be two triangles in which AC is equal to PQ, CD to QR, and ACD to PQR.

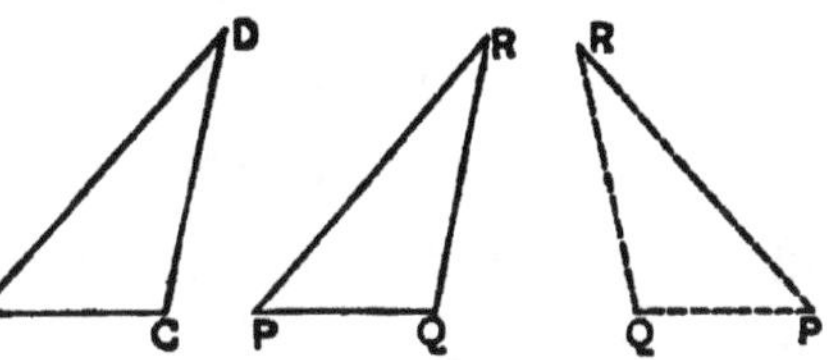

Let the triangle PQR be placed on ACD so that the angle Q shall coincide with C, QP falling on CA, and QR on CD; then, because R coincides with D, and P with A, RP coincides with DA, and consequently the triangles coincide throughout their whole extent; they are therefore equal in all their parts, *which was to be proved.*

Scho. 1. If PQR is situated in the manner indicated by the triangle on the right, it must be **reversed**, that is, turned completely over in its own plane, before it can be superposed on ACD.

Scho. 2. If two triangles are equal in all their parts, equal angles lie opposite the equal sides, and the reverse.

PROPOSITION IX. THEOREM.

If two sides of one triangle are equal to two sides of another triangle, each to each, and if the included angle in the first triangle is greater than the included angle in the second, the third side of the first is greater than the third side of the second, and conversely.

1st. Let ACD and PQR be two triangles in which AC is equal to PQ, CD to QR and $ACD > PQR$. From C draw CS equal to QR, and making ACS equal to PQR; draw SA. The triangles PQR and ACS have AC equal to PQ, CS to QR and ACS to PQR; they are therefore equal in all their parts, and consequently SA is equal to RP, (P. 8). In comparing SA with DA, there may be three cases: the vertex S may fall without the triangle ACD, it may fall within ACD, or it may fall on the side DA.

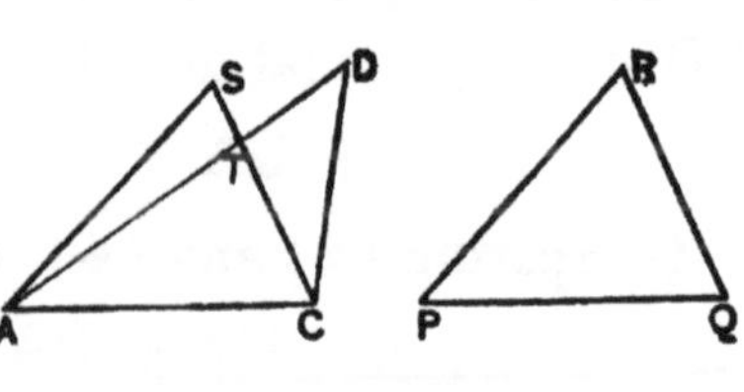

1°. If S falls without ADC; from P. 6, we have,

$$DT + CT > CD, \ldots\ldots\ldots (1)$$

and,

$$TA + TS > SA. \ldots\ldots\ldots (2)$$

adding (1) and (2), remembering that $DT + TA$ is equal to DA, and $CT + TS$ to CS, we have,

$$DA + CS > CD + SA; \ldots\ldots\ldots (3)$$

subtracting CS from the first member of (3), and its equal CD from the second member, we have,

$$DA > SA, \text{ or } DA > RP,$$

which was to be proved.

2°. If S falls within ADC: we have from P. 7,

$$CD + DA > CS + SA; \ldots (1)$$

subtracting CD from the first member of (1), and its equal CS from the second member, we have,

$$DA > SA, \text{ or } DA > RP,$$

which was to be proved.

3°. If S falls on DA: we have, from Ax. 6°,

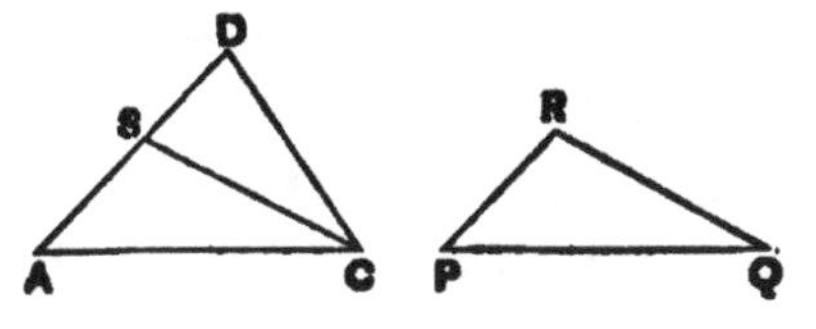

$$DA > SA, \text{ or } DA > RP,$$

which was to be proved.

Hence, the theorem is true in all cases.

2dly. *Conversely*, if DA is greater than RP, the angle ACD is greater than PQR. For, if ACD is less shan PQR, DA is less than RP, from what has just been shown; if ACD is equal to PQR, DA is equal to RP from P. 8; hence, ACD is neither less than, nor equal to PQR; it must therefore be greater than PQR, *which was to be proved.*

PROPOSITION X. THEOREM.

If two angles and their included side in one triangle are equal to two angles and their included side in another triangle, each to each, the triangles are equal in all their parts.

Let ACD and PQR be two triangles in which the angle A is equal to P, the angle C to Q, and the side AC to PQ.

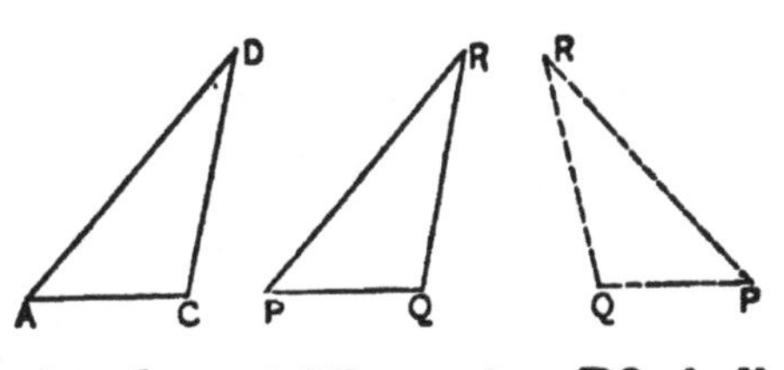

Let the triangle PQR be placed on ACD, so that PQ shall coincide with its equal AC, the angle P with its equal A, and the angle Q with its equal C; then will PR take the direction AD, QR the direction CD, and consequently R will fall on D; the triangles, therefore, coincide throughout; hence, they are equal in all their parts, *which was to be proved.*

PROPOSITION XI. THEOREM.

If the sides of one triangle are equal to the sides of another triangle, each to each, the triangles are equal in all their parts.

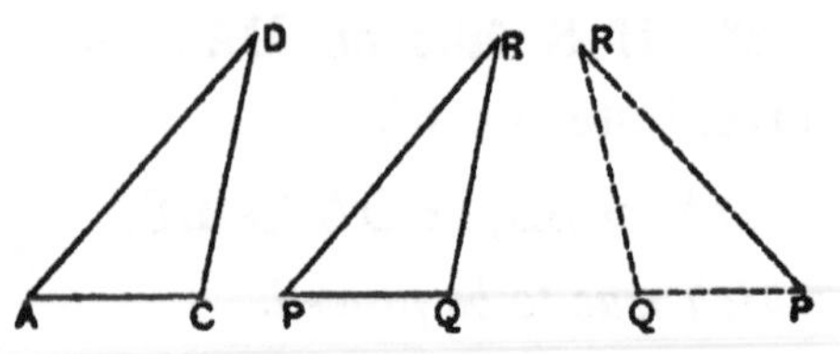

Let ACD and PQR be two triangles in which AC is equal to PQ, CD to QR, and DA to RP.

The sides that include the angle A are equal to those that include the angle P, each to each; if therefore A is greater than P, CD is greater than QR, or if A is less than P, CD is less than QR, (P. 9); but CD is neither greater nor less than QR; consequently A is neither greater nor less than P; hence, A is equal to P. The two triangles have therefore two sides and their included angle in one, equal to two sides and their included angle in the other, each to each; hence, the triangles are equal in all their parts, (P. 8), *which was to be proved.*

PROPOSITION XII. THEOREM.

In an isosceles triangle the angles opposite the equal sides are equal.

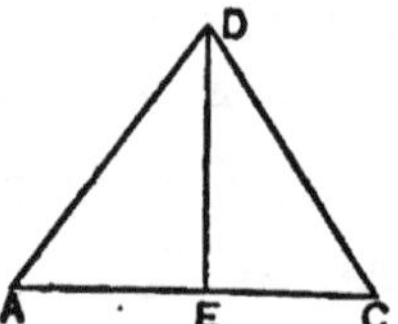

Let ACD be an isosceles triangle in which CD is equal to DA. Let the base AC be bisected in E and from the vertex D draw DE.

The triangles AED and CED have the side AE equal to CE, the side DE common, and the side DA equal to DC; they are therefore equal in all their parts, (P. 11). Hence, the angle A is equal to C, *which was to be proved.*

Cor. 1. An equilateral triangle is also equiangular.

Cor. 2. A line drawn from the *vertex* of an isosceles triangle to the middle of the *base* bisects the angle at the vertex, and is perpendicular to the base. For, from what has just been proved, the angle EDA is equal to EDC, and the angle DEA to DEC.

Proposition XIII. Theorem.

If two angles of a triangle are equal, their opposite sides are equal.

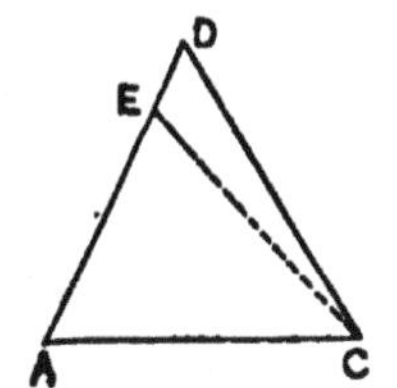

In the triangle ACD, let the angle A be equal to C.

If CD is not equal to DA, suppose DA to be the greater; on DA lay off AE equal to CD and draw CE. The triangles ACD and CAE, have the side AC common, the side CD equal to AE, by construction, and the angle ACD equal to EAC, by hypothesis; they are therefore equal in all their parts, (P. 8); but a part cannot be equal to the whole, (Ax. 6°); hence, the hypothesis that CD is not equal to DA is false; the two sides CD and DA are therefore equal, *which was to be proved.*

Cor. An equiangular triangle is equilateral.

Proposition XIV. Theorem.

If two angles of a triangle are unequal the sides opposite to them are unequal, and the greater side lies opposite the greater angle, and conversely.

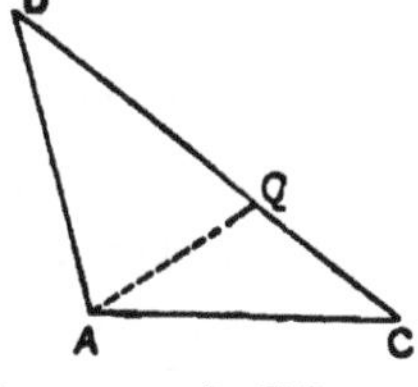

1°. In the triangle ACD, let the angle A be greater than the angle C. Through A draw AQ making the angle QAC equal to ACQ; then is AQ = CQ, (P. 13). From the triangle AQD, we have, (P. 6), $AQ + QD > DA$, or substituting for AQ its equal CQ, remembering that $CQ + QD = CD$, we have, $CD > DA$, *which was to be proved.*

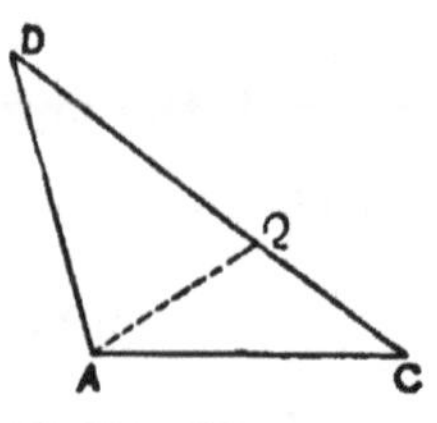

2°. *Conversely,* if CD is greater than DA, the angle A is greater than C. The angle A cannot be less than C, for in that case CD would be less than DA from what has just been shown; the angle A cannot be equal to C because in that case we should have CD equal to DA, (P. 13); since A is neither less than C nor equal to C it must be greater than C, *which was to be proved.*

PROPOSITION XV. THEOREM.

If from a point without a line a perpendicular is drawn to that line and also oblique lines:

1°. *Any two oblique lines that meet the given line at equal distances from the foot of the perpendicular are equal;*

2°. *Of two oblique lines that meet the given line at unequal distances from the foot of the perpendicular, that which meets it at the greater distance is the greater;*

3°. *The perpendicular is less than any oblique line.*

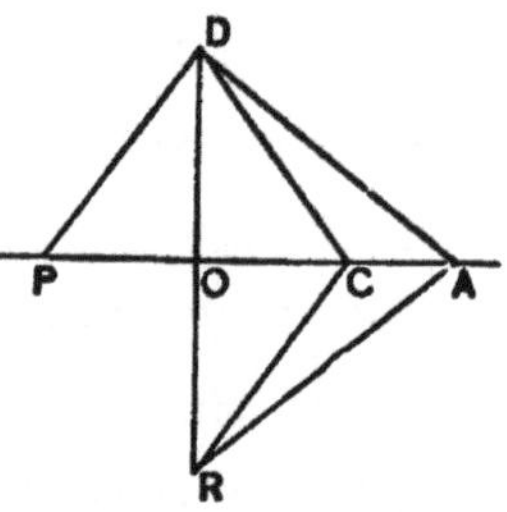

Let D be a point lying without PA; let DO be perpendicular to PA and let DP, DC, DA, be oblique lines, such that $OP = OC$, and $OA > OC$. Prolong DO, making OR equal to DO; draw RC and RA.

1°. The triangles POD, DOC, have the side DO common, OP equal to OC by hypothesis, and the angles POD and DOC equal, because DO is perpendicular to PA; they are therefore equal in all their parts, (P. 8); hence, DP is equal to DC, *which was to be proved.*

2°. Because AO is perpendicular to DR and OD equal to OR, we have, from what has just been shown, DC $=$ RC, and DA $=$ RA; but from P. 7, we have,

$$DA + RA > DC + RC;$$

substituting DA for its equal RA and DC for its equal RC, we have,

$$2DA > 2DC; \quad \text{or,} \quad DA > DC,$$

which was to be proved.

3°. From P. 6, we have,

$$DO + OR < DC + CR;$$

substituting DO for its equal OR, and DC for its equal CR, we have,

$$2DO < 2DC; \quad \text{or} \quad DO < DC,$$

which was to be proved.

Cor. 1. If DP is equal to DC, OP is equal to OC. For, if OP and OC were unequal, DP and DC would be unequal, from what has just been shown.

Cor. 2. Only two equal lines can be drawn from D to PA, and these lie on opposite sides of the perpendicular.

Cor. 3. If DA is greater than DC, OA is greater than OC. For if OA were equal to OC, DA would be equal to DC, and if OA were less than OC, DA would be less than DC, both of which conclusions are contrary to the hypothesis.

Cor. 4. In a right-angled triangle DOC, the hypothenuse DC is greater than either of the other sides, OC, OD; consequently the right angle O is greater than either of the other angles, that is, both the other angles are *acute.*

Cor. 5. The shortest line from D to PA is perpendicular to PA.

Scho. When the distance from a point to a line is spoken of, the *shortest* distance is always meant.

PROPOSITION XVI. THEOREM.

If one line is perpendicular to another at its middle point, any point of the first line is equally distant from the extremities of the second, and any point not on the first line is unequally distant from the extremities of the second.

1°. Let DR be perpendicular to PA at its middle point O, and let D be any point of the first line. Draw DP and DA. Then, because DR is perpendicular to PA, and because OP is equal to OA, we have DP equal to DA, (P. 15), *which was to be proved.*

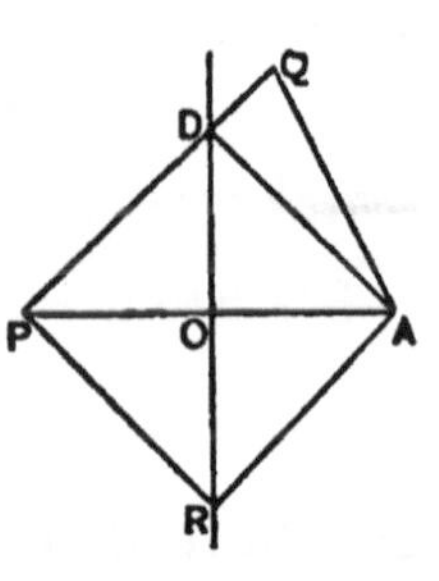

2°. Let Q be any point not on DR. Draw QP cutting DR at D; also draw QA and DA. Then will DP be equal to DA from what has just been proved.

From P. 6, we have

$$QA < QD + DA;$$

Substituting DP for its equal DA, we have,

$$QA < QD + DP, \text{ or } QA < QP,$$

that is, QA and QP are unequal, *which was to be proved.*

Cor. 1. The line DR contains every point that is equally distant from P and A.

Cor. 2. If each point of a line is equally distant from the extremities of a second line, the first line is perpendicular to the second, at its middle point.

PROPOSITION XVII. THEOREM.

If two right-angled triangles have the hypothenuse and an acute angle of one, equal to the hypothenuse and an acute angle of the other, each to each, the triangles are equal in all their parts.

Let ACD and PQR be two triangles right-angled at C and Q, in which the hypothenuse AD is equal to PR and the angle A to P. Let PQR be placed on ACD so that the angle P shall coincide with A, PQ taking the direction AC and PR the direction AD; then because PR is equal to AD, the vertex R will fall at D and RQ will coincide with the perpendicular from D to AC, that is, with the side DC (P. 5); hence, the vertex Q will fall at C, and the two triangles will coincide throughout their whole extent; they are therefore equal in all their parts, *which was to be proved.*

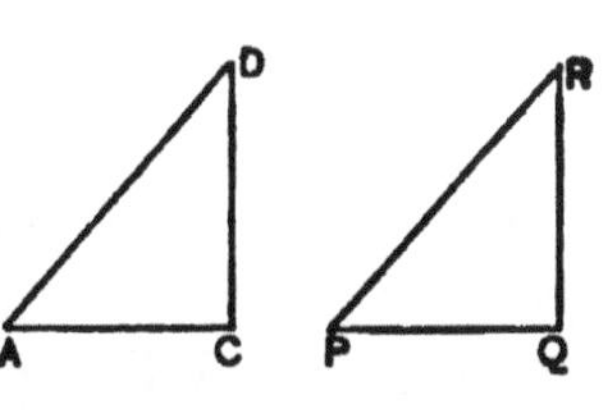

Definition.

38. The **bisectrix** of an angle is a line that divides it into two equal parts.

PROPOSITION XVIII. THEOREM.

Any point of the bisectrix of an angle is equally distant from the sides of that angle.

Let ACD be an angle whose bisectrix is CE, and let P be any point of CE. Draw PQ perpendicular to CA, and PR perpendicular to CD; these will be the distances from P to the sides of the angle. The right-angled triangles CQP and CRP have a common hypothenuse CP, and the acute angle QCP in the first is equal to the acute angle RCP in the second; hence, they are equal in all their parts (P. 17), and consequently PQ is equal to PR, *which was to be proved.*

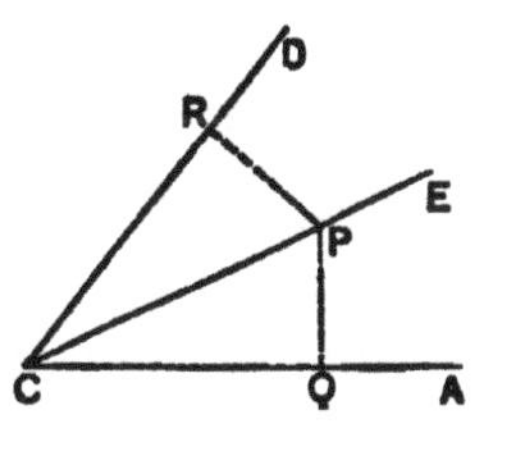

Definition.

39. Two lines lying in the same plane are **parallel**, when they cannot meet how far soever both may be prolonged.

If two lines are not parallel, they will meet if sufficiently prolonged.

PROPOSITION XIX. THEOREM.

If two lines are perpendicular to a third line, they are parallel to each other.

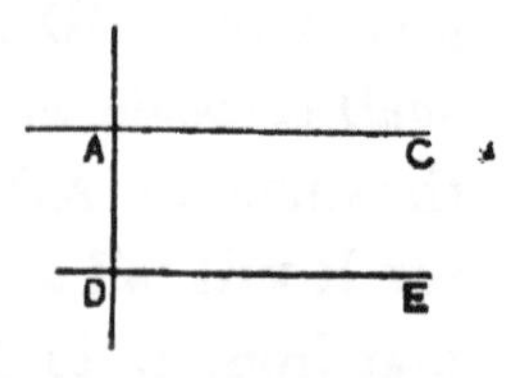

Let AC and DE be perpendicular to AD. If AC and DE are not parallel, they will meet if prolonged, and from the point in which they meet we shall have two perpendiculars to the line AD, which is impossible (P. 5); hence, the lines cannot meet; they are therefore parallel, *which was to be proved.*

Cor. 1. A line can be drawn through any point, A, lying without DE, that will be parallel to DE. For, draw AD perpendicular to DE, and at A draw AC perpendicular to AD; then will AC be parallel to DE.

It is assumed that only one parallel to a given line can be drawn through a given point; hence, if a line is perpendicular to one of two parallels, it is also perpendicular to the other.

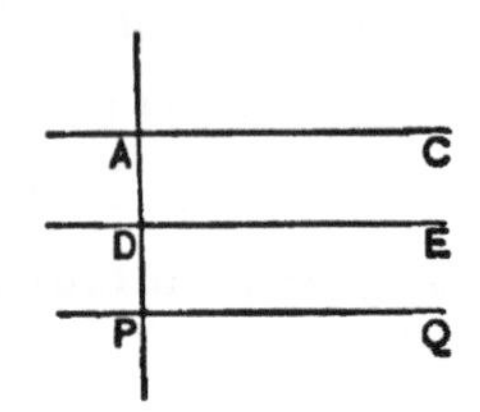

Cor. 2. If two lines AC and DE are parallel to a third, PQ, they are parallel to each other. For, let PA be perpendicular to PQ; it will also be perpendicular to both AC and DE; hence, AC and DE are parallel to each other.

Definitions.

40. A **secant** is a line that intersects or cuts another line. If two lines are cut by the same secant, the angles about the points of intersection, taken in pairs, receive different names according to their relative positions:

1°. Two angles that lie on the same side of the secant and between the other lines are called **interior angles on the same side**, as EQP, QPC;

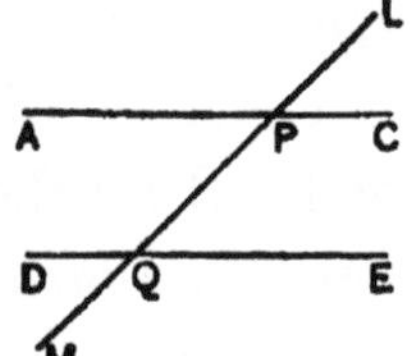

2°. Two angles that lie on the same side of the secant and without the other lines are called **exterior angles on the same side,** as CPL, MQE;

3°. Two angles that lie on opposite sides of the secant and within the other lines, but not adjacent, are called **alternate interior angles,** as APQ, PQE;

4°. Two angles that lie on opposite sides of the secant, and without the other lines, but not adjacent, are called **alternate exterior angles,** as CPL, DQM;

5°. Two angles that lie on the same side of the secant, one without and one within the other lines, but not adjacent, are called **opposite exterior and interior angles,** as CPL, EQP.

Proposition XX. Theorem.

If two lines are cut by a secant, making the sum of the interior angles on the same side equal to two right angles, the two lines are parallel.

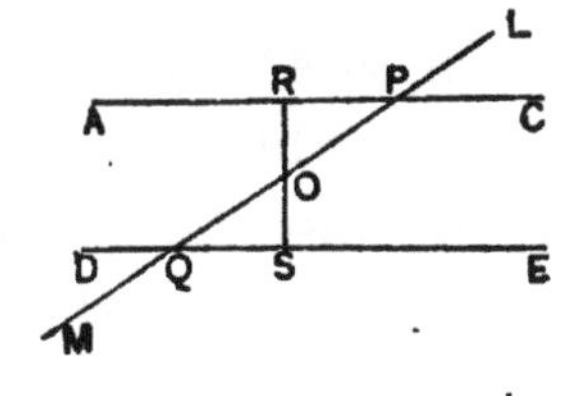

Let the lines AC and DE be cut by the secant LM, making the sum of the angles EQP and QPC equal to two right angles. Through O, the middle point of PQ, draw RS perpendicular to DE.

The angle RPO is the supplement of OPC, by P. 3, and the angle SQO is the supplement of OPC, by hypothesis; these angles are therefore equal; the angles POR and QOS are equal, because they are opposite angles (P. 1); and OP and OQ are equal, by hypothesis. Hence, the triangles OPR and OQS, have two angles and the included side in one equal to two angles and the included side in the other, each to each, and are therefore equal; consequently, the angle ORP is equal to OSQ, that is, it is a right angle; the lines AC and DE are therefore perpendicular to RS; hence, they are parallel, *which was to be proved.*

Cor. From the relations among the angles about P and Q, we infer that AC and DE are parallel:

1°. If the sum of the exterior angles on the same side is equal to two right angles;

2°. If the alternate interior angles are equal;

3°. If the alternate exterior angles are equal;

4°. If the opposite exterior and interior angles are equal.

Proposition XXI. Theorem.

If two parallel lines are cut by any secant, the sum of the interior angles on the same side is equal to two right angles.

Let AC and DE be parallel, and let LM cut them at P and Q. Through O, the middle of PQ, draw RS perpendicular to DE; it will also be perpendicular to AC (P. 19, *Cor.* 1).

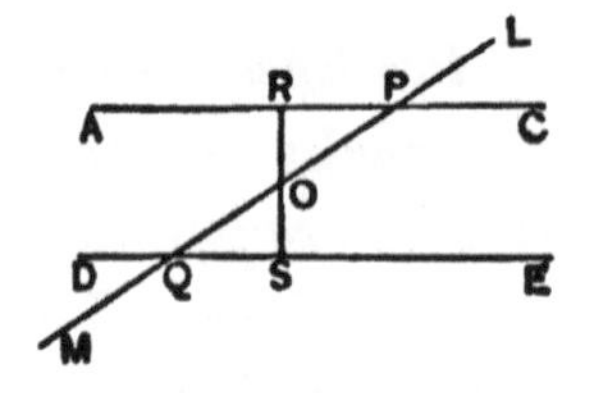

The right-angled triangles PRO and QSO have the hypothenuse OP equal to OQ, and the angles at O equal; the triangles are therefore equal in all their parts (P. 17), and consequently the angle RPO is equal to SQO. But the sum of RPO and OPC is equal to two right angles; hence, the sum of SQO and OPC is also equal to two right angles, *which was to be proved.*

Cor. 1. From the relations existing among the angles about P and Q, we infer that if AC and DE are parallel:

1°. The sum of the exterior angles on the same side is equal to two right angles;

2°. The alternate interior angles are equal;

3°. The alternate exterior angles are equal;

4°. The opposite exterior and interior angles are equal.

Cor. 2. If one of the interior angles on the same side is a right angle, the other one is also a right angle.

PROPOSITION XXII. THEOREM.

Two parallel lines are everywhere equally distant.

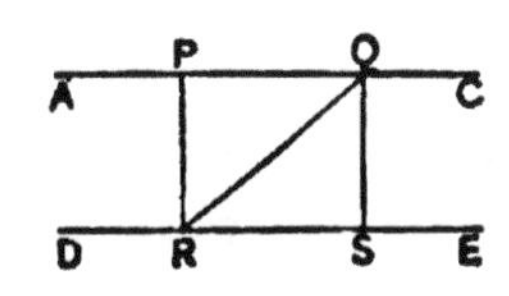

Let AC and DE be parallel; from any two points of AC, as P and Q, draw PR and QS perpendicular to DE; these lines will also be perpendicular to AC, and will therefore be the distances between AC and DE at those points. Draw the line QR. The right-angled triangles QPR and RSQ have the hypothenuse QR common and the acute angles PQR and SRQ equal (P. 21, *Cor.* 1); they are therefore equal in all their parts. Hence, PR and QS are equal, *which was to be proved.*

PROPOSITION XXIII. THEOREM.

If the corresponding sides of two angles are parallel and lie in the same, or in opposite directions, the two angles are equal.

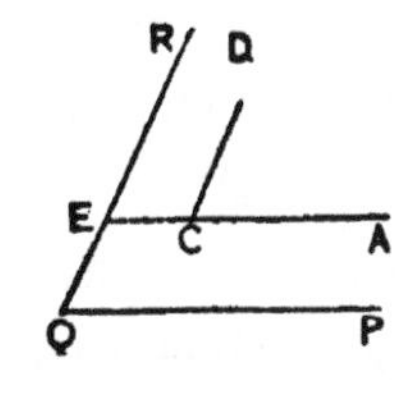

1°. Let ACD and PQR be two angles in which CA is parallel to QP, and CD to QR, and let the corresponding sides lie in the same direction. Prolong AC till it meets QR at E.

Because EA and QP are parallel and are cut by QR, the angle PQR is equal to AER; because QR and CD are parallel and are cut by EA, the angle AER is equal to ACD; hence, PQR is equal to ACD, *which was to be proved.*

2°. Let ACD and PQR be two angles in which AC is parallel to PQ and CD to QR, and let the corresponding sides lie in opposite directions. Prolong CA in the opposite direction to E, and CD in the opposite direction to S. The sides of the angle ECS are then parallel to the sides of PQR and lie in the same direction; the angles are therefore equal. Now, PQR and ECS are equal from what has just been proved, and ECS and ACD are equal because they are opposite angles; hence, ACD and PQR are equal, *which was to be proved.*

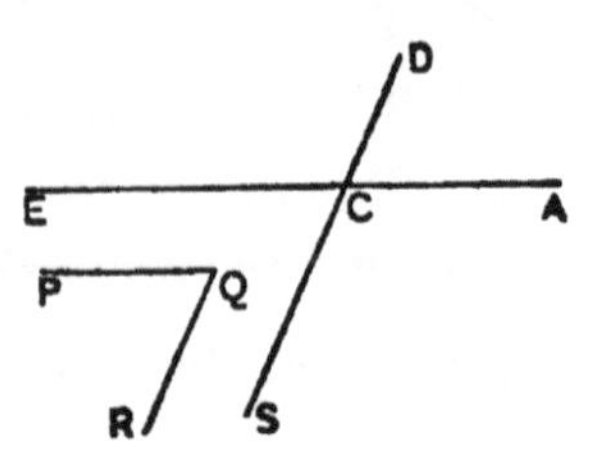

Cor. 1. If two corresponding sides of two angles are parallel and lie in the same direction, and if the other sides are parallel and lie in opposite directions, the angles are supplementary. Thus, DCE and RQP are supplementary. For, DCE is the supplement of ACD and consequently of PQR.

Cor. 2. If the corresponding sides of two angles are perpendicular, each to each, the angles are either equal or supplementary. For, let SQP be perpendicular to CA and QR to CD. Then let the angle PQR be revolved about Q in the direction indicated by the arrows and through a quarter of a complete revolution. The line SQP will take the position S′QP′ parallel to CA, and QR will take the position QR′ parallel to CD; from what has been shown we have P′QR′, or its equal PQR, equal to ACD; R′QS′, or its equal RQS, is the supplement of ACD. If the angles in question are both acute or both obtuse, they are equal; if one is acute and the other obtuse, they are supplementary.

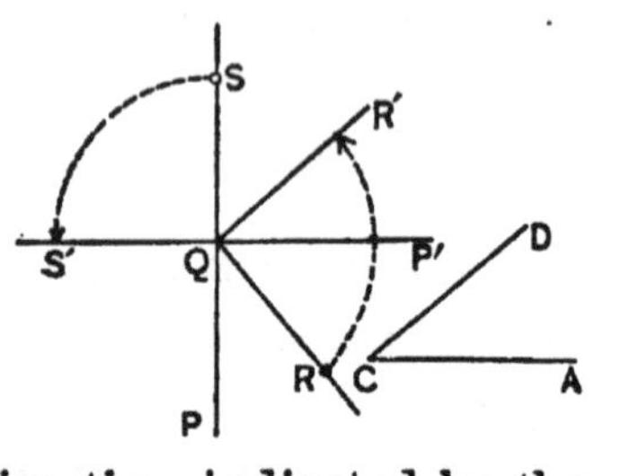

Cor. 3. Two lines will intersect when sufficiently prolonged, if they are respectively perpendicular to two lines that intersect.

PROPOSITION XXIV. THEOREM.

The sum of the angles of any triangle is equal to two right angles.

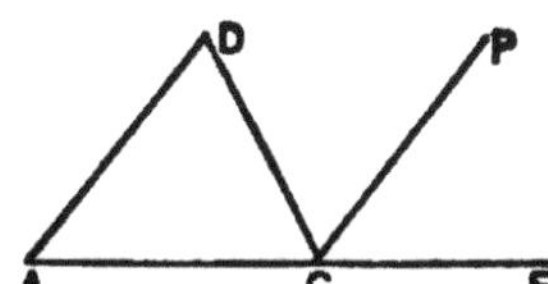

Let ACD be any triangle. Prolong the side AC in the direction CE and draw CP parallel to AD.

Because CP is parallel to AD and AC cuts them, the angles ECP and CAD are equal (P. 21, *Cor.* 1); because CP is parallel to AD and CD cuts them, the angles PCD and CDA are equal. Hence, the sum of the angles of the triangle is equal to the sum of the angles ECP, PCD, and DCA; but this sum is equal to two right angles, (P. 3, *Cor.* 2.); hence, the sum of the angles of the triangle is equal to two right angles, *which was to be proved.*

Cor. 1. The sum of any two angles of a triangle is equal to the supplement of the third angle, and conversely.

Cor. 2. Any **exterior angle of a triangle**, that is, the angle between the prolongation of any side and the side next in order, is equal to the sum of the **opposite interior angles**. Thus, DCE = CDA + DAC.

Cor. 3. The acute angles of a right-angled triangle are *complementary.*

Cor. 4. Each angle of an equiangular triangle is equal to one-third of two right angles, or to two-thirds of one right angle.

Definitions.

41. A **salient polygon** is a polygon such that no straight line can cut its perimeter in more than two points; as, ACDE. All other polygons are called **re-entrant polygons**, as PQRST.

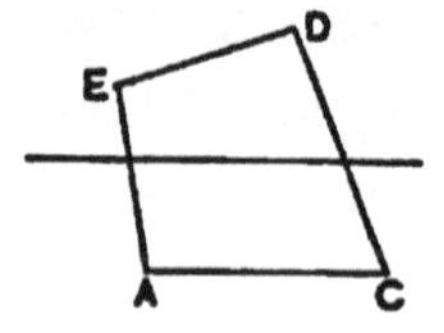

The polygons treated of in this work are supposed to be salient.

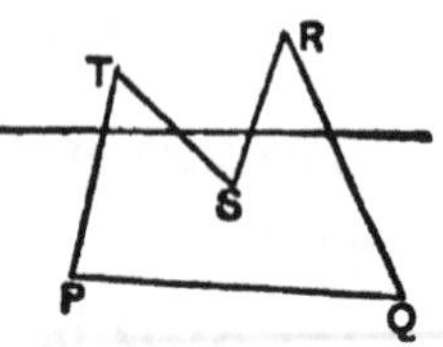

PROPOSITION XXV. THEOREM.

The sum of the angles of a polygon is equal to twice as many right angles as the polygon has sides, less four right angles.

Let ACDEG be a polygon and let O be a point within it. From O draw lines to each of the vertices of the polygon; these will divide the polygon into as many triangles as the polygon has sides.

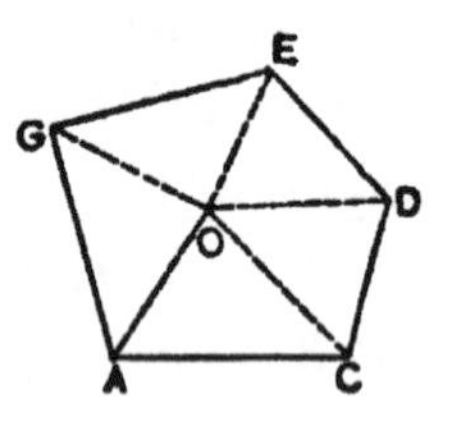

The sum of the angles of each triangle is two right angles; hence, the sum of all the angles of all the triangles is equal to twice as many right angles as the polygon has sides. But, the sum of the angles of the polygon is equal to the sum of all the angles of all the triangles *diminished* by the sum of the angles about O; and the sum of all angles about O is equal to four right angles (P. 3, *Cor.* 3). Hence, the sum of the angles of the polygon is equal to twice as many right angles as the polygon has sides, *less* four right angles, *which was to be proved.*

Cor. The sum of the angles of a quadrilateral is equal to *four* right angles; the sum of the angles of a pentagon is equal to *six* right angles; the sum of the angles of a hexagon is equal to *eight* right angles; and so on.

Scho. Any angle of an equiangular polygon is equal to the sum of its angles divided by the number of its angles. Thus, any angle of an equiangular pentagon is equal to $\frac{1}{5}$ of 6 right angles, or to $\frac{6}{5}$ of a right angle.

Definitions.

42. A **parallelogram** is a quadrilateral whose opposite sides are parallel; as ACDE.

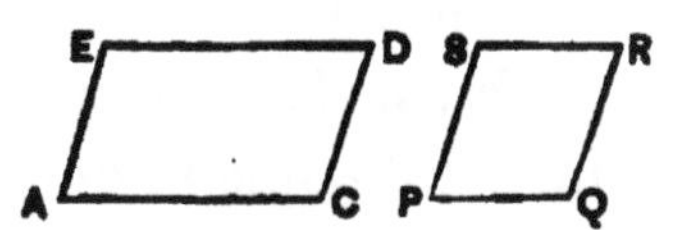

43. A **rhombus** is an equilateral parallelogram whose angles are oblique; as PQRS.

44. A **rectangle** is a parallelogram whose angles are right angles; as ACDE.

45. A **square** is an equilateral rectangle; as PQRS.

46. A **trapezoid** is a quadrilateral, two of whose side are parallel and the other two not parallel; as ACDE.

The parallel sides are called respectively the *upper* and the *lower bases.* Thus, ED is the upper base and AC the lower base of the trapezoid ACDE.

Proposition XXVI. Theorem.

The opposite sides of a parallelogram are equal, each to each.

Let ACDE be a parallelogram, and let CE be one of its diagonals.

Because AC and ED are parallel and CE cuts them, the angle ACE is equal to DEC (P. 21, *Cor.*); and because AE and CD are parallel and CE cuts them, the angles CEA and ECD are equal. The triangles ACE and DEC have the side CE common, the angle ACE equal to DEC, and the angle CEA equal to DCE; they are therefore equal in all their parts (P. 10); hence, AC is equal to ED, and AE to CD, *which was to be proved.*

Cor. 1. Parallels included between parallels are equal.

Cor. 2. Either diagonal of a parallelogram divides the parallelogram into equal triangles.

Cor. 3. The opposite angles of a parallelogram are equal, each to each.

Cor. 4. If two parallelograms have two sides and their included angle in one, equal to two sides and their included angle in the other, each to each, they are equal.

PROPOSITION XXVII. THEOREM.

If the opposite sides of a quadrilateral are equal, each to each, the quadrilateral is a parallelogram.

Let ACDE be a quadrilateral in which AC is equal to ED and AE to CD. Draw CE.

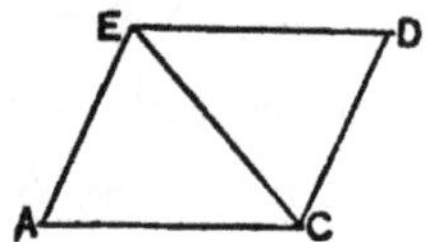

The triangles ACE and DEC have the side CE common, AC equal to DE, and EA to CD; they are therefore equal in all their parts (P. 11); consequently, the angle ACE is equal to DEC, and CEA to ECD. Because the alternate interior angles ACE and DEC are equal, AC and ED are parallel (P. 20, *Cor.*), and because the alternate angles CEA and ECD are equal, AE and CD are parallel. The quadrilateral ACDE is therefore a parallelogram, *which was to be proved.*

PROPOSITION XXVIII. THEOREM.

If two sides of a quadrilateral are parallel and equal, the quadrilateral is a parallelogram.

Let ACDE be a quadrilateral in which AC and ED are parallel and equal. Draw CE.

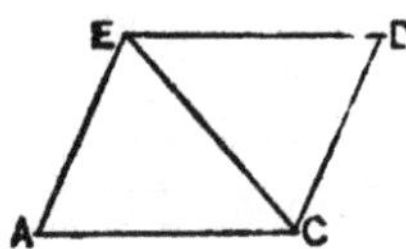

Because AC and ED are parallel and CE

cuts them, the angle ACE is equal to DEC. The triangles ACE and DEC have the side CE common, the side AC equal to ED, and the included angle ACE equal to the included angle DEC; they are therefore equal in all their parts (P. 8); consequently AE is equal to CD. Hence, the opposite sides of ACDE are equal each to each, and the figure is a parallelogram (P. 27), *which was to be proved.*

Cor. If two points on the same side of a line are equally distant from that line, the line that joins them is parallel to the given line.

PROPOSITION XXIX. THEOREM.

The diagonals of a parallelogram mutually bisect each other.

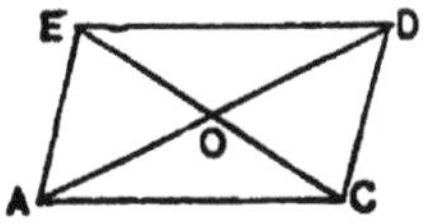

Let ACDE be a parallelogram whose diagonals intersect at O. The triangles ACO and DEO, have the side AC equal to DE, the angle ACO equal to DEO, and the angle OAC equal to ODE; they are therefore equal in all their parts. Hence, OA is equal to OD, and OC to OE, *which was to be proved.*

Cor. The diagonals of a rhombus are perpendicular to each other. For, if CD is equal to DE, the triangle CDE is isosceles, and because DO is drawn from the vertex D to the middle of the base CE, the line DA is perpendicular to CE.

EXERCISES ON BOOK I.

Let the student demonstrate the following theorems:

1°. *If the sum of two adjacent angles is equal to two right angles, the bisectrices of these angles are perpendicular to each other.*

2°. *The bisectrices of two opposite angles of a parallelogram are parallel to each other.*

3°. *The bisectrices of two consecutive angles of a parallelogram are perpendicular to each other.*

4°. *If two triangles ACE and ACD, (Fig. of Art. 31), have the side AC common and the sides CE and AD intersecting each other, then will the sum of the sides CE and AD, which intersect, be greater than the sum of the sides AE and CD, which do not intersect.*

5°. *If two lines mutually bisect each other, their extremities are the vertices of a parallelogram.*

6°. *If the opposite angles of a quadrilateral are equal, each to each, the quadrilateral is a parallelogram.*

7°. *If the diagonals of a parallelogram are equal, the parallelogram is a rectangle.*

8°. *The diagonals of a rhombus are the bisectrices of the angles of the rhombus.*

9°. *If a line included between two parallels is bisected and if a second line is drawn through the point of bisection and limited by the parallels, then will the second line be bisected at the same point.*

BOOK II.

THE CIRCLE, AND THE MEASURE OF ANGLES.

Definitions.

47. A **Circle** is a portion of a plane bounded by a curve all of whose points are equally distant from a point within, called the **centre**. The bounding line is called the **circumference**.

The centre of a circle is also the centre of its circumference.

48. A **radius** is a line drawn from the centre to any point of the circumference, as OA.

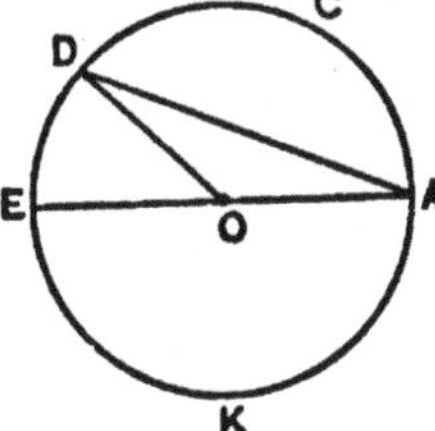

All radii of the same circle, or of equal circles, are equal.

49. A **diameter** is a line drawn through the centre and limited by the circumference, as AE.

A diameter is double the radius; all diameters of the same circle, or of equal circles, are equal.

50. An **arc** is a part of a circumference, as ACD. The line that joins the extremities of an arc is called a **chord**; thus, AD is the chord of the arc ACD and also of the arc AKD.

A chord is said to **subtend** the arc to which it belongs; thus, AD subtends the arc ACD and also the arc AKD.

51. An **angle at the centre** is an angle formed by two radii, as AOD.

The intercepted arc ACD is said to subtend the angle AOD.

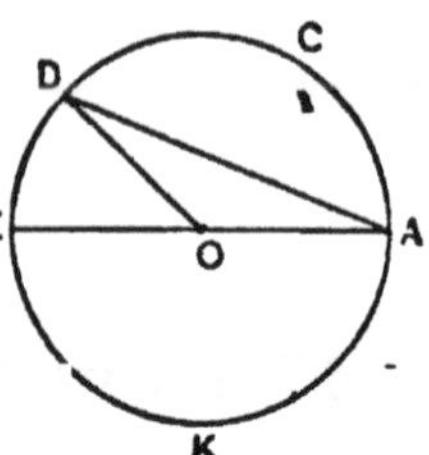

52. A **sector** is a part of a circle bounded by two radii and their intercepted arc, as AODC.

53. A **segment** is a part of a circle bounded by an arc and its chord, as ADC.

54. **Similar arcs, similar sectors, and similar segments,** are those which correspond to equal angles at the centre.

Two arcs cannot be made to coincide unless they belong to the same circumference, or to equal circumferences.

A circle, or a circumference, may be designated by two letters, one at the centre and the other at the extremity of any radius; thus, the circle whose centre is O and whose circumference is ACK is called the circle OA.

PROPOSITION I. THEOREM.

Any diameter divides the circle to which it belongs into two equal parts.

Let AE be any diameter of the circle OA: then are the parts ACE and AKE equal to each other.

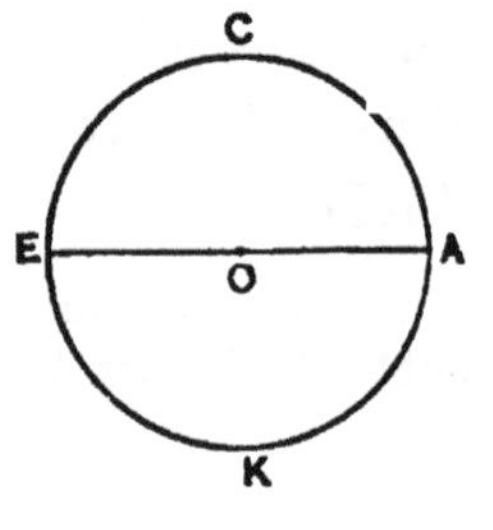

Let the part of the plane above AE be revolved about that line till it falls on the part below; each point of the curve ACE will fall on some point of the curve AKE, because all the points of both are equally distant from O; hence, the two parts of the circle coincide throughout; they are therefore equal, *which was to be proved.*

Cor. Any diameter bisects a circumference.

Scho. The segments formed by a diameter are *semi-circles*, and the arcs subtended by a diameter are *semi-circumferences*.

PROPOSITION II. THEOREM.

A diameter is greater than any other chord.

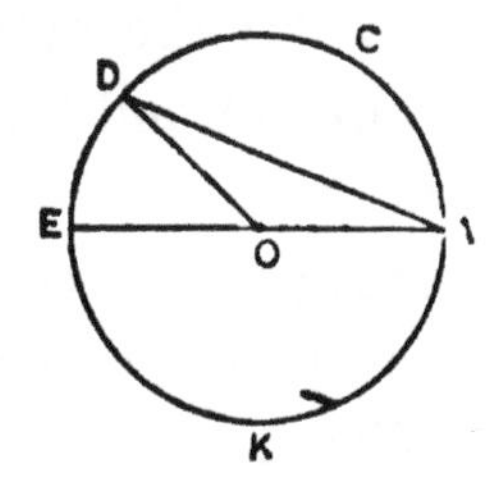

Let AD be any chord, not a diameter, of the circle OA, and let AE be a diameter through A: then is AE greater than AD.

Draw the radius OD.

The sum of the radii OA and OD is greater than AD (P. 6, B. 1); but the sum of OA and OD is equal to AE; hence, AE is greater than AD, *which was to be proved.*

Scho. The segments ACD and AKD are unequal; the arcs ACD and AKD, subtended by AD, are also unequal.

PROPOSITION III. THEOREM.

Equal angles at the centres of equal circles are subtended by equal arcs; and conversely, equal arcs subtend equal angles at the centre.

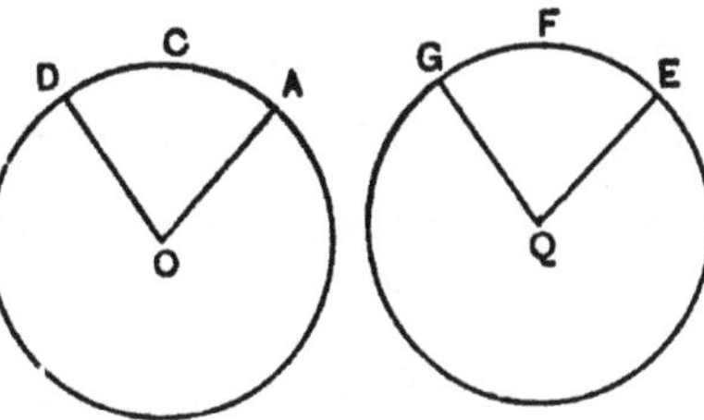

1°. Let AOD and EQG be equal angles at the centres of the equal circles OA and QE: then are the arcs ACD and EFG equal.

Let the sector EQG be placed on the sector AOD so that the angle Q shall coincide with the angle O; the point E will fall on A and the point G on D, and since all the points of the arcs EFG and ACD are equally distant from O, the two arcs will coincide throughout; they are therefore equal, *which was to be proved.*

2°. *Conversely,* if the arcs ACD and EFG are equal, the angles AOD and EQG are equal.

Let the sector EQG be placed on the sector AOD so that the arc EFG shall fall on ACD; the radius QE will fall on OA and the radius QG will fall on OD, and consequently the angles EQG and AOD coincide; they are therefore equal, *which was to be proved.*

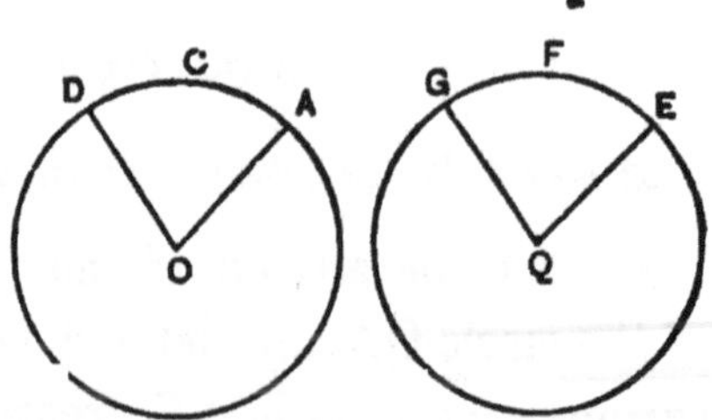

Cor. 1. The radius that bisects an angle at the centre bisects the corresponding arc, and *conversely,* the radius that bisects an arc bisects the corresponding angle at the centre.

Cor. 2. If radii are drawn dividing an angle at the centre into any number of equal parts, they will divide the corresponding arc into the same number of equal parts, and conversely, the radii that divide an arc into any number of equal parts divide the corresponding angle at the centre into the same number of equal parts.

Cor. 3. Two diameters, perpendicular to each other, form four angles at the centre, each equal to a right angle; they also divide the circumference into four arcs each equal to a **quadrant**, that is, to a quarter of a circumference; hence, a right angle at the centre is subtended by a quadrant.

Scho. The preceding proposition holds good if the angles and arcs belong to the same circle.

PROPOSITION IV. THEOREM.

In equal circles, equal arcs are subtended by equal chords; and conversely, equal chords subtend equal arcs.

1°. In the equal circles OA and QE let the arcs ACD and EFG be equal: then are the chords AD and EG equal.

Draw the radii OA, OD, QE, and QG.

Because the arcs ACD and EFG are equal, the angles O and Q are equal (P. 3), and consequently the triangles AOD and EGQ have two sides and their included angle equal, each to each; they are therefore equal in all their parts (P. 8, B. 1); hence, AD and EG are equal, *which was to be proved.*

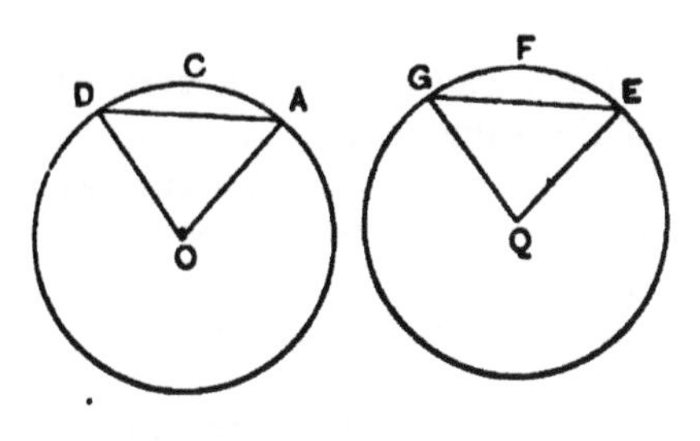

2°. *Conversely,* if the chords AD and EG are equal, the arcs ACD and EFG are equal.

The triangles AOD and EQG have their sides equal each to each; they are therefore equal in all their parts (P. 11, B. 1); the angle O is therefore equal to the angle Q, and consequently their subtending arcs ACD and EFG are equal, *which was to be proved.*

Scho. This proposition holds good when the arcs and their chords belong to the same circle.

Proposition V. Theorem.

The radius that is perpendicular to a chord bisects that chord.

Let AD be any chord in the circle OA, and let OC be perpendicular to it at H: then is AH equal to DH.

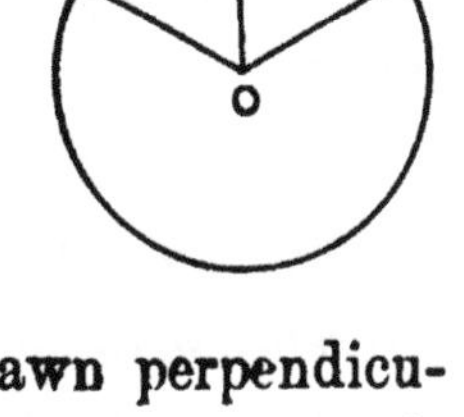

Draw the radii OA and OD.

The point O lies without the line AD, the oblique lines OA and OD are equal, being radii of the same circle, and OH is drawn perpendicular to AD; hence, AH is equal to DH (P. 15, *Cor.* 1, B. 1), *which was to be proved.*

Cor. The radius OC bisects the angle AOD, and also the corresponding arc ACD.

Scho. The centre of a circle, the middle of any chord, and the middle of the subtended arc lie in a line perpendicular to the chord; hence, a line that passes through any two of these points will pass through the other one and be perpendicular to the chord, and a line that passes through any one of the points and is perpendicular to the chord will pass through the other two.

PROPOSITION VI. THEOREM.

In equal circles, equal chords are equally distant from the centres.

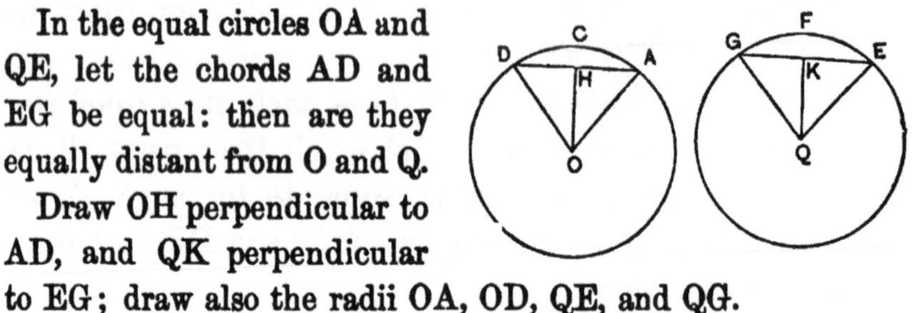

In the equal circles OA and QE, let the chords AD and EG be equal: then are they equally distant from O and Q.

Draw OH perpendicular to AD, and QK perpendicular to EG; draw also the radii OA, OD, QE, and QG.

In the triangles ADO and EGQ, the side AD is equal to EG, DO to GQ, and OD to QE; hence, the angle OAH is equal to QEK (P. 11, B. 1). In the right-angled triangles AHO and EKQ, the hypothenuse OA is equal to QE, and the angle OAH to QEK; hence, OH is equal to QK (P. 16, B. 1). But OH is the distance of AD from O, and QK is the distance of EG from Q; hence, their distances are equal, *which was to be proved.*

Scho. This proposition holds true when the chords belong to the same circle.

PROPOSITION VII. THEOREM.

Through three points, not in a straight line, one circumference may be made to pass, and but one.

1°. Let A, C, and D be three points, not in a straight line: then can one circumference be made to pass through them, and only one.

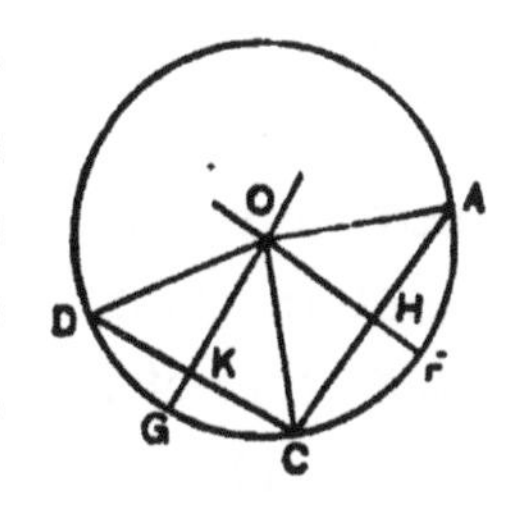

Draw AC and CD. Let FH be perpendicular to AC at its middle point, and let GK be perpendicular to CD at its middle point; these lines will intersect at some point, O (P. 23, *Cor.* 3, B. 1). Draw OA, OC, and OD.

Each point of FO is equally distant from A and C (P. 16, B. 1); hence, OA is equal to OC. In like manner it may be shown that OC is equal to OD. Hence, O is equally distant from A, C, and D; consequently, a circumference whose centre is O and whose radius is OA will pass through A, C, and D, *which was to be proved.*

2°. The line FO contains all the points that are equally distant from A and C, and GO contains all the points that are equally distant from C and D; hence, O is the only point that is equally distant from A, C, and D; the circumference just described is therefore the only one that can pass through these points, *which was to be proved.*

Cor. Two circumferences cannot intersect in more than two points; for if they could intersect in three points, two circumferences could pass through three points, which is impossible.

Definitions.

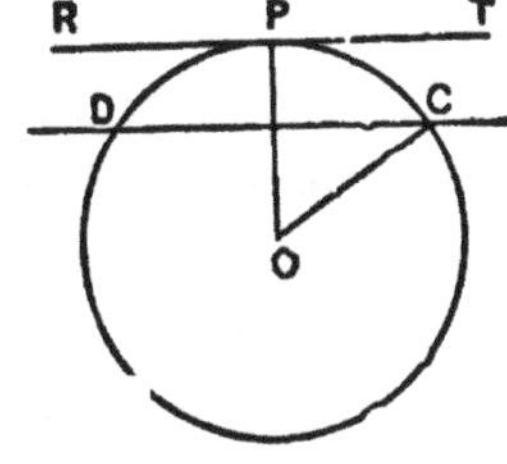

55. A **secant** is a line that cuts a circumference in two points, as DC. A **tangent** is a line that touches a circumference in one point and has no other point in common with it, as RPT. The point P, at which it touches the circumference, is called the **point of contact**, or the **point of tangency**.

The circumference OC is also said to be tangent to RPT.

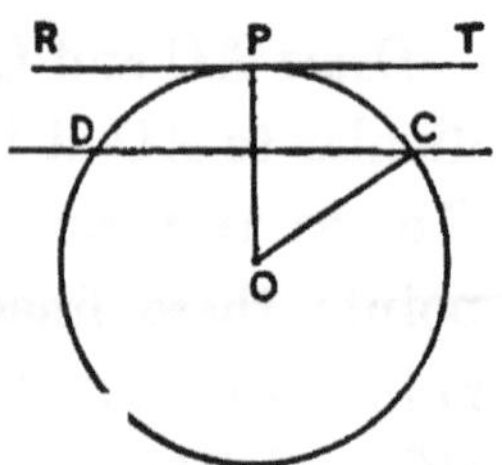

56. If two circumferences intersect, the corresponding circles are called **secant circles**; if they have but one point in common they are **tangent** to each other at that point.

57. The line passing through the centres of two circles is called their **line of centres.** The distance between the centres of two circles is called their **central distance.**

PROPOSITION VIII. THEOREM.

The perpendicular to a radius at its extremity is tangent to the circumference at that point; and conversely, a tangent to a circumference is perpendicular to the radius drawn to the point of contact.

1°. Let TS be perpendicular to the radius OP at P; then is it tangent to the circumference OP at P.

Let Q be any point of TS, except P, and draw OQ.

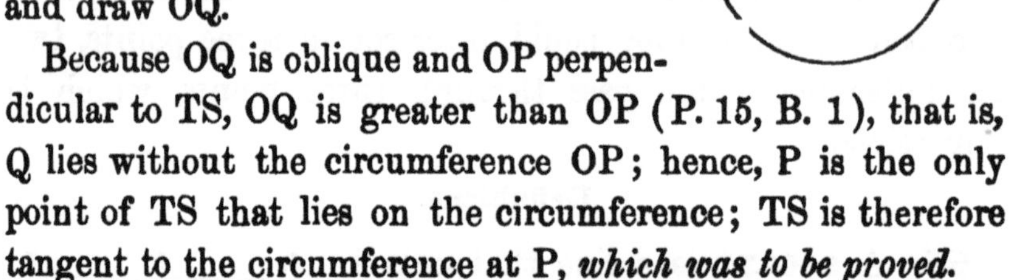

Because OQ is oblique and OP perpendicular to TS, OQ is greater than OP (P. 15, B. 1), that is, Q lies without the circumference OP; hence, P is the only point of TS that lies on the circumference; TS is therefore tangent to the circumference at P, *which was to be proved.*

2°. *Conversely,* let TS be tangent to the circumference OP at P: then is TS perpendicular to OP at P.

Because TS is a tangent, all its points except P lie without the circumference; hence, OP is the shortest line that can be drawn from O to TS; it is therefore perpendicular to TS at P (P. 15, *Cor.* 5, B. 1), *which was to be proved.*

Cor. 1. Only one tangent can be drawn to a circumference at the same point.

Cor. 2. Tangents at the extremities of the same diameter are parallel; and conversely, if two tangents are parallel the line joining their points of contact is a diameter.

Cor. 3. Parallel tangents intercept equal arcs of a circumference, each being a semi-circumference.

PROPOSITION IX. THEOREM.

Parallel secants intercept equal arcs of a circumference.

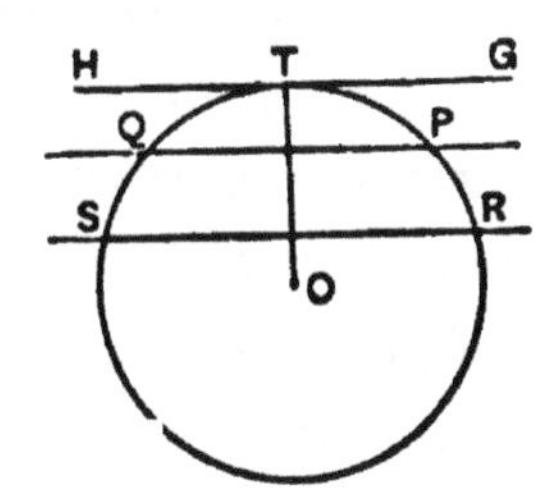

Let PQ and RS be parallel secants: then are the arcs PR and QS equal.

Draw a radius OT perpendicular to RS; it will also be perpendicular to PQ.

Because OT is perpendicular to RS and PQ it bisects the arcs RTS and PTQ (P. 5, *Cor.*); hence, the difference between the arcs TR and TP is equal to the difference between TS and TQ, that is, PR and QS are equal, *which was to be proved.*

Cor. The arcs intercepted by a secant and a parallel tangent are equal. For if the tangent GH is parallel to PQ the perpendicular radius OT passes through the point of contact, T, and consequently QT and PT are equal.

PROPOSITION X. THEOREM.

The line of centres of two secant circles is perpendicular to their common chord at its middle point.

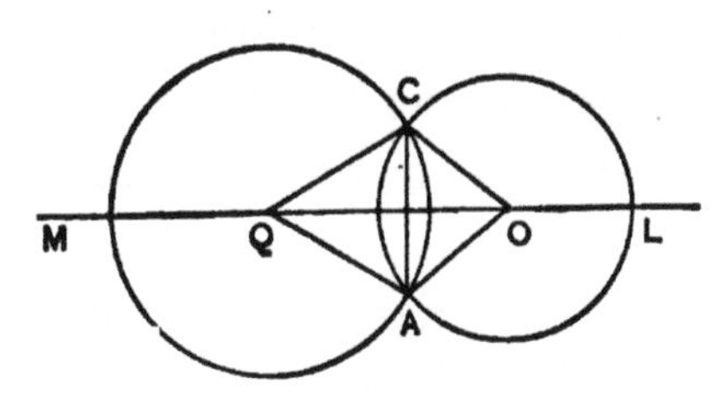

Let the circumferences OA and QA intersect at A and C, and let AC be their common chord: then is the line of centres LM perpendicular to AC at its middle point.

Because A and C are on the circumference OA they are

equally distant from O, and because they are on the circumference QA they are equally distant from Q; hence, LM has two of its points equally distant from A and C, it is therefore perpendicular to AC at its middle point (P. 16, *Cor.* 2, B. 1), *which was to be proved.*

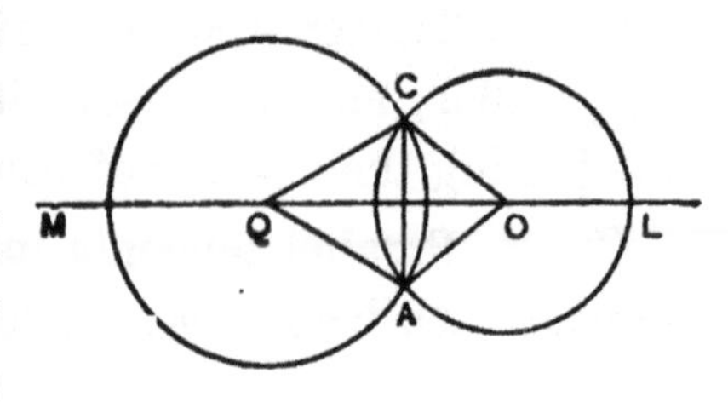

Cor. If two circumferences are tangent to each other their point of contact is on their line of centres. For if the circumference whose centre is Q is moved either to the right or left, its centre remaining on LM, the points of intersection will always be equally distant from the line of centres, and if the motion is sufficiently continued these points will meet on the line of centres, and the two circumferences will be tangent to each other at that point.

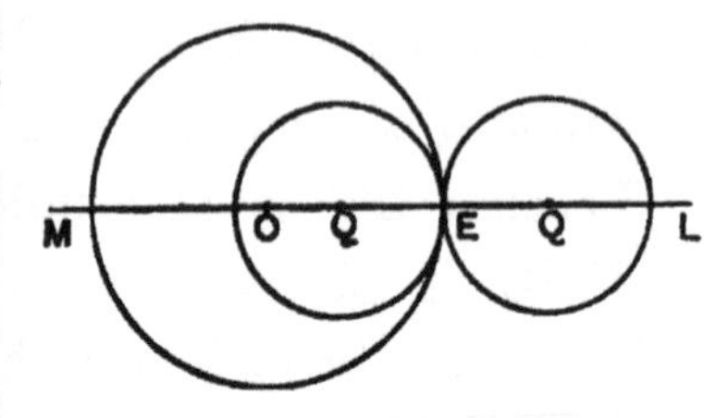

If the circumference is moved towards M the circumferences become tangent externally; if moved towards L they become tangent internally.

PROPOSITION XI. THEOREM.

If two circumferences intersect, their central distance is less than the sum, and greater than the difference of their radii.

Let the circumferences OC and QC intersect, and let OC and QC be radii drawn to one of the points of intersection.

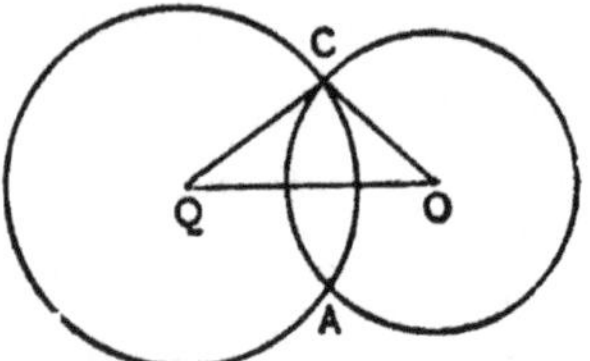

In the triangle OQC the side OQ is less than the sum, and greater than the difference, of QC and OC (P. 6, B. 1), *which was to be proved.*

PROPOSITION XII. THEOREM.

If two circles are tangent externally, their central distance is equal to the sum of their radii; if tangent internally their central distance is equal to the difference of their radii.

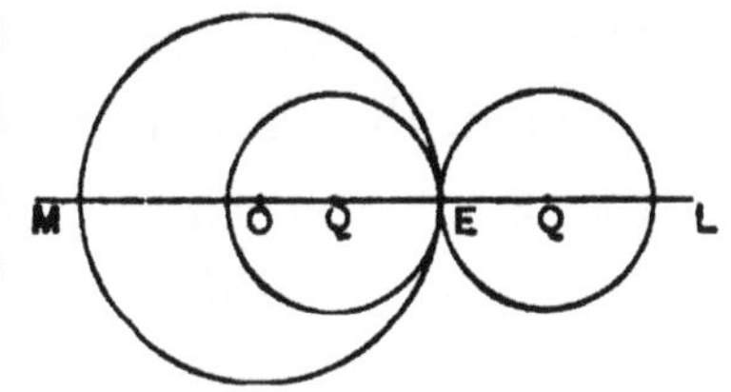

1°. If the circumferences are tangent externally at E, it is obvious that OQ is equal to the sum of OE and QE, *which was to be proved.*

2°. If the circumferences are tangent internally at E, it is obvious that OQ is equal to the difference of OE and QE, *which was to be proved.*

PROPOSITION XIII. THEOREM.

If one circumference lies wholly without another, their central distance is greater than the sum of their radii; if one circumference lies wholly within another, their central distance is less than the difference of their radii.

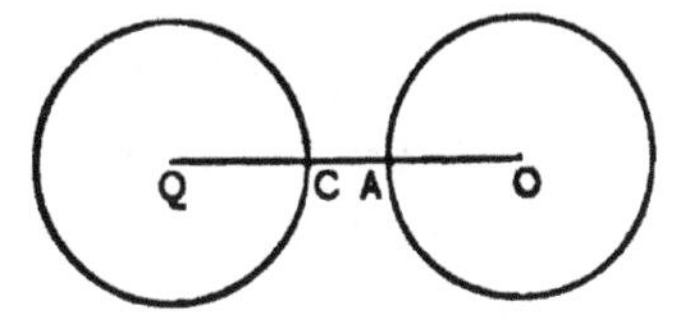

1°. Let the circumference QC lie wholly without the circumference OA, and let the line of centres cut the two circumferences at C and A. It is obvious that OQ is greater than the sum of OA, and QC, *which was to be proved.*

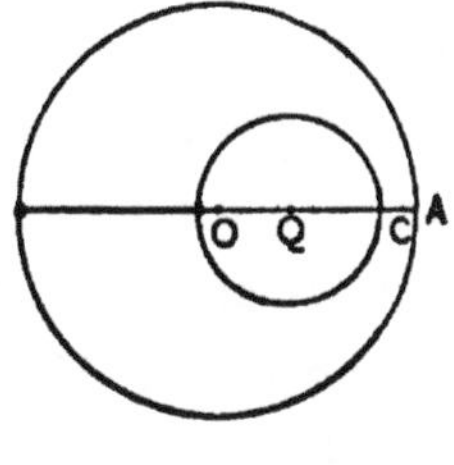

2°. Let the circumference QC lie wholly within the circumference OA, and let the line of centres cut the circumferences at C and A. It is obvious that OQ is less than the difference of OA and QC, *which was to be proved.*

From this and the two preceding propositions we deduce

the following conclusions relating to the positions of two circles:

Cor. 1. If the central distance of two circles is greater than the sum of their radii, one circle lies wholly without the other.

Cor. 2. If the central distance is equal to the sum of their radii, the circumferences are tangent externally.

Cor. 3. If the central distance is less than the sum and greater than the difference of their radii, the circumferences intersect in two points.

Cor. 4. If the central distance is equal to the difference of their radii, the circumferences are tangent internally.

Cor. 5. If the central distance is less than the difference of their radii, one circumference is wholly within the other.

Two circles that have a common centre are said to be concentric.

Measure of angles.

58. To **measure** a thing is to find its *numerical* value in terms of something of the same kind taken as a *unit.* The *primary* unit of angular measure is the *right angle.* The right angle is divided into 90 equal parts called **degrees**; the degree is divided into 60 equal parts called **minutes**; the minute is divided into 60 equal parts called **seconds.** Seconds are divided into **tenths**, tenths into **hundredths**, and so on indefinitely. A right angle at the centre of a circle is subtended by a quadrant, and the radii that divide the right angle into equal parts also divide the quadrant into the same number of equal parts (P. 3, *Cor.* 2); these parts of the arc are called by the *same names* as the corresponding parts of the angle. Hence, the *numerical value* of an arc is the same as the numerical value of the angle which it subtends, that is, the arc contains the same number of degrees and parts of a degree as the angle. We may therefore take the ***numerical value*** of an arc as the ***numerical value*** of the corresponding angle. In this sense, and in this sense only, we say that ***an angle at the centre is measured by the intercepted arc.***

Because units of arc are definite portions of the circumferences to which they belong, the measure of an arc is independent of the length of its radius, in the same manner that the measure of an angle is independent of the length of its sides.

Degrees, minutes, and *seconds* are denoted by the symbols °, ′, ″; thus, the expression 22°, 25′, 31″.34 is read 22 *degrees,* 25 *minutes,* 31.34 *seconds.*

Definitions.

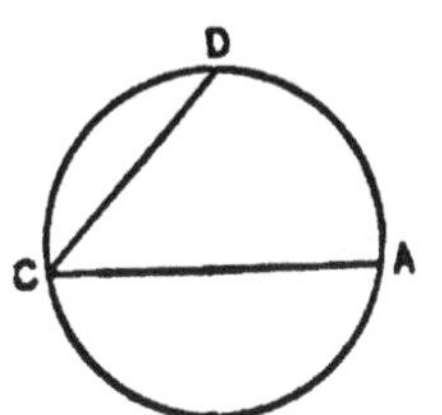

59. An **inscribed angle** is an angle whose vertex is on the circumference and whose sides are chords; as ACD.

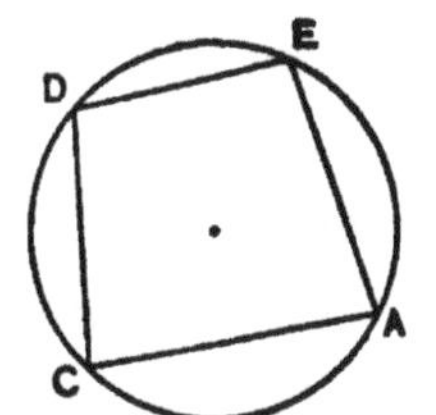

60. A polygon is said to be **inscribed** in a circle when all its sides are chords of the circumference; as ACDE.

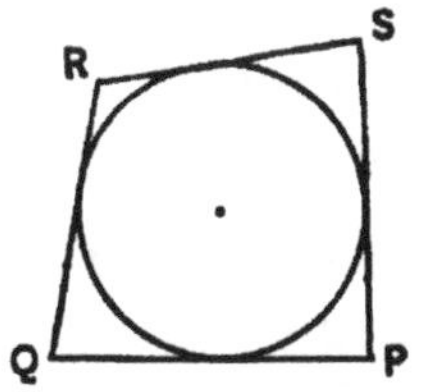

61. A polygon is said to be **circumscribed** about a circumference when all its sides are tangent to the circumference; as PQRS.

62. A circle is **circumscribed** about a polygon when the circumference of the circle passes through all the vertices of the polygon.

63. A circle is **inscribed** in a polygon when the circumference of the circle is tangent to all the sides of the polygon.

PROPOSITION XIV. THEOREM.

An inscribed angle is measured by half of the intercepted arc.

There may be three cases: the centre of the circle may lie on one of the sides of the angle; it may lie within the angle; or, it may lie without the angle.

1°. Let the centre O lie on the side AC of the inscribed angle ACD.

Draw the radius OD.

The angle AOD is an exterior angle of the triangle OCD and is therefore equal to the sum of the opposite interior angles OCD and ODC, (P. 24, *Cor.* 2, B. 1); but OCD and ODC are equal, because they lie opposite the equal sides OD and OC (P. 12, B. 1); hence, ACD is half the angle AOD; but AOD is an angle at the centre, and consequently is measured by the arc AD; hence, the angle ACD is measured by half the arc AD, *which was to be proved.*

2°. Let the centre O lie within the inscribed angle ACD.

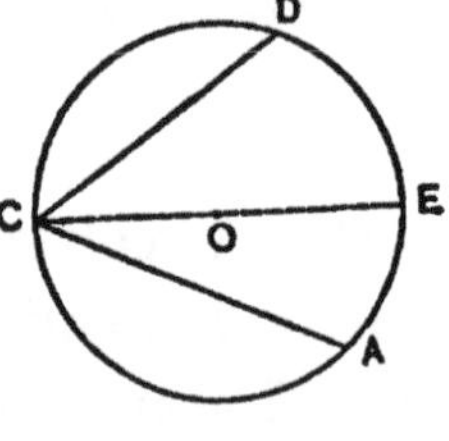

Draw the diameter CE.

From what has just been proved the angle ACE is measured by half the arc AE, and ECD by half the arc ED; hence, the angle ACD is measured by half the sum of the arcs AE and ED, that is, by half the arc AD, *which was to be proved.*

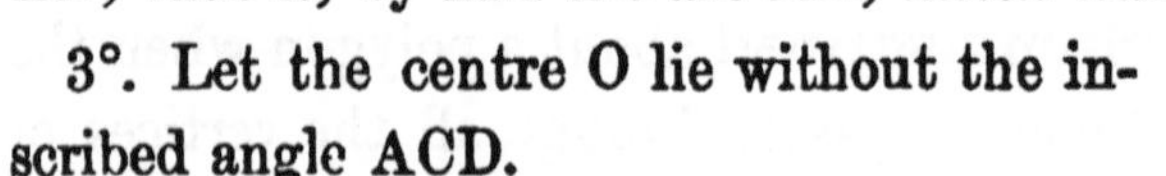

3°. Let the centre O lie without the inscribed angle ACD.

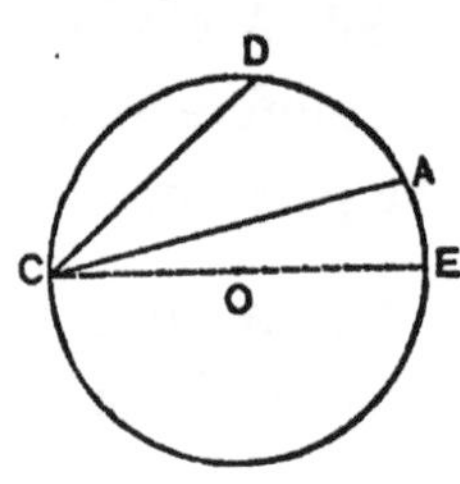

Draw the diameter CE.

From what precedes, the angle ECD is measured by half the arc ED, and ECA by half the arc EA; hence, ACD is measured by half the difference of ED and EA, that is, by half the arc AD, *which was to be proved.*

Cor. 1. Angles, as ACD, AED, inscribed in the same segment ACD are equal, for each is measured by half the arc AD.

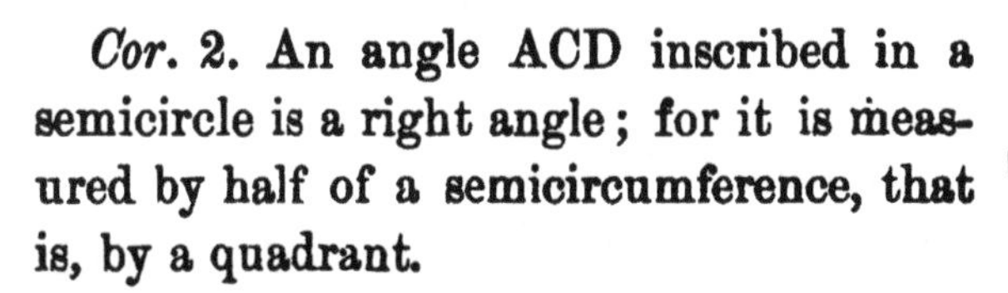

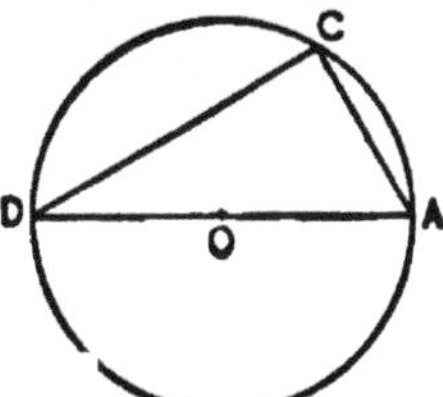

Cor. 2. An angle ACD inscribed in a semicircle is a right angle; for it is measured by half of a semicircumference, that is, by a quadrant.

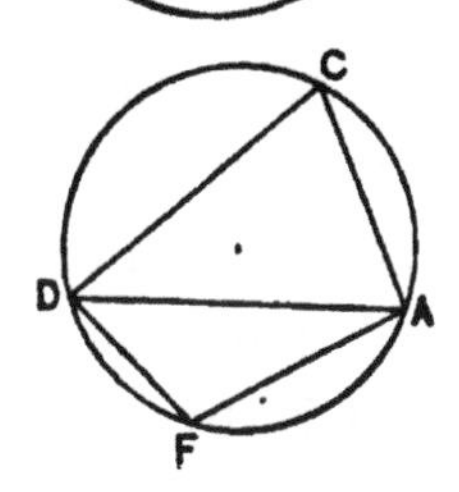

Cor. 3. An angle ACD inscribed in a segment greater than a semicircle is acute, and an angle AFD inscribed in a segment less than a semicircle is obtuse: for the former is measured by an arc less than a quadrant and the latter by an arc greater than a quadrant.

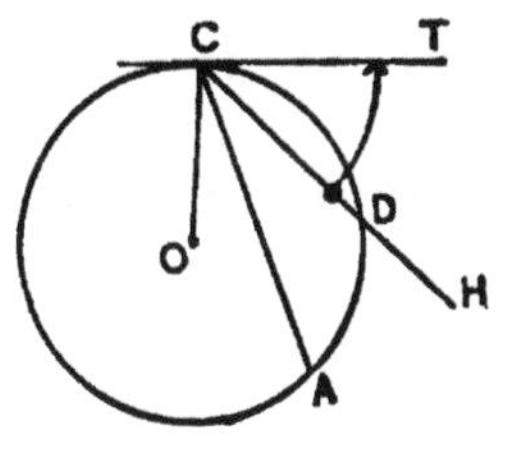

Cor. 4. The angle between a tangent and a chord through the point of contact is measured by half the intercepted arc. For, let CT be tangent to the circumference OC at C, and let CA be a chord. Then draw any secant CH through C cutting the circumference at D. The angle ACH is measused by half the arc AD, whatever may be the position of CH. Now let CH be revolved about C in the indicated direction; the point D will approach C, and the arc AD will become more nearly equal to the arc AC; finally when CH coincides with CT the point D coincides with C and the arc AD with the arc AC; hence, the angle ACT is measured by half the arc AC.

PROPOSITION XV. THEOREM.

If two secants intersect within a circumference their included angle is measured by half the sum of the intercepted arcs; if they intersect without the circumference their included angle is measured by half the difference of the intercepted arcs.

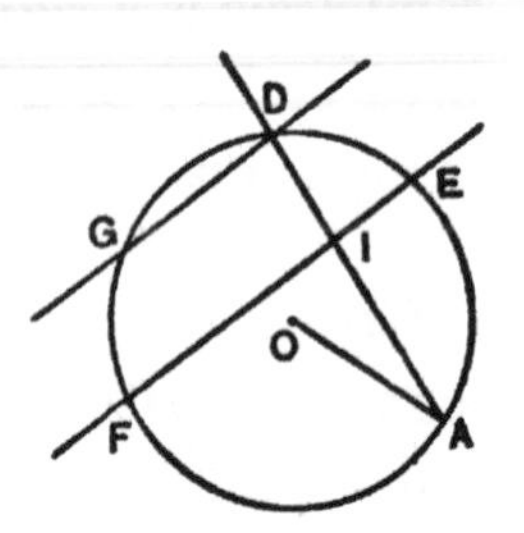

1°. Let the secants AD and EF intersect at a point I within the circumference OA, and let DG be drawn parallel to EF.

The angle AIF is equal to ADG, because they are opposite exterior and interior angles (P. 21, *Cor.*, B. I); but ADG, and consequently AIF is measured by half the arc AFG, that is, by half the sum of the arcs AF and FG (P. 14); but the arc FG is equal to DE (P. 9); hence, the angle AIF is measured by half the sum of the arcs AF and DE, *which was to be proved.*

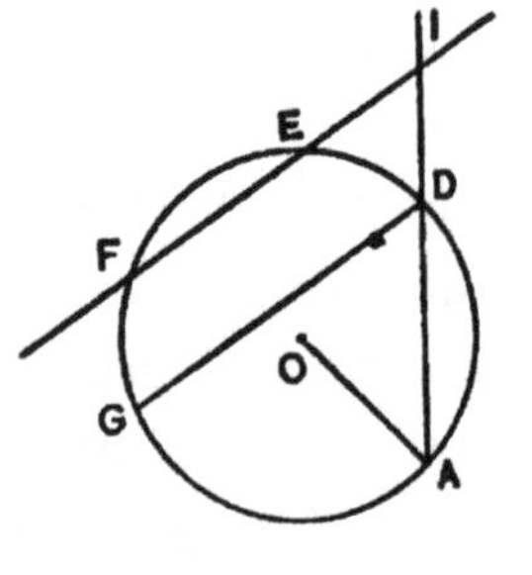

2°. Let the secants AD and EF intersect at a point I without the circumference OA, and let DG be drawn parallel to EF.

The angle I is equal to ADG because they are opposite exterior and interior angles; the angle ADG, and consequently the angle I is measured by half the arc AG, that is, by half the difference of the arcs AF and FG; but FG is equal to DE because they lie between the parallels DG and EF; hence, the angle I is measured by half the difference of the arcs AF and DE, *which was to be proved.*

Exercises on Book II.

Let the student demonstrate the following theorems:

1°. *A straight line cannot cut a circumference in more than two points.*

2°. *In equal circles, the greater of two chords subtends the greater arc, and conversely.*

3°. *In equal circles, the less of two chords is at a greater distance from the centre.*

4°. *The angle between two tangents that intersect is measured by half the difference of the intercepted arcs.*

5°. *The sum of two opposite angles of an inscribed quadrilateral is equal to two right angles.*

6°. *An inscribed parallelogram is a rectangle.*

7°. *An inscribed quadrilateral formed by joining the extremities of two perpendicular diameters is a square.*

8°. *If circumferences are constructed on the equal sides of an isosceles triangle as diameters, they will intersect at the middle of the base.*

9°. *If two intersecting chords make equal angles with the diameter through their common point, they are equal.*

APPLICATION OF PRECEDING PRINCIPLES TO THE SOLUTION OF PROBLEMS.

64. The problems treated of in Elementary Geometry are those whose elements are the straight line and the circle, and the only instruments actually required for their solution are the *ruler*, and the *compasses* or *dividers*. The ruler is used to guide a pencil, or pen, in drawing a straight line, and the dividers are used for guiding a pencil or pen in drawing a circle. The student will find it convenient to have a *scale of equal parts*, a *triangular ruler*, and a *semicircular protractor.* It is assumed that the pupil understands the methods of drawing a straight line through two points, of laying off a line equal to a given line, and of drawing a circumference from a given centre with a given radius.

PROBLEM I.

To bisect a given line by a perpendicular.

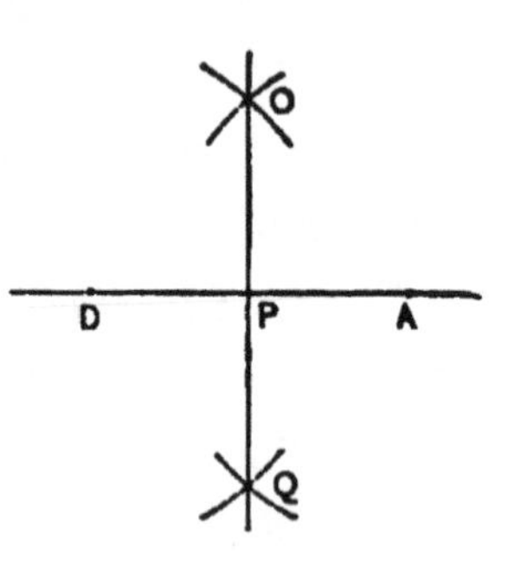

Let AD be the given line. From A and D as centres, and with a radius greater than half of AD, draw two arcs intersecting at O and Q; then draw the line OQ, cutting AD at P. The line OQ is perpendicular to AD at its middle point P, because two of its points, O and Q, are each equally distant from A and D (P. 16, *Cor.* 2, B. 1).

PROBLEM II.

To erect a perpendicular to a line at a given point.

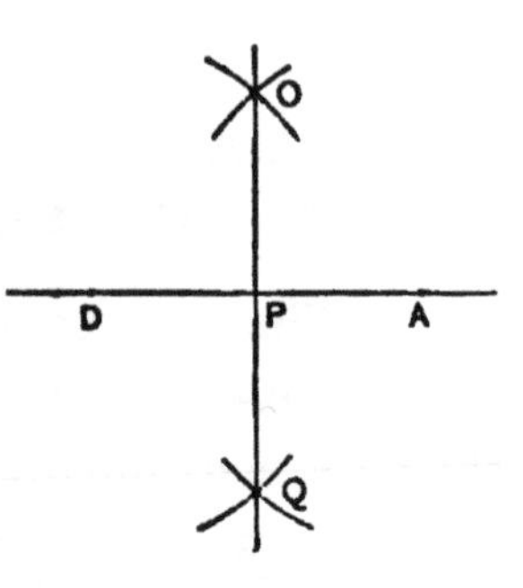

1°. Let P be any point of the given line. Assume a point A, at a convenient distance from P, and lay off PD equal to PA; then bisect AD by the line OQ, as in Prob. I, and OQ will be the required perpendicular.

2°. Let P be near one extremity of the line MN. Take a point O so that

OP shall make an acute angle with NM; with O as a centre and OP as a radius draw an arc meeting MN at A and the prolongation of AO at D; then draw PD and it will be the required perpendicular. For, the angle APD is inscribed in a semicircle and is consequently a right angle (P. 14, *Cor.* 2).

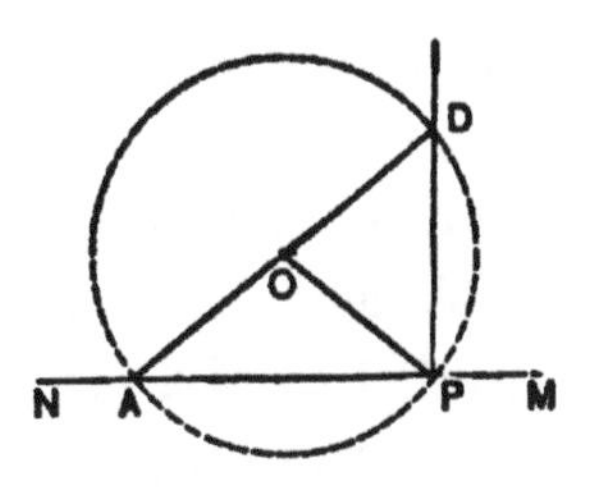

Problem III.

To draw a perpendicular to a given line from a point without.

Let P be a point lying without the line MN. From P as a centre, with a radius greater than the shortest distance from P to MN, draw an arc cutting MN at A and D; from A and D as centres, with equal radii, draw two arcs intersecting at Q, and join P and Q. The line PQ having two of its points each equally distant from A and D, is perpendicular to MN (P. 16, *Cor.* 2, B. 1).

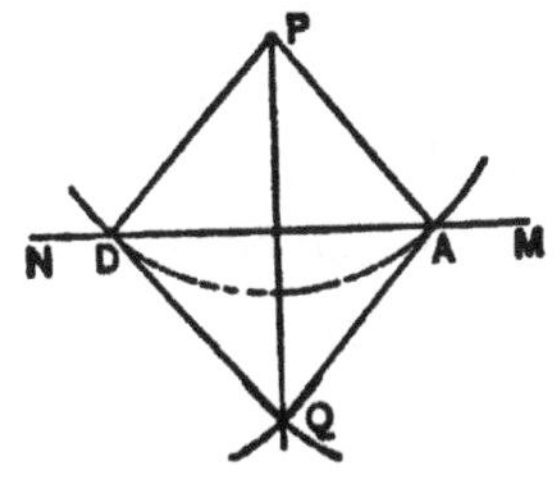

Problem IV.

To draw a line through a given point parallel to a given line.

1°. Let P be the given point and MN the given line. From P as a centre, with a radius greater than the shortest distance from P to MN, draw an indefinite arc AL, and from A as a centre with the same radius draw the arc PD; draw also the chord of PD; then from A as a centre, with a radius equal to DP, draw an arc intersecting AL at Q, and draw PQ; this will be parallel to MN. For, the arcs AQ and PD are equal, because they are arcs of equal circles subtended by equal chords (P. 4); the angles QPA and PAD are therefore equal (P. 3); hence, PQ is parallel to MN (P. 20, *Cor.*, B. 1).

2°. The problem may be solved by means of the triangular ruler as follows:

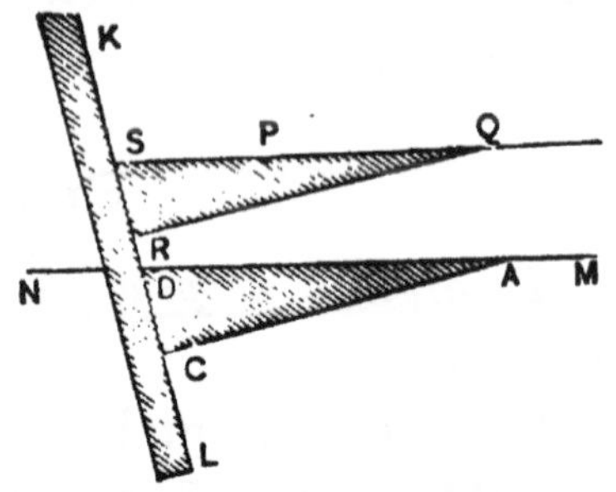

Let KL be an ordinary ruler and ACD a triangular ruler, right-angled at C. Place the edge DA of the triangular ruler so as to coincide with MN, and press the ruler KL against it so as to touch the edge CD; then hold the ruler KL fast and move the triangular ruler along it till the edge passes through P; the line SP is parallel to MN, for the angles QSR and ADC are equal (P. 20, *Cor.*, B. 1).

Problem V.

To construct an angle at a given point equal to a given angle.

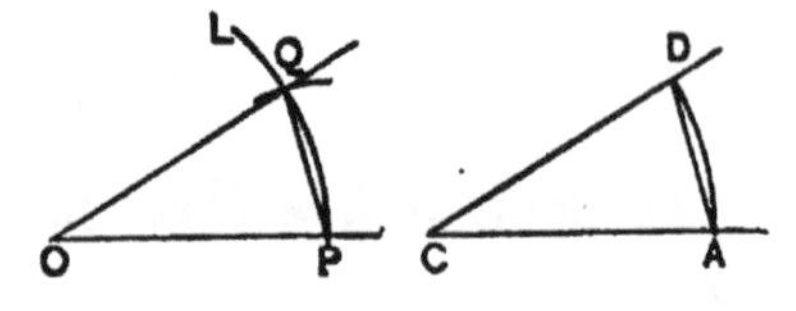

Let ACD be the given angle, and OP be an indefinite line through the given point O. With C as a centre, and with any convenient radius, CA, draw the arc AD, also draw its chord; then from O as a centre, with the same radius, draw the indefinite arc PL; from P as a centre, with AD as a radius, draw an arc cutting PL at Q; draw OQ and also the chord PQ. Then is the angle POQ equal to the angle ACD; for, the arcs PQ and AD are equal (P. 4), and consequently the corresponding angles are equal (P. 3).

PROBLEM VI.

To bisect a given angle.

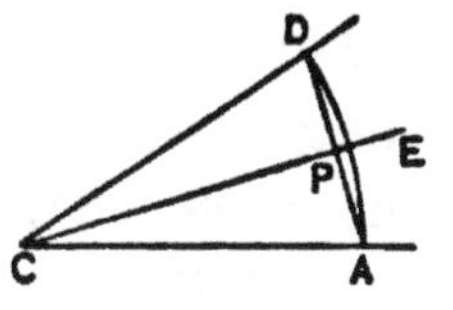

Let ACD be the given angle. From C as a centre, with any radius, CA, draw the arc AD; also draw the chord AD; then bisect the chord AD by the perpendicular EP (Prob. 1); EP will pass through C and will bisect the angle ACD (P. 12, *Cor.* 2, B. 1).

PROBLEM VII.

To find the third angle of a triangle when the other angles are known.

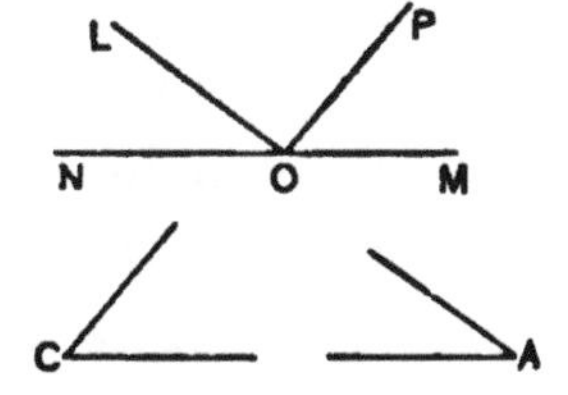

Let A, and C, be two angles of a triangle. At any point, O, of the indefinite line MN, construct the angle NOL equal to A, and MOP equal to C; then will POL be the required angle. For POL is the supplement of the sum of the angles NOL and MOP (P. 3, *Cor.* 2, B. 1), and consequently it is the supplement of the sum of the angles A and C; it is therefore the required angle (P. 24, *Cor.* 1, B. 1).

PROBLEM VIII.

To construct a triangle when its sides are given.

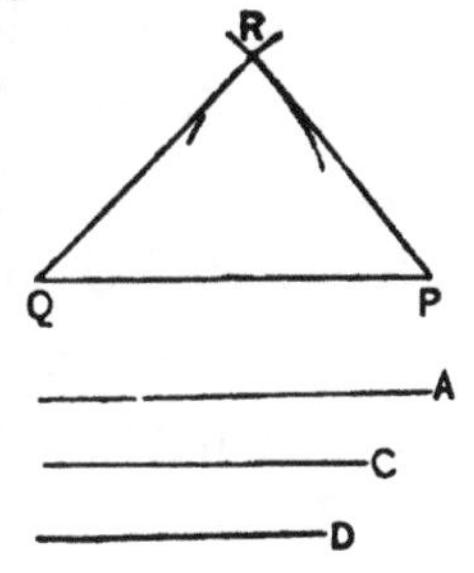

Let A, C, and D be the given sides. On an indefinite line lay off PQ equal to A; from P and Q as centres, with radii equal to C, and D, draw two arcs intersecting at R; then draw QR and PR, the triangle PQR is the required triangle.

PROBLEM IX.

To construct a triangle when two sides and their included angle are given.

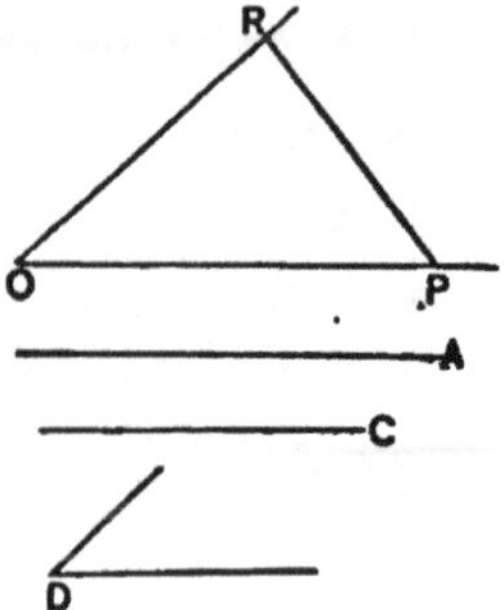

Let A, and C, be the given sides and D, the given angle. At any point, O, construct the angle POR equal to D, on one side lay off OP equal to A and on the other side lay off OR equal to C, and draw RP; the triangle OPR, is the required triangle.

PROBLEM X.

To construct a triangle when two angles and their included side are given.

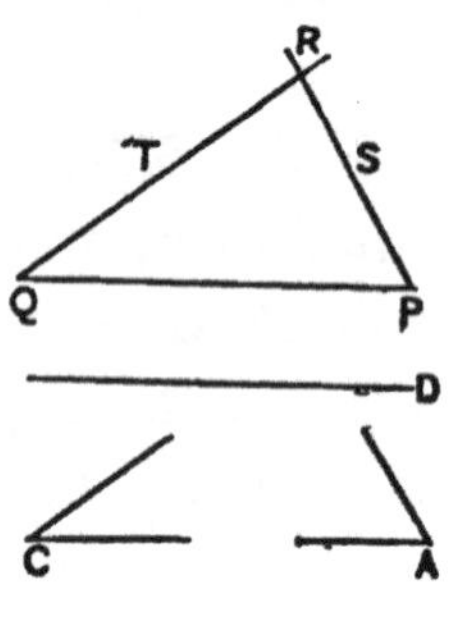

Let A, and C, be the given angles and let D be their included side. On an indefinite line lay off PQ equal to D; at P and Q construct angles respectively equal to A and C and prolong their sides till they meet at R; the triangle PQR is the required triangle.

PROBLEM XI.

To construct a triangle when two sides and the angle opposite one are given.

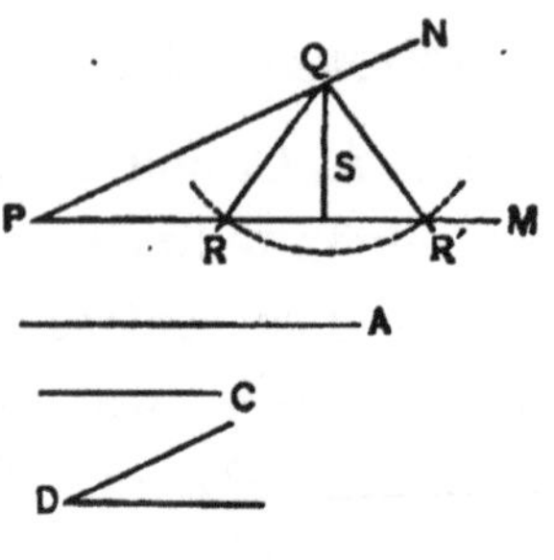

Let A and C be the given sides and let D be the angle opposite C.

Draw the indefinite line PM and at P construct an angle equal to D; on PN lay off PQ equal to A; from Q as a centre with a radius equal C, draw an arc cutting PM at R and R′; then join R and Q and also R′ and Q.

There may be several cases:

1°. The angle D may be acute and the side C may be less than the side A; in this case denote the perpendicular distance from Q to PM by S. Then if C is greater than S, there will be two solutions as in the figure; if C is equal to S, there will be but one solution; if C is less than S, there will be no solution, that is, the problem is impossible.

2°. The angle D may be acute and C may be greater than A; in this case one point of intersection will fall on one side of P and one on the other side of P, that is, on the prolongation of MP, and there will, therefore, be but one solution.

3°. The angle D may be a right angle, or an obtuse angle; in this case if C is greater than A there will be one solution, but if C is equal to, or less than, A, there will be no solution.

PROBLEM XII.

To draw a tangent to a circumference at a given point.

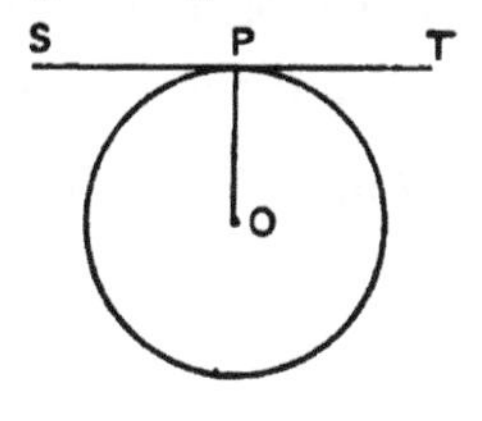

Let P be any point of the circumference OP, and let OP be the radius through that point. Draw ST perpendicular to OP at P; then will ST be the required tangent (P. 8).

PROBLEM XIII.

To draw a tangent to a circumference and parallel to a given line.

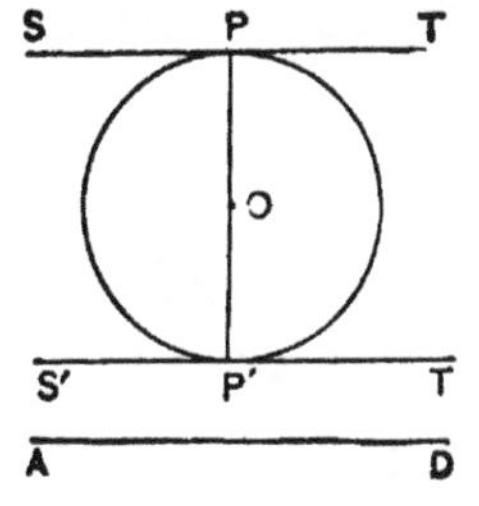

Let AD be the given line, and let the circumference OP, be the given circumfer-ference. Draw the diameter PP′ perpendicular to AD, and at the points in which this meets the circumference draw ST and S′T′ perpendicular to PP′; these will be parallel to AD (P. 19, B. 1) and tangent to the circumference OP (P. 8).

There are two solutions.

PROBLEM XIV.

To draw a tangent to a given circumference from a point without the circumference.

Let P be a point without the circumference OM. Draw PO, and on it, as a diameter, construct a circumference cutting the given circumference at M and N; then draw PM and PN; they will pass through P and be tangent to the circumference OM. For draw the radii OM and ON; the angles PMO and PNO are right angles because they are inscribed in semicircles; hence, PM and PN are perpendicular to the radii OM, and ON, at M, and N; they are therefore tangent to the circumference OM; the problem has *two* solutions.

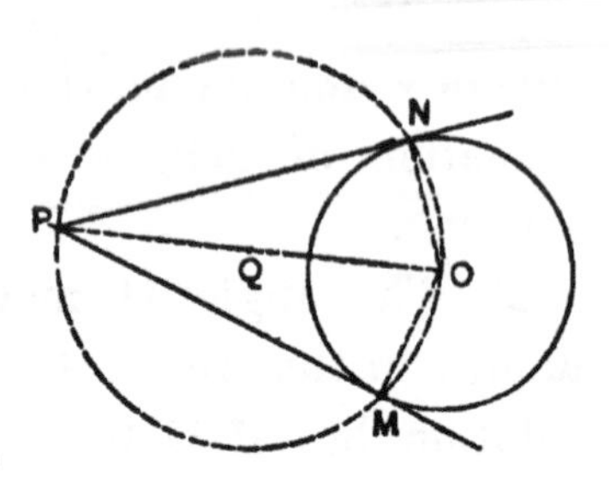

Cor. The right-angled triangles PNO and PMO have a common hypothenuse PO, and the angles OPN, and OPM are equal, because they are measured by halves of equal arcs ON and OM; the triangles are therefore equal in all their parts (P. 17, B. 1); hence, the tangents PN and PM are equal, and the line PO bisects the angle between them, also the angle NOM, between the radii to the point of contact.

PROBLEM XV.

To inscribe a circle in a given triangle.

Let ACD be the given triangle. Bisect the angles A and C, and prolong their bisectrices till they meet at O. The point

O is equally distant from the three sides of the triangle. For draw OM, ON, and OP, respectively perpendicular to AC, CD, and DA; because AO is the bisectrix of the angle A, OM and OP are equal, and because CO is the bisectrix of the angle C, OM and ON are equal (P. 18, B. 1); hence, the three perpendiculars are all equal. From O as a centre, with OM as a radius, describe a circumference and it will be tangent to all the sides of the given triangle; hence, the circle OM is inscribed in the triangle ACD.

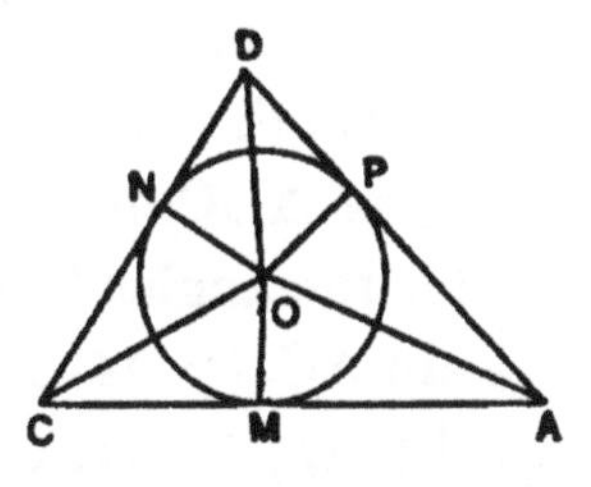

PROBLEM XVI.

To circumscribe a circle about a given triangle.

Let ACD be the given triangle. Bisect AC and AD by perpendiculars (Prob. 1), and prolong them till they meet at O; from O as a centre with the radius OA describe a circumference and it will pass through all the vertices of the given triangle (P. 7); hence, the circle OA is circumscribed about the given triangle.

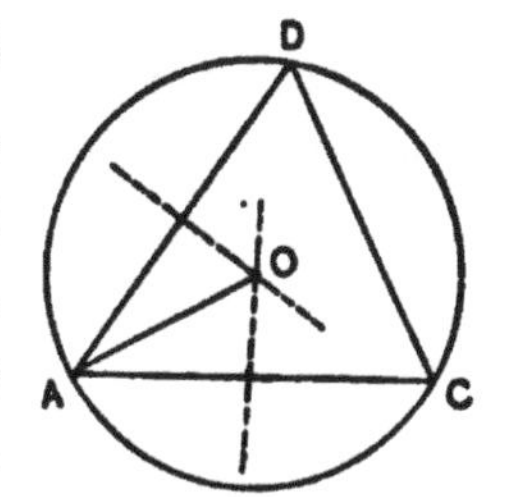

PROBLEM XVII.

To construct a segment on a given line that shall contain a given angle.

Let AC be the given line and Q the given angle. At A construct the angle CAT equal to Q, and draw AM perpendicular to AT at A; bisect AC by a perpendicular and prolong it till it meets AM at O; with O

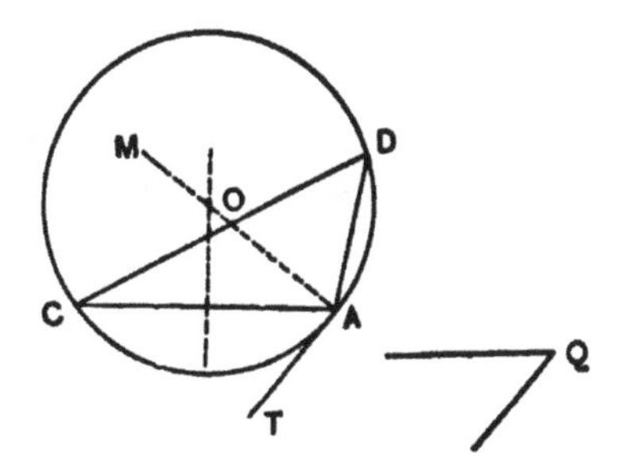

as a centre and OA as a radius, draw a circumference; it will pass through C and be tangent to AT at A. Then will any angle, as ADC, inscribed in the segment CDA be equal to Q; for the angles CDA and CAT are each measured by half the arc AC (P. 14, and P. 14, *Cor.* 4); but the angle CAT is equal to Q; hence, CDA is also equal to Q.

ADDITIONAL PROBLEMS.

Let the pupil solve the following additional problems, using the annexed diagrams as guides to their solution.

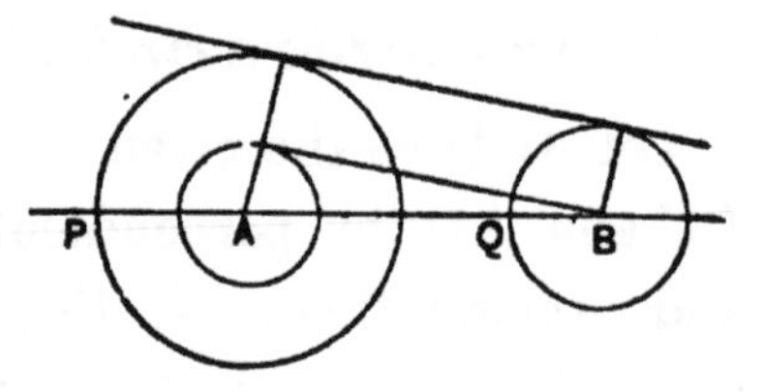

PROB. XVIII. *Draw a line tangent to two given circumferences* AP *and* BQ.

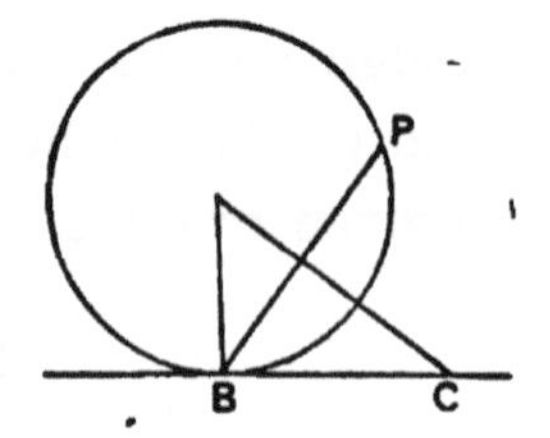

PROB. XIX. *Through a given point* P *draw a circumference tangent to a given line* CB *at a given point* B.

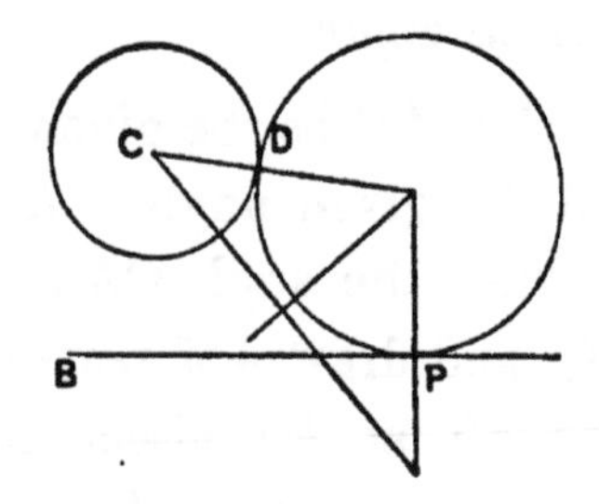

PROB. XX. *Draw a circumference tangent to a given circumference* CD *and to a given line* BP *at a given point* P.

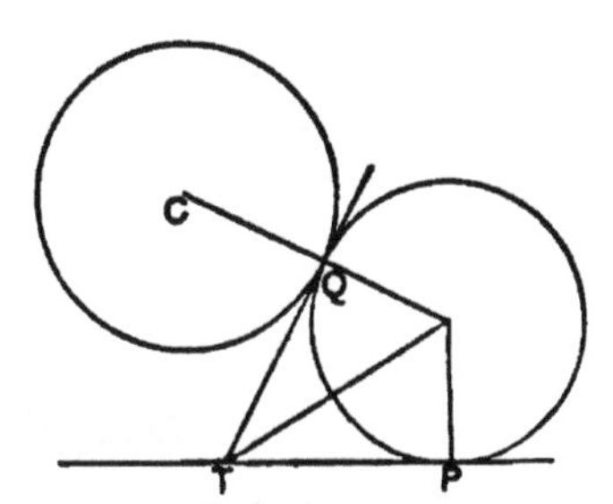

PROB. XXI. *Draw a circumference tangent to a given line* TP *and to a given circumference* CQ *at a given point* Q.

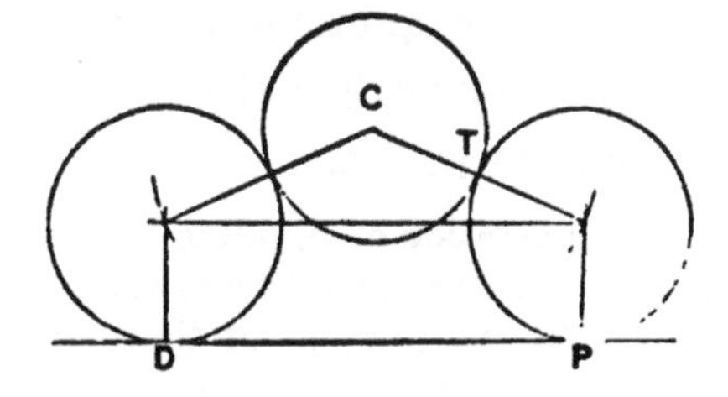

PROB. XXII. *Draw a circumference with a given radius tangent to a given line* DP *and to a given circumference* CT.

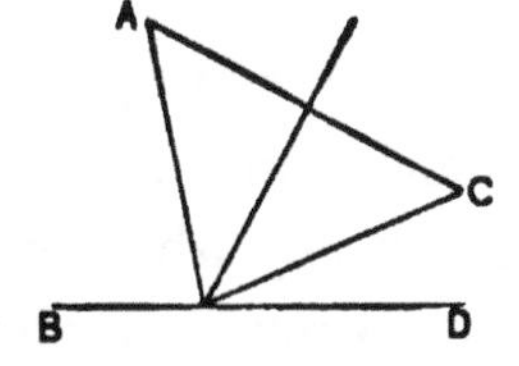

PROB. XXIII. *From two given points* A *and* C *draw two lines meeting on a given line* BD *and equal to each other.*

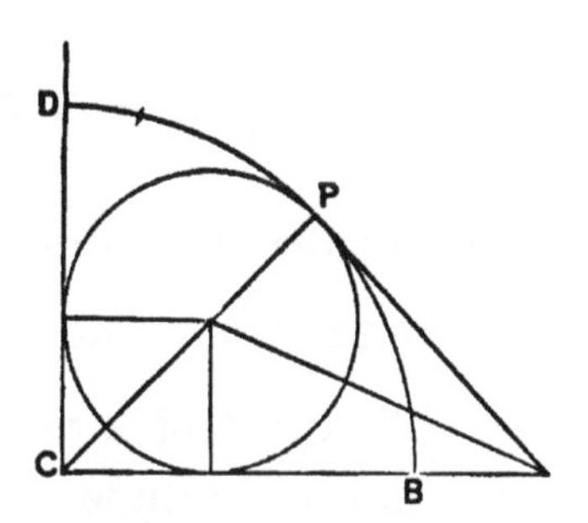

PROB. XXIV. *Inscribe a circle in a given quadrant* BCDP.

NOTE.—The figures in the margin are intended to suggest the methods of solution.

BOOK III.

RATIOS, PROPORTIONS, AND COMPARISON OF LINES.

Definitions and Principles.

65. The **ratio** of one magnitude to another of the same kind is the quotient of the *second* by the *first.* The first magnitude is called the **antecedent** and the second is called the **consequent.**

The operation of dividing one magnitude by another consists in dividing the *number of times* that any assumed unit is contained in the former by the number of times it is contained in the latter. If the magnitudes are commensurable the ratio is *exact;* if not, it is only *approximate.*

Let P and Q be two incommensurable magnitudes of the same kind. Suppose P to be divided into m equal parts, m being any whole number, and let Q contain n of these parts with a remainder less than one of them; then will,

$$\frac{Q}{P}=\frac{n}{m} \text{ to within less than } \frac{1}{m}.$$

The greater the value of m, the more nearly will $\frac{n}{m}$ approach to the true value of the ratio of P to Q.

66. The **limit** of a varying magnitude, is a value *towards* which that magnitude may approach, *with* which it may coincide, but *beyond* which it cannot pass.

It is assumed that whatever is true of all values of a varying magnitude is true of its limit.

67. Two quantities that differ by less than any assignable quantity are said to be equal to each other.

PROPOSITION I. THEOREM.

If the ratio of one magnitude to another of the same kind is equal to $\frac{n}{m}$, *to within less than* $\frac{1}{m}$, *for all values of* m, *the true value of the ratio is the limit of* $\frac{n}{m}$.

Let $\frac{Q}{P} = \frac{n}{m}$, *to within less than* $\frac{1}{m}$, for all values of m.

As m increases, $\frac{1}{m}$ diminishes and the value of $\frac{n}{m}$ becomes more nearly equal to $\frac{Q}{P}$; when m becomes greater than any assignable number, or *infinite*, $\frac{1}{m}$ becomes less than any assignable number or *zero*. Hence the true value of $\frac{Q}{P}$ is the limiting value of $\frac{n}{m}$, *which was to be proved.*

Cor. If each of the two ratios, $\frac{Q}{P}$ and $\frac{S}{R}$, is equal to $\frac{n}{m}$, *to within less than* $\frac{1}{m}$, for all values of m, the two ratios are equal; for, each is equal to the limiting value of $\frac{n}{m}$.

Additional Definitions and Principles.

68. **A proportion** is an expression of equality between two ratios.

A proportion may be written in either of two ways. Thus, if the ratio of a to b is equal to the ratio of c to d, the equality may be indicated by either of the following expressions: $\frac{b}{a} = \frac{d}{c}$, or $a : b :: c : d$; . . (1) either of these expressions may be read *a is to b as c is to d.*

The first and third terms of a proportion are **antecedents**; the second and fourth terms are **consequents**; thus in (1) a and c are antecedents, b and d are consequents.

The first and fourth terms are **extremes**; the second and third terms are **means**; thus, a and d are extremes, b and c are means.

The first ratio is called the **first couplet**; the second ratio is called the **second couplet**; thus, $a : b$ is the first, $c : d$, the second couplet.

The fourth term is said to be a **fourth proportional** to the other three; thus, d is a fourth proportional to a, b, and c.

If the second term is equal to the third term, either is a **mean proportional** between the other two; in this case, there are but three different terms, and the last is a **third proportional** to the other two; thus in the proportion

$$a : b :: b : c,$$

b is a mean proportional between a and c, and c is a third proportional to a and b.

69. Two varying quantities are **reciprocally proportional** when their product is equal to a fixed quantity; thus, if $xy = m$, x and y are reciprocally proportional.

70. Equimultiples of two quantities are the results obtained by multiplying both by the same quantity; thus, ma and mb are equimultiples of a and b, no matter what may be the value of m.

71. A continued proportion is an expression of continued equality among several ratios; thus,

$$a : b :: c : d :: e : f,$$

is a continued proportion. It is read, a is to b, as c is to d, as e is to f. This proportion may also be written,

$$a : c : e :: b : d : f,$$

in which case it is to be read as before.

Transformations.

72. If antecedents are made consequents and consequents antecedents, a proportion is said to be transformed **by inversion.**

73. If antecedent is compared with antecedent and consequent with consequent, the proportion is said to be transformed **by alternation.**

74. If the sum of antecedent and consequent, in each couplet, is compared with the corresponding antecedent, or with the corresponding consequent, the proportion is said to be transformed **by composition.**

75. If the difference of antecedent and consequent, in each couplet, is compared with the corresponding antecedent, or with the corresponding consequent, the proportion is said to be transformed **by division.**

PROPOSITION II. THEOREM.

If four quantities are in proportion, the product of the means is equal to the product of the extremes.

Assume the proportion,

$$a : b :: c : d, \text{ whence } \frac{b}{a} = \frac{d}{c}; \quad . \quad . \quad . \quad (1)$$

Multiplying both members of (1) by ac, and reducing, we have

$$bc = ad, \quad . \quad . \quad . \quad . \quad . \quad . \quad . \quad . \quad . \quad (2)$$

which was to be proved.

Cor. If $b = c$, there are but three different terms in the proportion, and we then have the square of the mean equal to the product of the extremes.

PROPOSITION III. THEOREM.

If the product of two quantities is equal to the product of two other quantities, the first two may be made the means and the other two the extremes of a proportion.

Assume the equation,

$$bc = ad, \quad . \quad . \quad . \quad . \quad . \quad . \quad . \quad . \quad . \quad (1)$$

dividing both members of (1) by ac, and reducing, we have,

$$\frac{b}{a} = \frac{d}{c}, \quad \text{or,} \quad a : b :: c : d, \quad . \quad . \quad . \quad . \quad (2)$$

which was to be proved.

Cor. 1. Proportion (2) may be transformed in any way that will satisfy equation (1); we may therefore write the following proportions:

$$b : a :: d : c, \quad \text{and} \quad a : c :: b : d;$$

hence, if four quantities are in proportion, they are also in proportion *by inversion*, and *by alternation.*

Cor. 2. If a is greater than b, c must be greater than d; if a is equal to b, c must be equal to d; if a is less than b, c must be less than d.

PROPOSITION IV. THEOREM.

If a couplet in each of two proportions is the same, the remaining couplets will form a proportion.

Assume the proportions,

$$a : b :: c : d, \quad \text{whence} \quad \frac{b}{a} = \frac{d}{c}, \quad . \quad . \quad . \quad (1)$$

and

$$a : b :: e : f, \quad \text{whence} \quad \frac{b}{a} = \frac{f}{e}; \quad . \quad . \quad . \quad (2)$$

placing the second members of (1) and (2) equal, we have,

$$\frac{d}{c} = \frac{f}{e}, \quad \text{or} \quad c : d :: e : f, \quad . \quad . \quad . \quad . \quad (3)$$

which was to be proved.

Cor. If the antecedents of two proportions are equal, each to each, the corresponding consequents will form a proportion; for, if we have,

$$a : c :: b : d, \quad \text{and} \quad a : e :: b : f,$$

we have by alternation,

$$a : b :: c : d, \quad \text{and} \quad a : b :: e : f,$$

and consequently,

$$c : d :: e : f.$$

PROPOSITION V. THEOREM.

If four quantities are in proportion, they are in proportion by composition.

Assume the proportion,

$$a : b :: c : d, \quad \text{whence} \quad \frac{b}{a} = \frac{d}{c}; \quad . \quad . \quad . \quad (1)$$

adding each member of the preceding equation to 1, we have

$$1 + \frac{b}{a} = 1 + \frac{d}{c}, \quad \text{or} \quad \frac{a+b}{a} = \frac{c+d}{c}; \quad . \; . \; . \quad (2)$$

whence the proportion,

$$a : a + b :: c : c + d. \quad . \; . \; . \; . \; . \; . \quad (3)$$

From (1) and (3) we have (P. 4, *Cor.*),

$$b : a + b :: d : c + d. \quad . \; . \; . \; . \; . \; . \quad (4)$$

From (3) and (4) we have, by inversion,

$$a + b : a :: c + d : c, \quad \text{and} \quad a + b : b :: c + d : d,$$

which was to be proved.

PROPOSITION VI. THEOREM.

If four quantities are in proportion, they are in proportion by division.

Assume the proportion,

$$a : b :: c : d, \quad \text{whence} \quad \frac{b}{a} = \frac{d}{c}, \quad . \; . \; . \quad (1)$$

subtracting each member of the preceding equation from 1, we have,

$$1 - \frac{b}{a} = 1 - \frac{d}{c}, \quad \text{or} \quad \frac{a-b}{a} = \frac{c-d}{c};$$

whence,

$$a : a - b :: c : c - d. \quad . \; . \; . \; . \; . \; . \quad (3)$$

From (1) and (3) we have (P. 4, *Cor.*),

$$b : a - b :: d : c - d. \quad . \; . \; . \; . \; . \; . \quad (4)$$

From (3) and (4) we have, by inversion,

$$a - b : a :: c - d : c, \quad \text{and} \quad a - b : b :: c - d : d,$$

which was to be proved.

PROPOSITION VII. THEOREM.

Equimultiples of two quantities are proportional to the quantities themselves.

Let ma and mb be any equimultiples of a and b; we have,

$$\frac{mb}{ma} = \frac{b}{a}, \quad \text{or} \quad ma : mb :: a : b,$$

which was to be proved.

Cor. Both terms of either couplet of a proportion may be multiplied or divided by the same quantity.

PROPOSITION VIII. THEOREM.

In a continued proportion, the sum of all the antecedents is to the sum of all the consequents as any antecedent is to the corresponding consequent.

Assume the continued proportion,

$$a : b :: c : d :: e : f, \quad \text{or,} \quad \frac{b}{a} = \frac{d}{c} = \frac{f}{e}; \quad . \; . \; (1)$$

we have

$$\frac{b}{a} = \frac{b}{a}, \quad \text{or,} \quad ab = ba, \quad . \; . \; . \; . \; (2)$$

$$\frac{b}{a} = \frac{d}{c}, \quad \text{or,} \quad cb = da, \quad . \; . \; . \; . \; (3)$$

$$\frac{b}{a} = \frac{f}{e}, \quad \text{or,} \quad eb = fa; \quad . \; . \; . \; . \; (4)$$

adding (2), (3), and (4), member to member, and factoring, we have,

$$(a + c + e)\,b = (b + d + f)\,a \quad . \; . \; . \; (5)$$

From (5), we have, (P. 3),

$$a + c + e : b + d + f :: a : b.$$

which was to be proved.

Scho. The demonstration may be extended to cases in which there is any number of couplets.

PROPOSITION IX. THEOREM.

If the corresponding terms of two proportions are multiplied together, the products will be in proportion.

Assume the proportions,

$$a : b :: c : d, \text{ or, } \frac{b}{a} = \frac{d}{c} \quad . \quad . \quad . \quad . \quad (1)$$

and

$$e : f :: g : h, \text{ or, } \frac{f}{e} = \frac{h}{g} \quad . \quad . \quad . \quad . \quad (2)$$

Multiplying (1) and (2), member by member, we have,

$$\frac{bf}{ae} = \frac{dh}{cg}, \text{ or, } ae : bf :: cg : dh,$$

which was to be proved.

Cor. If four terms are in proportion, their squares are in proportion.

Scho. The principle above demonstrated may be extended to any number of proportions.

PROPOSITION X. THEOREM.

If lines parallel to any side of a triangle divide one of the other sides into equal parts, they will also divide the remaining side into equal parts.

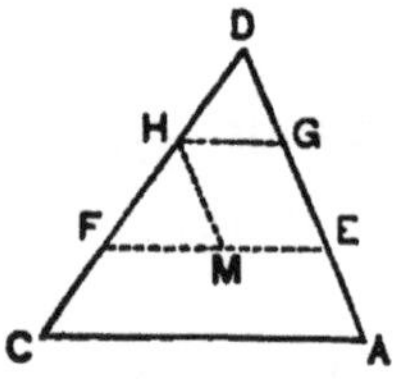

Let EF and GH, parallel to the side AC of the triangle ACD, divide AD into equal parts, AE, EG, and GD; then will they divide CD into equal parts, CF, FH, and HD.

Through H draw HM parallel to DA.

Because MH and EG are parallels between parallels they are equal (P. 26, *Cor.* 1, B. 1); but EG is equal to GD; hence, MH is equal to GD. Because MH and GD are parallel, the angles FHM and HDG are opposite exterior and interior angles; they are therefore equal. Because the angles HMF and DGH have their sides parallel and lying in the same direction they are equal (P. 23, B. 1). The triangles MFH and GHD have therefore two angles and their included side equal, each to each; consequently the triangles are equal in all their parts (P. 10, B. 1): hence, FH is equal to HD. In like manner, it may be shown that CF is equal to HD; the parts CF, FH, and HD are therefore equal to each other, *which was to be proved.*

Scho. The demonstration may be extended to the case in which parallels divide AD into any number of equal parts.

PROPOSITION XI. THEOREM.

A line parallel to any side of a triangle divides the other sides into proportional parts.

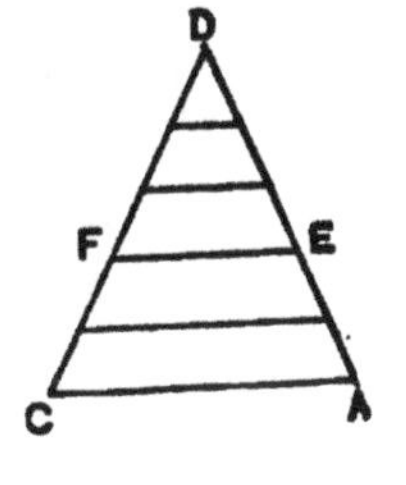

Let EF be parallel to AC, in the triangle ACD; then will AE and ED be proportional to CF and FD.

1°. Let AE and ED be to each other as two whole numbers, say as 2 is to 3. Divide AE into two, and ED in three equal parts; these will all be equal to each other; then through each point of division draw a line parallel to AC. These lines will divide CF into two and FD into three parts, all of which will be equal (P. 10, *Scho.*); hence,

$$CF : FD :: 2 : 3;$$

but,

$$AE : ED :: 2 : 3;$$

hence (P. 4), we have,

$$AE : ED :: CF : FD,$$

which was to be proved.

2°. Let AE and ED be incommensurable. Divide AE into m equal parts, m being any whole number; let one of these parts be applied to ED and suppose that it is contained in ED n times with a remainder less than that part; then will

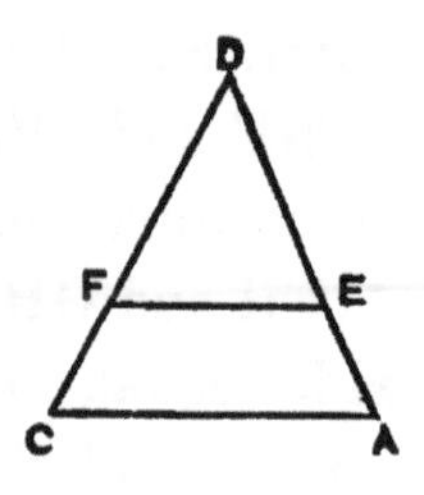

$$\frac{ED}{AE} = \frac{n}{m} \text{ to within less than } \frac{1}{m}. \quad . \quad . \quad . \quad (1)$$

Through the points of division on AD draw lines parallel to AC; the lines between A and E will divide CF into m equal parts and those between E and D will divide FD into n equal parts with a remainder less than one of these parts; hence,

$$\frac{FD}{CF} = \frac{n}{m} \text{ to within less than } \frac{1}{m}. \quad . \quad . \quad . \quad (2)$$

Because m is any whole number the true values of the ratios $\frac{ED}{AE}$ and $\frac{FD}{CF}$ are equal to each other (P. 1, *Cor.*); hence,

$$\frac{ED}{AE} = \frac{FD}{CF}, \quad \text{or,} \quad AE : ED :: CF : FD \quad . \quad . \quad (3)$$

which was to be proved.

Cor. 1. Transforming proportion (3) by composition, remembering that AE + ED = AD and CF + FD = CD, we have

$$AD : ED :: CD : FD, \quad . \quad . \quad . \quad . \quad (4)$$

and

$$AD : AE :: CD : CF; \quad . \quad . \quad . \quad . \quad (5)$$

whence, by alternation,

$$AD : CD :: ED : FD, \quad . \quad . \quad . \quad . \quad (6)$$

and

$$AD : CD :: AE : CF. \quad . \quad . \quad . \quad . \quad (7)$$

Cor. 2. The lines AC and EF are proportional to AD and

ED, and consequently to CD and FD. For, draw EK parallel to DC: then will KC be equal to EF, and we shall have (*Cor.* 1),

AD : ED :: AC : KC, or EF;

but, AD : ED :: CD : FD;

hence, AC : EF :: CD : FD.

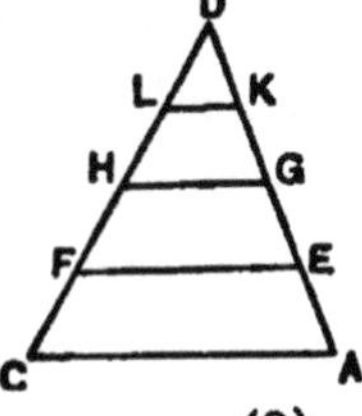

Cor. 3. If EF, GH, and KL are parallel to AC, they divide AD and CD proportionally. For, from (3), taken by inversion, we have,

ED : AE :: FD : CF, ... (8)

and from *Cor.* 2, we have,

ED : EG :: FD : FH; . . . . (9)

From (8) and (9), we have (P. 4, *Cor.*),

AE : EG :: CF : FH.

In like manner, we have,

EG : GK :: FH : HL,

and so on.

PROPOSITION XII. THEOREM.

If a line cuts two sides of a triangle proportionally it is parallel to the third side.

Let EF cut the sides AD and CD, of the triangle ACD, so that

AE : ED :: CF : FD; . . (1)

then is EF parallel to AC.

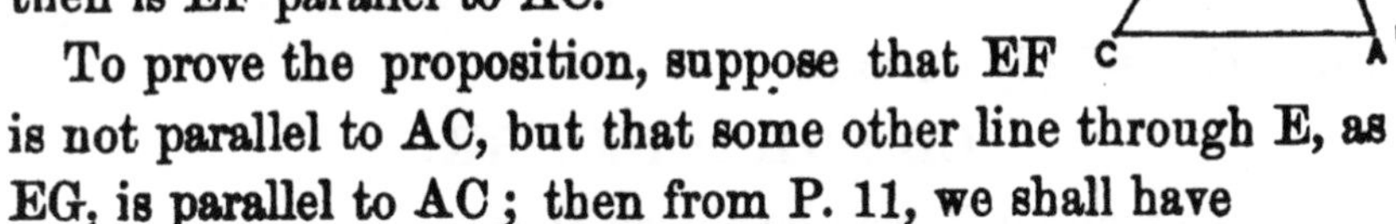

To prove the proposition, suppose that EF is not parallel to AC, but that some other line through E, as EG, is parallel to AC; then from P. 11, we shall have

AE : ED :: CG : GD. . . . . . (2)

Comparing (1) and (2), (P. 4), we have

CF : FD :: CG : GD.

Whence, by alternation,

CF : CG :: FD : GD. . . . . . . (3)

In proportion (3), CF is greater than CG and FD is less than GD, which is impossible (P. 3, *Cor.* 2); hence, the supposition that led to proportion (3) is false; its contradictory must therefore be true, that is, EF is parallel to AC, *which was to be proved.*

Definition.

76. If two lines cut a third line in two points, the part of the third line included between the first two is called an **intercept.**

PROPOSITION XIII. THEOREM.

If several lines radiating from a point cut two parallel lines the corresponding intercepts are proportional.

Let the lines AC, AD, AE, and AT, radiating from A, cut the parallel lines CT and GL; then we shall have

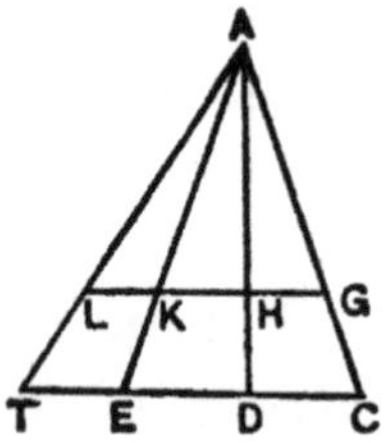

CD : DE : ET :: GH : HK : KL.

From the triangles CDA and GHA, we have (P. 11, *Cor.* 2),

DA : HA :: CD : GH, ... (1)

and from the triangles DEA and HKA we have

DA : HA :: DE : HK. . . . . . . (2)

From (1) and (2) we have

CD : GH :: DE : HK.

In like manner it may be shown that

DE : HK :: ET : KL.

Hence,

CD : DE : ET :: GH : HK : KL,

which was to be proved.

Scho. The demonstration holds good where there is any number of radiating lines.

PROPOSITION XIV. THEOREM.

The bisectrix of any angle of a triangle divides the opposite side into segments proportional to the adjacent sides.

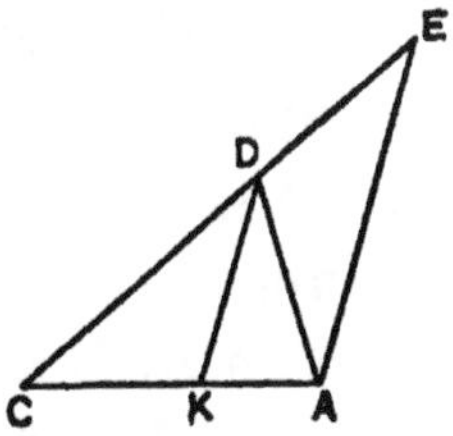

Let DK bisect the angle CDA of the triangle ACD; then

AK : KC :: AD : CD.

Prolong CD till DE is equal to DA, and draw EA. Because DE and DA are equal, the triangle ADE is isosceles and consequently the angles EAD and DEA are equal. But, CDA being an exterior angle of the triangle ADE is equal to the sum of the opposite interior angles EAD and DEA; consequently half of CDA, or CDK, is equal to the angle DEA; hence, DK is parallel to EA. We have, therefore, from P. 11,

AK : KC :: ED : DC.

Substituting AD for ED, we have,

AK : KC :: AD : CD,

which was to be proved.

Scho. The converse of this proposition may be proved in a manner entirely similar to that followed in P. 12.

Definitions.

77. Two polygons are **similar** when their corresponding angles are equal and when the sides about these angles taken in the same order are proportional. Corresponding lines of two similar polygons are called **homologous** lines; corresponding angles are called **homologous angles**.

PROPOSITION XV. THEOREM.

If the corresponding angles of two triangles are equal the triangles are similar.

In the triangles ACD and POR let the angles P, O, and R be respectively equal to the angles A, C, and D; then are the triangles similar.

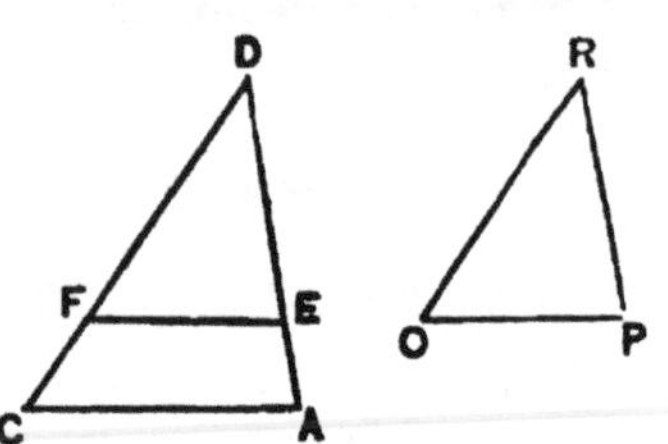

Because the corresponding angles are equal, it only remains to be shown that the sides taken in the same order are proportional.

Suppose DA to be greater than RP. Place the triangle POR on ACD so that the angle R shall coincide with D, P falling at E and O at F. Then because the angle DEF is equal to the angle A, EF is parallel to AC, and from P. 11, *Cor.* 2, we have

$$AC : CD : DA :: EF : FD : DE,$$

or, $$AC : CD : DA :: PO : OR : RP,$$

which was to be proved.

Cor. 1. If two triangles have two angles of the one respectively equal to two angles of the other, the triangles are similar.

Cor. 2. If two triangles have their corresponding sides parallel or perpendicular, each to each, the triangles are similar.

PROPOSITION XVI. THEOREM.

If two triangles have an angle in each equal and the including sides proportional, the triangles are similar.

In the triangles ACD and POR let the angles R and D be equal, and let the sides CD and DA be proportional to OR and RP; then are the triangles similar.

Place the triangle POR on the

triangle ACD so that the angle R shall coincide with D, the vertex P falling at E and the vertex O at F. Then, because DC and DA are proportional to DF and DE, EF is parallel to AC (P. 12); hence, the angle EFD, or O, is equal to C, and DEF, or P, is equal to A; the given triangles are therefore mutually equiangular, and consequently similar, *which was to be proved.*

Proposition XVII. Theorem.

If the three sides of one triangle are proportional to the three sides of another triangle, the triangles are similar.

In the triangles ACD and POR, let AC, CD, and DA be proportional to PO, OR, and RP; then are the triangles similar.

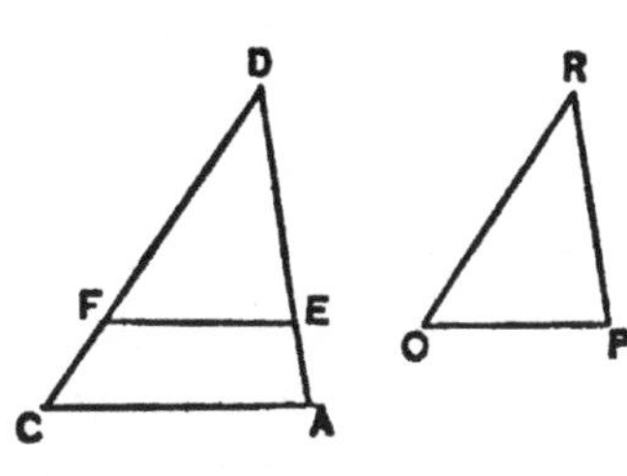

Lay off DE equal to RP, DF equal to RO, and draw EF. Then, because DA and DC are proportional to DE and DF, the line EF is parallel to AC (P. 12), and the triangles ACD and EFD are similar (P. 16). From P. 11, *Cor.* 2, we have

$$AC : CD :: EF : FD; \quad . \quad . \quad . \quad . \quad . \quad (1)$$

but by hypothesis, we have,

$$AC : CD :: PO : OR, \text{ or } FD. \quad . \quad . \quad . \quad (2)$$

Because proportions (1) and (2) have three corresponding terms equal, each to each, their other terms must be equal, that is, EF is equal to PO; hence, the triangles POR and EFD have their sides respectively equal; they are therefore equal in all their parts: but ACD and EFD are similar; hence, ACD and POR are similar, *which was to be proved.*

PROPOSITION XVIII. THEOREM.

If a line is drawn from the vertex of the right angle of a right-angled triangle perpendicular to the hypothenuse:

1°. *It will divide the triangle into two triangles that are similar to the given triangle and to each other;*

2°. *It will be a mean proportional between the segments into which it divides the hypothenuse;*

3°. *Either side about the right angle will be a mean proportional between the entire hypothenuse and the adjacent segment.*

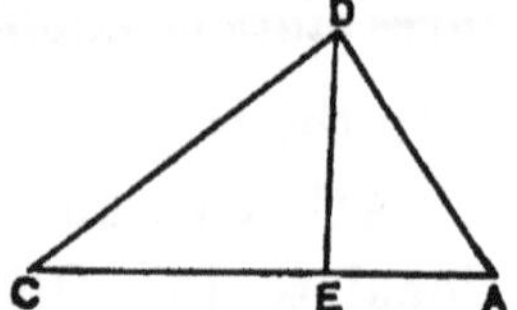

Let ACD be a triangle, right-angled at D, and let DE be perpendicular to the hypothenuse AC.

1°. The triangles DCE and ACD have the angle C common, and the angles CED and CDA equal, because both are right angles; hence, DCE is similar to ACD (P. 15, *Cor.* 1). For like reasons, the triangle ADE is similar to ACD; and because both DCE and ADE are similar to ACD, they are similar to each other, *which was to be proved.*

2°. Because DCE and ADE are similar, the angles DCE and ADE being homologous, we have,

$$AE : DE :: DE : CE,$$

which was to be proved.

3°. Because the triangles ACD and DCE are similar, the angle C being common, we have,

$$AC : CD :: CD : CE;$$

and because ACD and ADE are similar, the angle A being common, we have

$$AC : AD :: AD : AE,$$

which was to be proved.

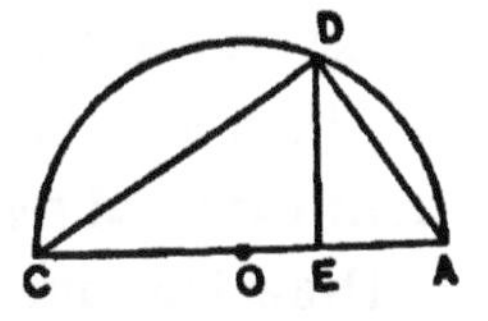

Cor. From any point D on a semicircumference let a perpendicular DE be drawn to the diameter, and also chords DA and DC to the extremities of the diameter; then will ACD be a right-angled triangle, and from what was proved above we infer that the perpendicular is a mean proportional between the segments into which it divides the diameter, and that each chord is a mean proportional between the whole diameter and the adjacent segment.

PROPOSITION XIX. THEOREM.

If two chords intersect within a circumference, they divide each other into segments that are inversely proportional.

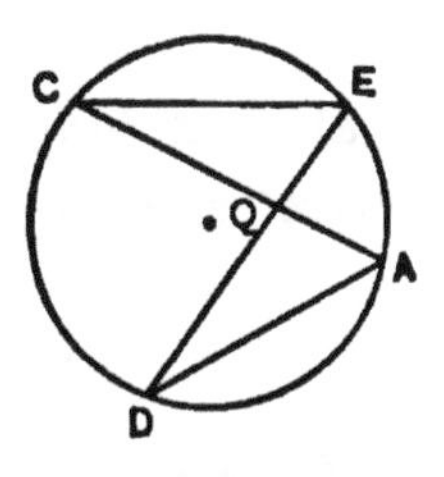

Let the chords AC and DE intersect at Q; then are their segments inversely proportional, that is, one segment of the first is to one segment of the second as the remaining segment of the second is to the remaining segment of the first.

Draw AD and CE.

The triangles ADQ and ECQ have the angles at Q equal because they are opposite or vertical angles, and the angles D and C equal because each is measured by half the arc AE; consequently, the triangles are similar (P. 15, *Cor.* 1); hence

$$AQ : DQ :: QE : QC,$$

which was to be proved.

PROPOSITION XX. THEOREM.

If two secants intersect without a circumference, the entire secants are inversely proportional to their external segments.

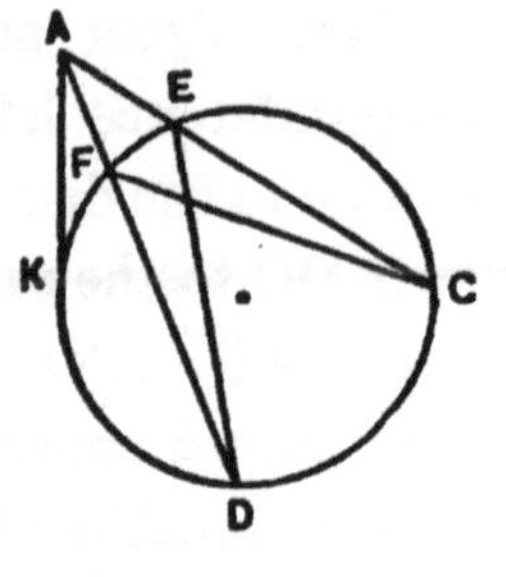

Let the secants AC and AD intersect at A; then are they inversely proportional to their external segments.

Draw CF and DE.

The triangles ACF and ADE have the angle A common, and the angles C and D equal, because each is measured by half the arc FE; these triangles are therefore similar: hence,

AC : AD :: AF : AE, ... (1)

which was to be proved.

Cor. If the secant AD is revolved about A, towards AK, the points D and F approach each other, and when AD becomes a tangent, the two points coincide at the point of contact K. In this case, proportion (1) becomes

AC : AK :: AK : AE;

that is, *a tangent is a mean proportional between the entire secant and its external segment.*

PROPOSITION XXI. THEOREM.

The perimeters of two similar polygons are to each other as any two homologous sides.

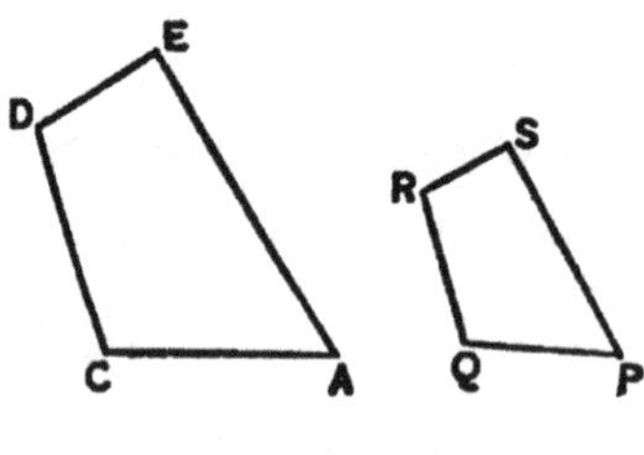

Let ACDE and PQRS be two similar polygons, in which the angles A, C, D, and E, are respectively equal to P, Q, R, and S; then from the definition of similar polygons, we have

AC : CD : DE : EA :: PQ : QR : RS : SP; . . . (1)

from (1), we have, by P. 8,

AC + CD + DE + EA : PQ + QR + RS + SP :: AC : PQ, &c.,

which was to be proved.

Cor. The perimeters of similar polygons are proportional to any homologous diagonals, or to any other homologous lines.

EXERCISES ON BOOK III.

Let the student demonstrate the following theorems:

1°. *If* DE *bisects the external angle* ADF *of the triangle* ACD, *show that*

EA : EC :: AD : CD.

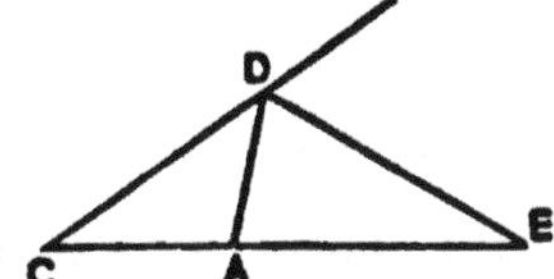

2°. *Show that the bases of two similar triangles are to each other as the distances of these bases from the opposite vertices.*

3°. *Show that homologous diagonals of similar polygons are proportional to any two homologous sides.*

4°. *If several parallels are cut by two lines intersecting at a point, show that the intercepts are proportional to their distances from the point of intersection.*

5°. *If a chord is drawn through a fixed point within a circumference, show that the product of its two segments is always the same, no matter what may be the direction of the chord.*

APPLICATION OF PRECEDING PRINCIPLES TO THE CONSTRUCTION OF PROBLEMS.

PROBLEM I.

To divide a given line into any number of equal parts.

Let it be required to divide AE into three equal parts. From A draw any other line, and on it lay off AK, KL, and LM, each equal to any assumed line; join

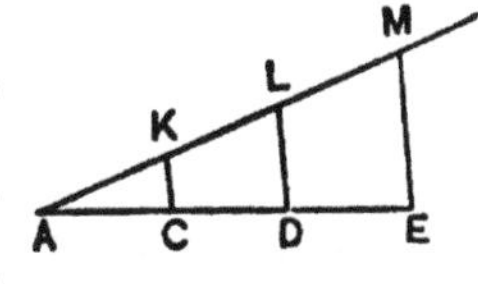

M and E, and draw LD and KC, parallel to ME: then will AC, CD, and DE be the required parts of AE. (P. 10).

Scho.—The method of proceeding is the same, when it is required to divide AE into any number of equal parts.

PROBLEM II.

To divide a given line into parts proportional to any given lines.

Let it be required to divide AE into parts proportional to the lines P, Q, and R. From A draw any line and on it lay off AK = P, KL = Q, and LM = R; join M and E, and draw LD and KC parallel to ME: then will AC, CD, and DE, be the required parts (P. 11, *Cor.* 3).

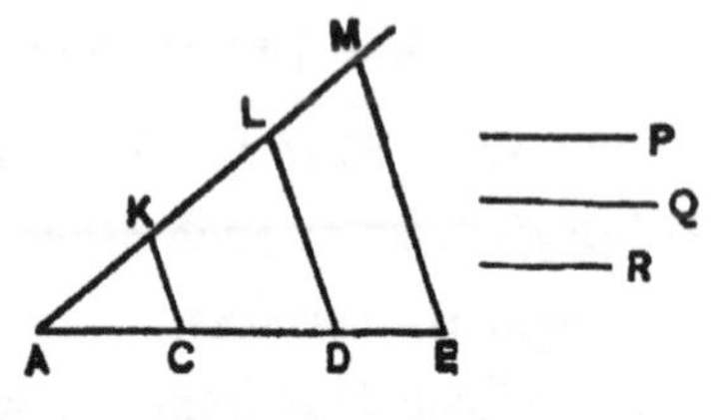

Scho.—The method is the same when there are any number of required parts.

PROBLEM III.

To find a fourth proportional to three given lines.

Let it be required to find a fourth proportional to P, Q, and R.

From A draw two lines, making any angle with each other; on these lay off AE = P, EC = Q, and AF = R; join E and F, and draw CD parallel to EF: then will FD be the required fourth proportional. For, from P. 11, we have,

AE : EC :: AF : FD,

or P : Q :: R : FD.

Cor. If Q = R, FD is a third proportional to P and Q.

PROBLEM IV.

To construct a mean proportional between two given lines.

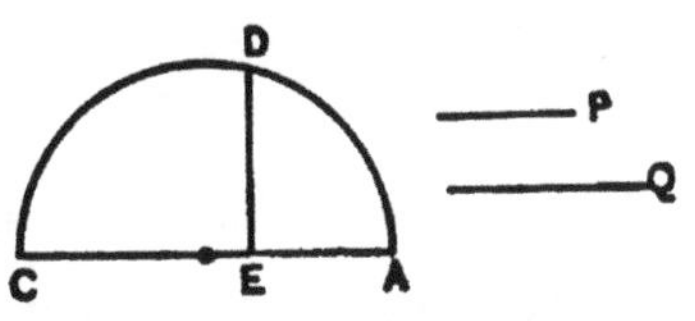

Let it be required to construct a mean proportional between P and Q. On an indefinite line AC lay off AE = P and EC = Q. On AC as a diameter construct a semi-circumference and from E draw a perpendicular to AC, meeting the semi-circumference at D; then is ED the required mean proportional. For, from P. 18, *Cor.*, we have

$$AE : ED :: ED : CE,$$

or

$$P : ED :: ED : Q.$$

PROBLEM V.

To divide a given line so that the greater part shall be a mean proportional between the whole line and the less part.

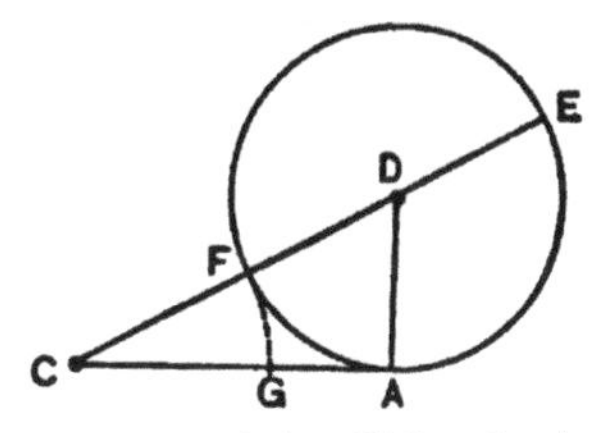

Let CA be the given line. At A draw AD perpendicular to AC and lay off AD equal to half of AC; from D as a centre with DA as a radius draw the circumference AEF, also draw the secant CDE; the circumference will be tangent to CA at A and its diameter FE will be equal to CA. From C as a centre with CF as a radius draw an arc cutting CA at G; then will G be the required point of division. For from P. 20, *Cor.*, we have

$$CE : CA :: CA : CF.$$

Whence, *by division*,

$$CE - CA : CA :: CA - CF : CF; \quad . \quad . \quad (1)$$

but CE − CA = CF, or CG, CA − CF = CA − CG = GA, and CF = CG; substituting these in (1), we have,

$$CG : CA :: GA : CG,$$

whence, *by inversion*,

$$CA : CG :: CG : GA.$$

The segment CG is the greater, because CA > CG.

Scho. The line CA is said to be divided in extreme and mean ratio at the point G.

PROBLEM VI.

To draw a line through a given point within a given angle so that the parts between that point and the sides of the angle shall be equal.

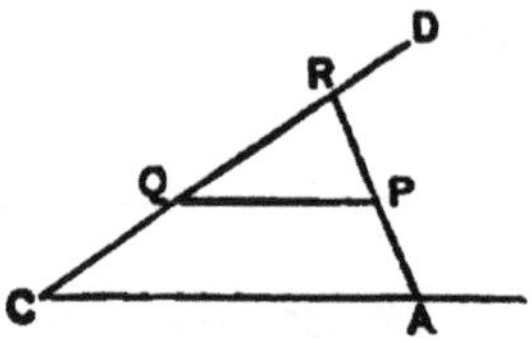

Let P be the given point and ACD the given angle.

Draw PQ parallel to AC cutting CD at Q; lay off QR equal to CQ and draw RPA; then will RA be the required line.

Note. Let the student show that RP = PA.

BOOK IV.

MEASURE AND COMPARISON OF AREAS.

Definitions and Explanations.

78. The **unit of measure** of a surface is a square described on the linear unit as a side; thus, if the linear unit is a *foot*, the unit of surface is a *square foot.*

A unit of surface is sometimes called a superficial unit.

79. The **area** of a surface is an expression for the surface in terms of a superficial unit.

80. The **product of two lines** is *the number of times* the linear unit is contained in one line, multiplied by *the number of times* the same unit is contained in the other line; the unit of the product is a superficial unit.

The operation of comparing two surfaces is purely *numerical.* In what follows, we shall regard two surfaces as *equal* when they have the same area; if they can be made to coincide throughout they are said to be *equal in all their parts.*

81. The **altitude of a triangle** is the perpendicular from the vertex of any angle of the triangle to the opposite side, or to the opposite side prolonged; the vertex considered is called the **vertex of the triangle,** and the opposite side is called the **base of the triangle.**

82. The **altitude of a parallelogram,** or **of a trapezoid,** is the perpendicular distance between two parallel sides; the sides considered are called **bases**, one being the **lower base** and the other the **upper base.**

In a parallelogram either pair of parallel sides may be taken as bases.

PROPOSITION I. THEOREM.

If two rectangles have equal altitudes they are to each other as their bases.

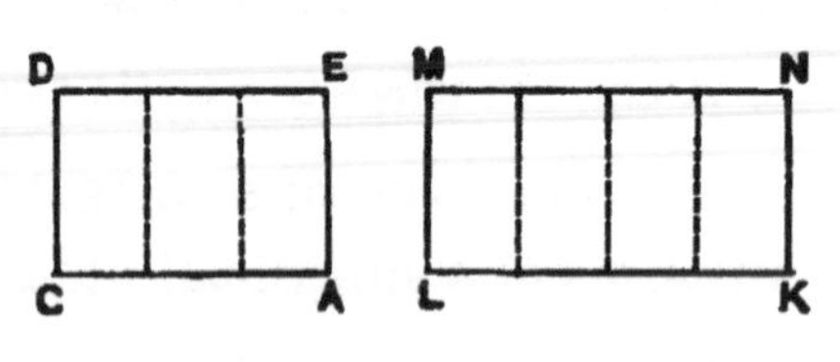

Let the altitudes CD and LM of the rectangles CE and LN be equal; then are the rectangles proportional to AC and KL.

1°. Let the bases be to each other as two whole numbers, say as 3 is to 4.

Divide AC into three equal parts and KL into four equal parts, then at the points of division draw lines perpendicular to the corresponding bases. These lines will divide the rectangle CE into three partial rectangles, and the rectangle LN into four partial rectangles, all of which are equal to each other (P. 26, *Cor.* 4, B. 1); *hence*

$$ACDE : KLMN :: 3 : 4, \quad . \quad . \quad . \quad . \quad (1)$$

but,

$$AC : KL :: 3 : 4; \quad . \quad . \quad . \quad . \quad . \quad (2)$$

hence, from proportions (1) and (2), (P. 4, B. 3), we have

$$ACDE : KLMN :: AC : KL,$$

which was to be proved.

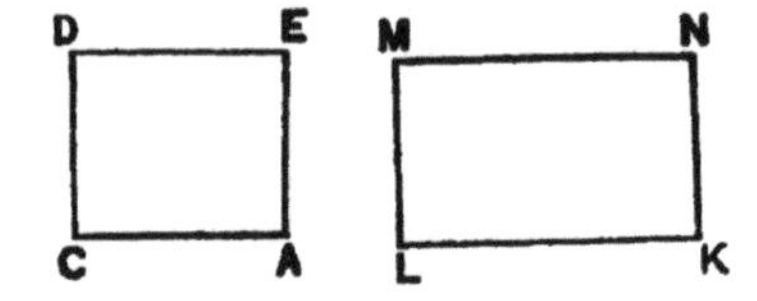

2°. Let the bases be incommensurable.

Let AC be divided into m equal parts; let one of these parts be applied to KL and suppose that it is contained in KL, n times, with a remainder less than that part; then will

$$\frac{KL}{AC} = \frac{n}{m} \text{ to within less than } \frac{1}{m}. \quad . \quad . \quad . \quad (1)$$

At each point of division on AC and KL, let lines be drawn respectively perpendicular to the bases. These will divide the rectangle CE into m equal rectangles, and the rectangle LN into n corresponding and equal rectangles, with a remainder less than one of these rectangles; hence,

$$\frac{KLMN}{ACDE} = \frac{n}{m} \text{ to within less than } \frac{1}{m}. \quad . \quad . \quad . \quad (2)$$

But m is any whole number; hence, the true values of the ratios given by equations (1) and (2) are equal to each other, (P. 1, *Cor.*, B. 3); that is,

$$\frac{KLMN}{ACDE} = \frac{KL}{AC}, \text{ or } ACDE : KLMN :: AC : KL,$$

which was to be proved.

PROPOSITION II. THEOREM.

The area of a rectangle is equal to the product of its base and altitude.

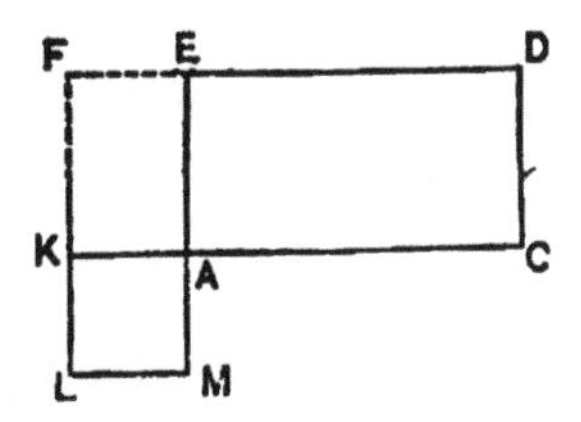

Let AD be a rectangle and AL the assumed superficial unit, that is, a square each of whose sides is equal to a linear unit; let the square AL be so placed that AK shall be on the prolongation of CA, and AM on the prolongation of EA. Prolong DE and LK till they meet at F, forming the rectangle AF. Now, if we take AC and AK as bases, the rectangles AD and AF will have the same altitude AE, and we shall have, (P. 1),

$$ACDE : AEFK :: AC : AK; \quad . \quad . \quad . \quad (1)$$

if we take AE and AM as bases, the rectangles AF and AL will have the same altitude AK, and we shall have

$$AEFK : AKLM :: AE : AM; \quad . \quad . \quad . \quad (2)$$

multiplying proportions (1) and (2) together, term by term (P. 9, B. 3), and dividing both terms of the first couplet of the resulting proportion by AEFK, we have

$$ACDE : AKLM :: AC \times AE : AK \times AM.. \quad . \quad (3)$$

But AK and AM are each equal to the linear unit, and AKLM is equal to the superficial unit. Passing to numerical values, proportion (3) becomes,

$$ACDE : 1 :: AC \times AE : 1, \text{ or } ACDE = AC \times AE,$$

that is, the number of superficial units in the rectangle is equal to the number of linear units in its base multiplied by the number of linear units in its altitude, *which was to be proved.*

Cor. Any two rectangles are to each other as the products of their bases and altitudes; if their bases are equal, they are to each other as their altitudes.

Scho. The product of two lines is often called the rectangle of the lines, because the product expresses the area of a rectangle whose adjacent sides are the given lines.

PROPOSITION III. THEOREM.

The area of a parallelogram is equal to the product of its base and altitude.

Let ACDE be a parallelogram; draw AL and CK perpendicular to AC meeting ED, and ED prolonged at L and K: then, is the area of the parallelogram equal to the product of AC and CK.

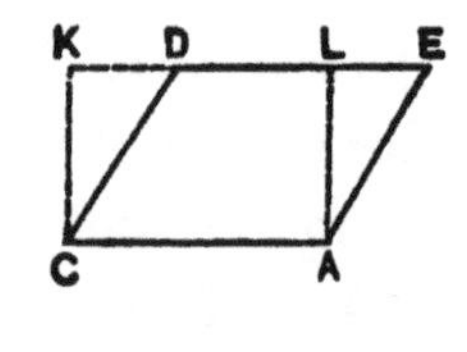

The triangles ALE and CKD, have AE parallel and equal to CD, AL parallel and equal to CK, and the included angle A equal to the included angle C; they are therefore equal in all their parts. If we now add to the figure ACDL the triangle CKD, we shall have the rectangle ACKL; if we add to the figure ACDL the equal triangle ALE we shall have the parallelogram ACDE; hence, the rectangle CL is equal to the parallelogram CE (Ax. 2). But the area of the rectangle is equal to the product of AC and CK; hence, the area of the parallelogram is also equal to the product of AC and CK, *which was to be proved.*

Cor. Any two parallelograms are to each other as the products of their bases and altitudes. If their bases are equal, they are to each other as their altitudes; if their altitudes are equal, they are to each other as their bases.

PROPOSITION IV. THEOREM.

The area of a triangle is equal to half the product of its base and altitude.

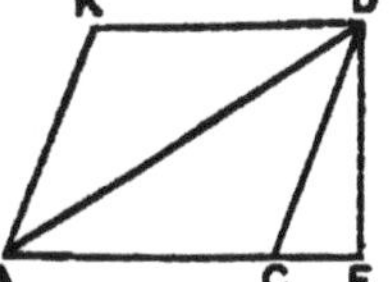

Let ACD be any triangle, and let DE be its altitude: then is the area of the triangle equal to ½ (AC × DE).

Through D draw DK parallel to CA and through A draw AK parallel to CD; the figure ACDK is then a parallelogram equal to twice the given triangle (P. 26, *Cor.* 2, B. 1). But, the area of the parallelogram is equal to AC × DE; hence, the area of the triangle is ½ (AC × DE), *which was to be proved.*

Cor. 1. Any two triangles are to each other as the products of their bases and altitudes; if their altitudes are equal, they are to each other as their bases; if their bases are equal, they are to each other as their altitudes.

Cor. 2. If two triangles have a common base and if their vertices are on a line parallel to that base they are equal.

Cor. 3. If a triangle has the same base as a parallelogram and if the vertex of the triangle is on the opposite base, or on that base produced, the triangle is equal to half the parallelogram.

PROPOSITION V. THEOREM.

The area of a trapezoid is equal to the product of its altitude, and the half sum of its parallel bases.

Let ACDE be any trapezoid, and let DK be its altitude: then is the area of the trapezoid equal to

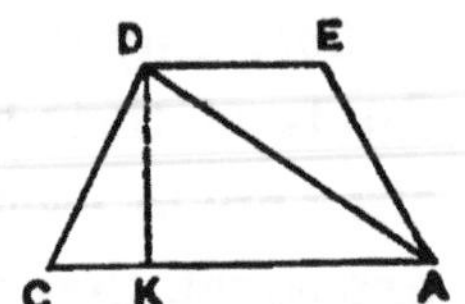

$$DK \times \tfrac{1}{2}(AC + DE).$$

Draw AD dividing the trapezoid into two triangles; the altitude of each triangle is equal to DK. Now, the area of ACD is equal to $\tfrac{1}{2}AC \times DK$, and the area of ADE is equal to $\tfrac{1}{2}DE \times DK$ (P. 4); hence, the area of the trapezoid is equal to

$$\tfrac{1}{2}AC \times DK + \tfrac{1}{2}DE \times DK, \text{ or to } DK \times \tfrac{1}{2}(AC + DE).$$

which was to be proved.

PROPOSITION VI. THEOREM.

The square of the sum of two lines is equal to the sum of the squares of the lines, increased by twice their rectangle.

Let AC and CD be any two lines, CD lying in the prolongation of AC: then is

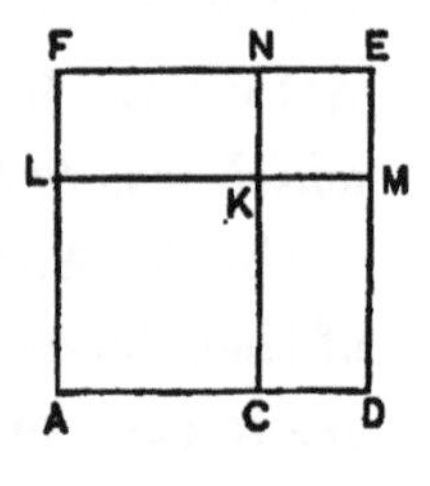

$$(AC + CD)^2 = \overline{AC}^2 + \overline{CD}^2 + 2AC \times CD.$$

On AD construct the square AE, and on AC construct the square AK; then prolong CK to N and LK to M.

Because CK and KL are each equal to AC, KN and KM are each equal to CD; hence, the measure of each of the rectangles CM and KF is $AC \times CD$, and the measure of the square KE is $\overline{CD}^2$. But the square AE is made up of the squares AK and KE, together with the rectangles CM and KF; hence,

$$(AC+CD)^2=\overline{AC}^2+\overline{CD}^2+2AC\times CD,$$

which was to be proved.

Cor. If AC is equal to CD the four parts into which AE is divided are equal squares; hence, the square of a given line is equal to four times the square of half that line.

Proposition VII. Theorem.

The square of the difference of two lines is equal to the sum of the squares of the lines, diminished by twice their rectangle.

Let AC and DC be two lines placed as shown in the figure; then is

$(AC-CD)^2=\overline{AC}^2+\overline{CD}^2-2AC\times CD.$

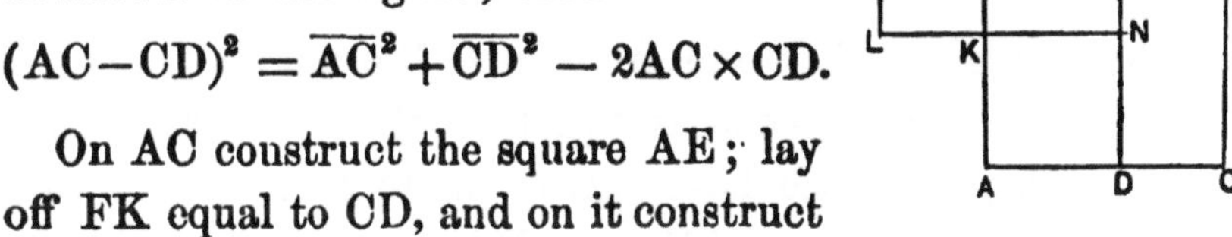

On AC construct the square AE; lay off FK equal to CD, and on it construct the square KM; draw DP parallel to CE and prolong LK to N.

Because LN is equal to AC and LM to CD, the rectangle NM is equal to the rectangle DE, and each is measured by AC × CD. If now from the entire figure CM, which is equal to the sum of the squares AE and KM, we take away the rectangles NM and DE, there will remain the square AN, that is, the square on AC — CD; hence,

$$(AC-CD)^2=\overline{AC}^2+\overline{CD}^2-2AC\times CD,$$

which was to be proved.

PROPOSITION VIII. THEOREM.

The square of the hypothenuse of a right-angled triangle is equal to the sum of the squares of the other two sides.

Let ACD be a right-angled triangle, AC being the hypothenuse and D the right angle; then is

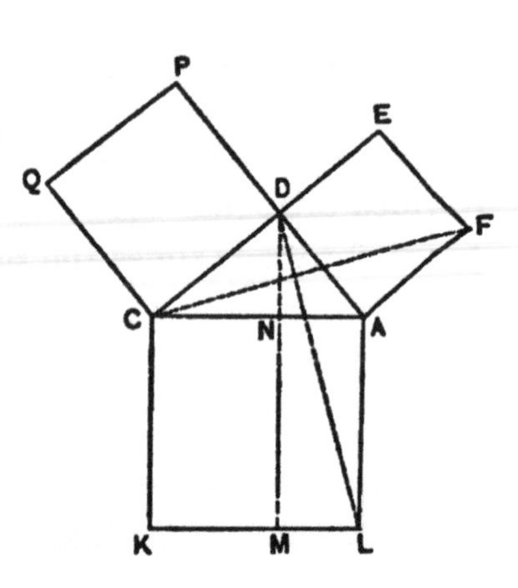

$$\overline{AC}^2 = \overline{CD}^2 + \overline{DA}^2.$$

Prolong CD till DE is equal to AD, and complete the square AE; prolong AD till DP is equal to CD, and complete the square DQ; also, on AC construct the square AK. Then draw DM perpendicular to KL; also draw CF and DL.

The triangles ALD and ACF have the angles DAL and FAC equal, because each is equal to DAC increased by a right angle; they also have AL equal to AC, because they are sides of the square AK, and DA equal to FA, because they are sides of the square AE; hence, the triangles are equal (P. 8, B. 1).

But the triangle ALD is half the rectangle AM, and the triangle ACF is half the square AE (P. 4, *Cor.* 3); hence, the rectangle AM is equal to the square AE. In like manner it may be shown that the rectangle CM is equal to the square DQ. But, the sum of the rectangles AM and CM is equal to the square AK; hence,

$$\overline{AC}^2 = \overline{CD}^2 + \overline{DA}^2, \quad . \quad . \quad . \quad . \quad (1)$$

which was to be proved.

Cor. 1. From equation (1), we have,

$$\overline{CD}^2 = \overline{AC}^2 - \overline{DA}^2 \text{ and } \overline{DA}^2 = \overline{AC}^2 - \overline{CD}^2.$$

Cor. 2. If CD = DA, they form the adjacent sides of a square whose diagonal is AC. In this case

$$\overline{AC}^2 = 2\overline{CD}^2 \quad \text{whence} \quad AC = \sqrt{2} \times CD.$$

Whence by P. 3, B. 3, we have,

$$AC : CD :: \sqrt{2} : 1 ;$$

that is, the diagonal of a square is to one of the sides as $\sqrt{2}$ is to 1.

PROPOSITION IX. THEOREM.

In any triangle, the square of a side opposite an acute angle is equal to the sum of the squares of the other sides, diminished by twice the rectangle of the base and the distance from the vertex of the acute angle to the foot of the perpendicular drawn from the vertex of the triangle to the base, or to the base produced.

In the triangle ACD, let AC be the base, D the vertex, and DE a perpendicular from D to AC, or to AC prolonged; also let A be an acute angle; then is

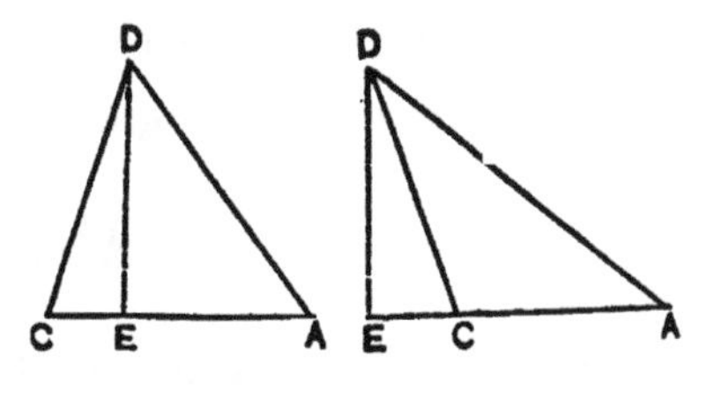

$$\overline{CD}^2 = \overline{DA}^2 + \overline{AC}^2 - 2AC \times AE.$$

Because CE is equal to the difference between AC and AE, we have (P. 7),

$$\overline{CE}^2 = \overline{AE}^2 + \overline{AC}^2 - 2AC \times AE; \quad . \quad . \quad (1)$$

adding $\overline{ED}^2$ to both members of (1), we have,

$$\overline{CE}^2 + \overline{ED}^2 = (\overline{AE}^2 + \overline{ED}^2) + \overline{AC}^2 - 2AC \times AE;$$

substituting for $\overline{CE}^2 + \overline{ED}^2$ its value $\overline{CD}^2$, and for $\overline{AE}^2 + \overline{ED}^2$ its value $\overline{DA}^2$ (P. 8), we have,

$$\overline{CD}^2 = \overline{DA}^2 + \overline{AC}^2 - 2AC \times AE,$$

which was to be proved.

PROPOSITION X. THEOREM.

In any obtuse-angled triangle, the square of the side opposite the obtuse angle is equal to the sum of the squares of the other two sides, increased by twice the rectangle of the base and the distance from the vertex of the obtuse angle to the foot of the perpendicular drawn from the vertex of the triangle to the base produced.

In the obtuse-angled triangle ACD, let AC be the base, D the vertex, and DE a perpendicular from D to CA produced; also let A be the obtuse angle; then is

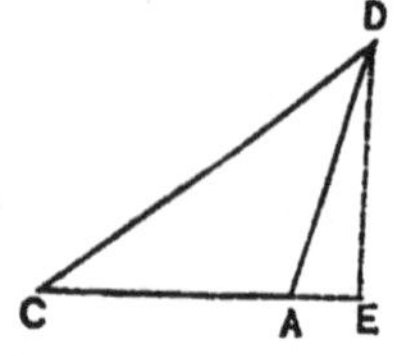

$$\overline{CD}^2 = \overline{DA}^2 + \overline{AC}^2 + 2AE \times AC.$$

Because CE is equal to the sum of AE and AC, we have (P. 6),

$$\overline{CE}^2 = \overline{AE}^2 + \overline{AC}^2 + 2AE \times AC; \quad . \ . \quad (1)$$

adding $\overline{ED}^2$ to both members of (1), we have,

$$\overline{CE}^2 + \overline{ED}^2 = (\overline{AE}^2 + \overline{ED}^2) + \overline{AC}^2 + 2AE \times AC;$$

substituting for $\overline{CE}^2 + \overline{ED}^2$ its value $\overline{CD}^2$, and for $\overline{AE}^2 + \overline{ED}^2$ its value $\overline{AD}^2$ (P. 8), we have

$$\overline{CD}^2 = \overline{DA}^2 + \overline{AC}^2 + 2AE \times AC,$$

which was to be proved.

Scho. The right-angled triangle is the only one in which the sum of the squares of two sides is equal to the square of the third side.

PROPOSITION XI. THEOREM.

If two triangles have an angle in each equal, the triangles are to each other as the products of the including sides.

Let the triangles ACD and PQR have the angles D and R equal: then

ACD : PQR :: CD × DA : QR × RP.

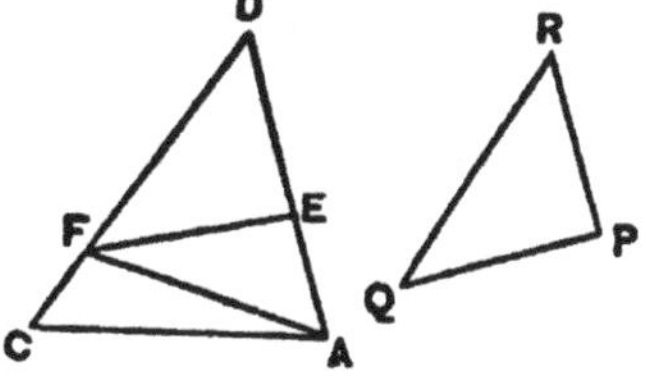

Place the triangle PQR on the triangle ACD so that the angle R shall coincide with the angle D, P falling at E, and Q at F; draw FA.

The triangles ACD and AFD, have their bases CD and FD in the same line and their vertices at the same point A; hence, they have the same altitude; they are therefore proportional to their bases (P. 4, *Cor.* 1), that is,

ACD : AFD :: CD : FD. . . . . (1)

The triangles AFD and EFD have their bases DA and DE in the same line and their vertices at the same point F; hence, they have the same altitude; they are therefore proportional to their bases, that is,

AFD : EFD :: DA : DE. . . . . (2)

Multiplying (1) by (2), term by term, (P. 9, B. 3), and then dividing both terms of the first couplet of the resulting proportion by AFD, (P. 7, *Cor.*, B. 3), we have,

$$ACD : EFD :: CD \times DA : FD \times DE,$$

$$\text{or, } ACD : PQR :: CD \times DA : QR \times RP, \quad . \quad . \quad (3)$$

which was to be proved.

Cor. 1. If CD and DA are proportional to QR and RP, the triangles ACD and PQR will be similar, and we shall have

$$DA : CD :: RP : QR, \text{ or } \frac{CD}{DA} = \frac{QR}{RP}. \quad . \quad . \quad (4)$$

Multiplying the first term of the second couplet of (3) by $\frac{CD}{DA}$, and the second term of that couplet by the equal quantity $\frac{QR}{RP}$, and reducing, we have

$$ACD : PQR :: \overline{CD}^2 : \overline{QR}^2.$$

But,

$$\overline{CD}^2 : \overline{QR}^2 :: \overline{DA}^2 : \overline{RP}^2 :: \overline{AC}^2 : \overline{PQ}^2.$$

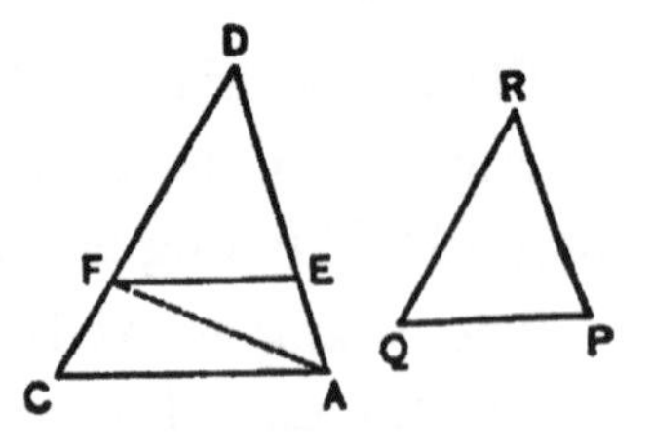

Hence, *two similar triangles are to each other as the squares of any two homologous sides.*

Cor. 2. If ACD and PQR are similar, ACD and EFD are also similar; consequently the second couplets of (1) and (2) are equal; hence,

$$ACD : AFD :: AFD : EFD,$$

that is, AFD is a mean proportional between ACD and EFD.

Proposition XII. Theorem.

Two similar polygons may be divided into the same number of similar triangles, similarly placed.

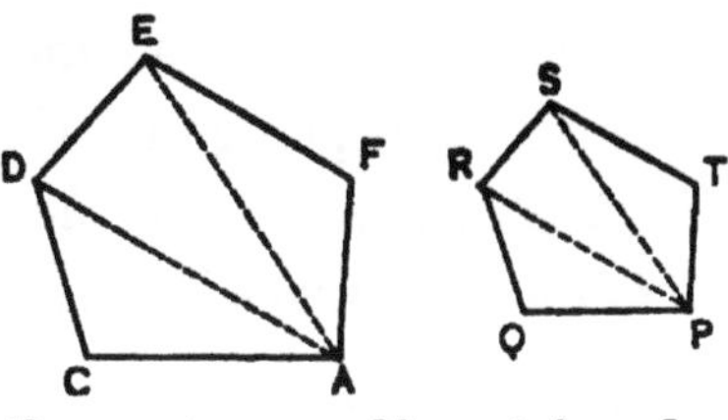

Let ACDEF and PQRST be similar polygons, the equal angles in both being taken in the order named. Draw the diagonals AD and AE, also the diagonals PR and PS; then will the corresponding triangles of the polygons be similar.

The angle C is equal to the angle Q, and from the definition of similar polygons AC and CD are proportional to PQ and QR; hence, the triangles ACD and PQR are similar (P. 16, B. 3).

Because ACD and PQR are similar, the angle CDA is equal to QRP, and the sides DA and RP are proportional to CD and QR, and consequently to DE and RS. If we take the equal angles CDA and QRP from the equal angles CDE and QRS, the remaining angles ADE and PRS are equal. Consequently, the triangles ADE and PRS have an angle in each equal, and their including sides proportional; hence, their triangles are similar. In like manner it may be shown that AEF and PST are similar. Hence, the corresponding triangles of the two polygons are similar, *which was to be proved.*

Cor. 1. Any two corresponding triangles are like parts of the polygons to which they belong. For, the triangles ACD and PQR, being similar, are proportional to $\overline{AD}^2$ and $\overline{PR}^2$, and the triangles ADE and PRS being similar are also proportional to $\overline{AD}^2$ and $\overline{PR}^2$; consequently ACD and PQR are proportional to ADE and PRS. In like manner it may be shown that ADE and PRS are proportional to AEF and PST. Hence,

ACD : ADE : AEF :: PQR : PRS : PST. . (1)

Now, the sum of the antecedents of (1) is equal to the first polygon, and the sum of the consequents of (1) is equal to the second polygon. Hence, from P. 8, B. 3, we infer that *the first polygon is to the second, as any triangle in the first is to the corresponding triangle in the second.*

Cor. 2. From the preceding corollary, it follows that two similar polygons are to each other as the squares of any two homologous sides of any two corresponding triangles; but, the homologous sides of the corresponding triangles are the homologous sides or the homologous diagonals of the polygons to which the triangles belong; *hence, two similar polygons are to each other as the squares of their homologous sides, or as the squares of their homologous diagonals.*

Exercises on Book IV.

Let the student demonstrate the following theorems:

1°. *The area of a triangle is equal to half the product of its perimeter and the radius of the inscribed circle.*

2°. *The rectangle of the sum and the difference of two lines is equal to the difference of their squares.*

3°. *The sum of the squares of two sides of a triangle is equal to twice the square of half the third side, increased by twice the square of the line drawn from the middle of the third side to the opposite vertex.*

4°. *The sum of the squares of the diagonals of a parallelogram is equal to the sum of the squares of its sides.*

5°. *If lines are drawn from any point to the vertices of a rectangle, the sum of the squares of the lines to the extremi-*

ties of one diagonal is equal to the sum of the squares of the lines to the extremities of the other diagonal.

6°. *If a line is drawn from the centre of a circle to any point of any chord, the square of this line, increased by the rectangle of the segments of the chord, is equal to the square of the radius.*

APPLICATION OF PRECEDING PRINCIPLES TO THE SOLUTION OF PROBLEMS.

Problem I.

To construct a square equal to the sum of two given squares; also, to construct a square equal to the difference of two given squares.

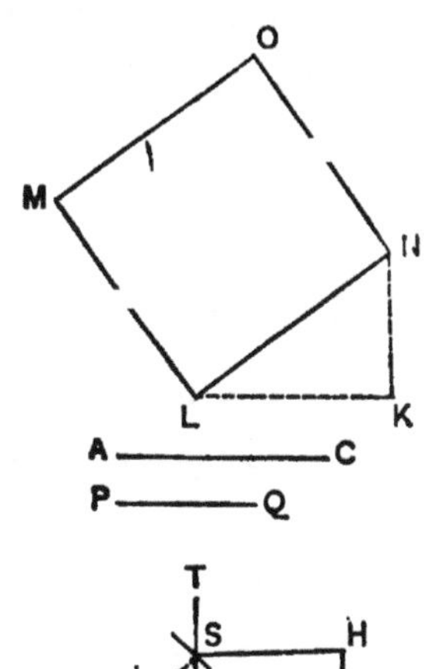

Let AC and PQ be sides of the given squares, AC being the greater.

1°. Draw two lines KL and KN perpendicular to each other at K; lay off KL equal to AC and KN equal to PQ; draw LN and on it, as a side, construct the square LO; it will be equal to the sum of the given squares (P. 8).

2°. Draw two lines KL and KT perpendicular to each other at K; lay off KL equal to PQ; from L as a centre, with a radius equal to AC, draw an arc cutting KT at S, and on KS as a side construct the square KH; it will be equal to the difference of the given squares (P. 8, *Cor.* 1).

Problem II.

To construct a triangle equal to a given polygon.

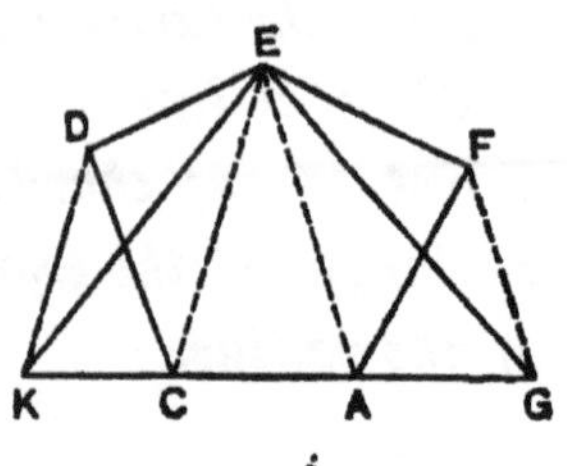

Let ACDEF be the given polygon.

Take AC as a base and prolong it in both directions; draw the diagonals AE and CE. Through F draw FG parallel to EA, and draw EG. Because the triangles AEF and AEG have the common base AE, and because their vertices F and G are on a line FG parallel to that base they are equal (P. 4, *Cor.* 2). If we take the triangle AEF from the given polygon and then add the equal triangle AEG, the resulting polygon GCDE will be equal to the given polygon. Again draw DK parallel to EC and join the points K and E. Because the triangles ECD and ECK have a common base EC, and because their vertices D and K are in the line DK parallel to that base, the triangles are equal. If we take from the polygon GCDE the triangle CDE, and add the equal triangle CKE, the resulting triangle KEG will be equal to the polygon GCDE, and consequently to the given polygon.

Problem III.

To construct a square equal to a given triangle.

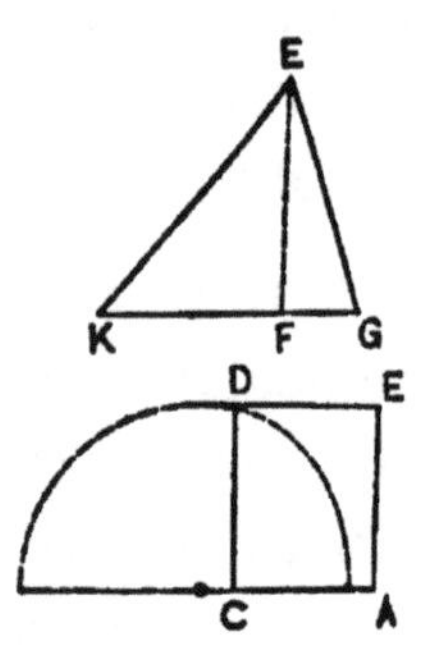

Let GKE be the given triangle, GK being its base and EF its altitude.

Construct a mean proportional CD between $\frac{1}{2}$GK and FE (Prob. 4, B. 3); then, on CD as a side construct the square AD: it will be the required square; for,

$\overline{AC}^2 = \frac{1}{2}GK \times EF$, (P. 2, and P. 4).

Scho. By means of Problems 2 and 3 we can construct a square equal to a given polygon, and by the aid of Problem 1 we can construct a square equal to the sum, or to the difference of any two polygons.

Problem IV.

To construct a polygon on a given line as a side, that shall be similar to a given polygon.

Let ACDEF be the given polygon and let PQ be the side which is to be homologous with AC; suppose PQ to be placed parallel to AC.

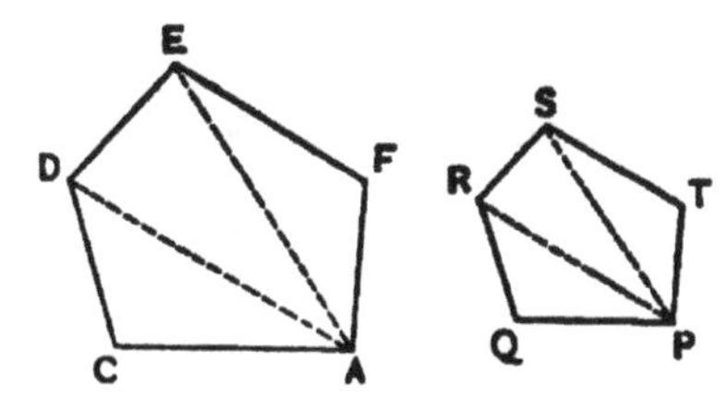

Draw AD and AE.

Draw QR parallel to CD, and PR parallel to AD, and let these lines intersect at R; then are the triangles ACD and PQR similar, (P. 15, *Cor.* 2, B. 3). Draw RS parallel to DE, and PS parallel to AE, and let these lines intersect at S; then are the triangles ADE and PRS similar. In like manner draw ST parallel to EF, and PT parallel to AF, and let these lines intersect at T; then are the triangles AEF and PST similar. From the similarity of the homologous triangles we infer the equality of the corresponding angles of the given polygon and of the polygon PQRST; we also infer the proportionality of their sides taken in the same order; hence, the polygon PQRST is similar to the given polygon.

BOOK V.

PROPERTIES AND MEASURE OF REGULAR POLYGONS AND CIRCLES.

Definition.

83. **A regular polygon** is a polygon that is both *equilateral* and *equiangular*.

PROPOSITION I. THEOREM.

If two regular polygons have the same number of sides they are similar.

For, their corresponding angles are equal, because any angle of either is equal to twice as many right angles as the figure has sides, less four right angles, *divided* by the number of angles (P. 25, B. 1); furthermore, their corresponding sides are proportional, because all the sides of each polygon are equal to each other; hence, the polygons are similar, *which was to be proved.*

Cor. If two regular polygons have the same number of sides, their perimeters are proportional to any homologous lines, and their areas are proportional to the squares of these lines.

PROPOSITION II. THEOREM.

A circle can be circumscribed about any regular polygon; a circle can also be inscribed within it.

Let ACDEF be a regular polygon. Bisect the angles A and C and suppose their bisectrices to meet at O; from O, draw lines to the vertices, D, E, and F; also draw OP, OQ, &c., perpendicular to the sides of the polygon.

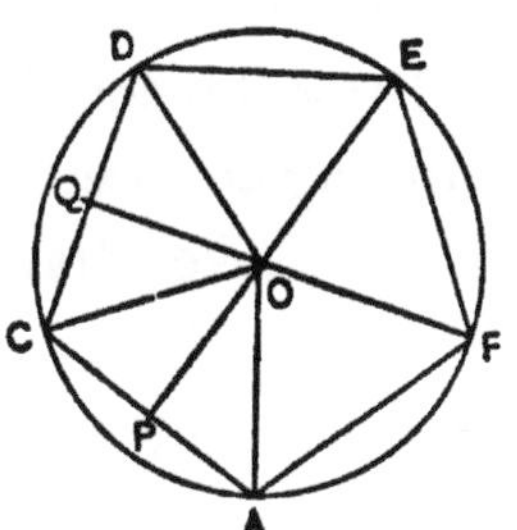

1°. Because the angles A and C are equal, their halves are equal, that is, CAO and ACO are equal; the triangle ACO is therefore isosceles; hence, OC is equal to OA. The triangles ACO and CDO are equal, because OA is equal to OC, AC to CD, and the angle OAC to OCD; hence, OD is equal to OC, and consequently to OA. In like manner it may be shown that OE and OF are each equal to OA. Hence, if a circumference is drawn from O as a centre and with OA as a radius it will pass through all the vertices of the polygon, that is, it will be circumscribed about the polygon, *which was to be proved.*

2°. Because the triangles ACO, CDO, &c., are equal, their altitudes OP, OQ, &c., are equal; hence, if a circumference is drawn from O as a centre, with a radius OP, it will be tangent to all the sides of the polygon, that is, it will be inscribed within the polygon, *which was to be proved.*

PROPOSITION III. THEOREM.

If a circumference is divided into equal arcs, the corresponding chords will form a regular inscribed polygon; and if tangents are drawn at the middle points of each of these arcs, they will form a regular circumscribed polygon.

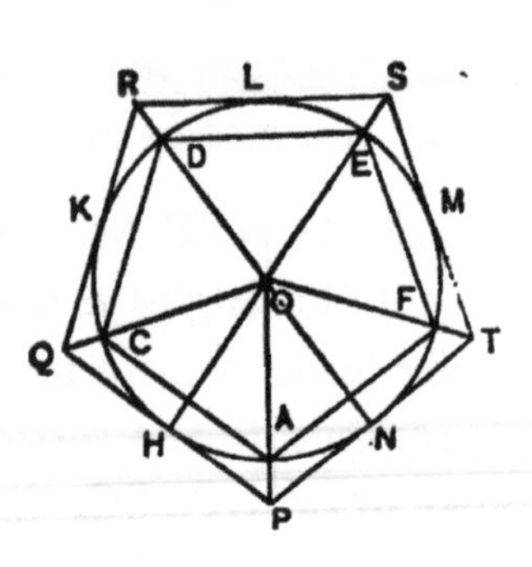

Let AHC, CKD, &c., be equal arcs of the circumference OA, and let H, K, &c., be their middle points.

Draw the chords AC, CD, DE, &c., forming an inscribed polygon, and at H, K, L, &c., draw tangents meeting at P, Q, R, &c., forming a circumscribed polygon.

1°. In the polygon ACDEF, the sides are equal, because they are chords of equal arcs; and the angles are equal because they are measured by halves of equal arcs; hence, the inscribed polygon is regular, *which was to be proved.*

2°. Because PN and PH are tangent to the circumference OA, the line OP bisects the angle NPH and also the angle NOH (Prob. 14, *Cor.*, B. 2); hence, OP passes through A, the middle of the arc NAH. In like manner it may be shown that each of the lines OQ, OR, &c., passes through the corresponding vertex of the inscribed polygon. In the similar triangles ACO and PQO, we have,

$$OC : OQ :: AC : PQ, \quad . \quad . \quad . \quad . \quad (1)$$

and in the similar triangles CDO and QRO, we have,

$$OC : OQ :: CD : QR; \quad . \quad . \quad . \quad . \quad (2)$$

from (1) and (2), we have,

$$AC : PQ :: CD : QR;$$

but, CD is equal to AC; hence, QR is equal to PQ. In like manner it may be shown that RS is equal to QR, and consequently to PQ; and so on; hence, the circumscribed polygon is equilateral. It is also equiangular; for each of its angles is equal to the corresponding angle of the inscribed polygon, because their sides are respectively parallel and lie in the same direction (P. 23, B. 1). Hence, the circumscribed polygon is both equilateral and equiangular; it is therefore regular, *which was to be proved.*

Cor. 1. The polygons ACDEF and PQRST are similar (P. 1).

Cor. 2. If a circumference is divided into equal arcs, the tangents drawn at the points of division, form a regular circumscribed polygon; hence, if tangents are drawn at the vertices of a regular inscribed polygon they will form a similar circumscribed polygon.

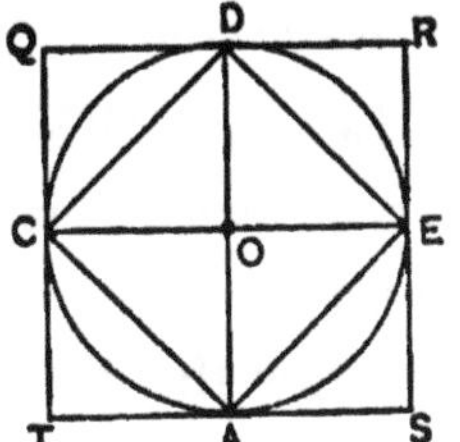

Cor. 3. If the arcs subtended by the sides of a regular inscribed polygon are bisected, the corresponding chords will form a regular inscribed polygon having double the number of sides.

Scho. If we have given a regular inscribed polygon we can always construct a similar circumscribed polygon. Then, by the methods already described, we may construct a regular inscribed and a regular circumscribed polygon, having double the number of sides. By repeating the process we may again double the number of sides, and so on, *ad infinitum.* This method of proceeding is called **the method of continued duplication.** Thus, if we start with an inscribed and a circumscribed square, we can find in succession, regular inscribed and circumscribed polygons, of 8, 16, 32, &c., sides.

Definitions.

84. The **centre of a regular polygon** is the common centre of the inscribed and circumscribed circles.

85. The **radius of a regular polygon** is a line drawn from the centre to any vertex.

86. The **apothem of a regular polygon** is perpendicular from the centre to any side.

87. The **angle at the centre of a regular polygon** is the angle between two radii drawn to the extremities of any side.

PROPOSITION IV. THEOREM.

The area of a regular polygon is equal to half the product of its perimeter and apothem.

Let ACDEF be a regular polygon, whose centre is O and whose apothem is OP.

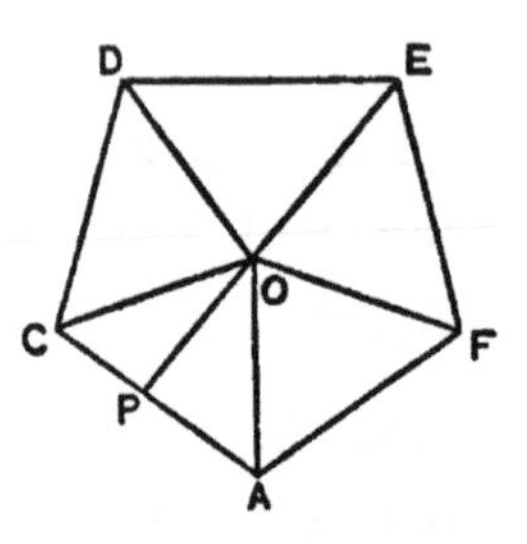

Draw the radii OA, OC, OD, OE, and OF; these will divide the polygon into equal triangles whose bases are the sides of the polygon and whose altitudes are each equal to the apothem (P. 2).

Now the area of any one of these triangles, as ACO, is equal to half the product of the side AC and the apothem OP; hence, the sum of the areas of the triangles, which is the area of the polygon, is equal to the sum of the sides, or the perimeter, multiplied by half the apothem, *which was to be proved.*

PROPOSITION V. THEOREM.

The circumference is the common limit of the perimeters of both the inscribed and the circumscribed regular polygons when the number of their sides is indefinitely increased by the method of continued duplication.

Let AC be a side of any regular polygon inscribed in the circle OA, and let DE be the corresponding side of the circumscribed polygon. Let OA and OF be the radius and apothem of the first polygon, and let OD and OG be the radius and apothem of the second polygon.

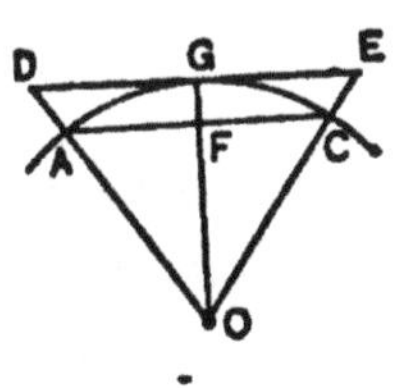

1°. In the triangle OAF, we have the relation,

$$OA - OF < AF,$$

that is, the difference between the radius and the apothem of the polygon is less than half of the side. If we now suppose the number of sides of the polygons to be indefinitely increased by the method of continued duplication, the value of AC will *decrease*, that of OF will *increase*, and the difference of OA and OF will *diminish* ; finally when AC becomes less than any assignable quantity, OA and OF will become equal, and the perimeter will coincide with the circumference, *which was to be proved.*

2°. Because OD and OG are proportional to OA and OF, OD will become equal to OG when OA becomes equal to OF, and the perimeter of the circumscribed polygon will then coincide with the circumference OA, *which was to be proved.*

Cor. 1. When the perimeters of the circumscribed and inscribed polygons coincide with the circumference OA, their areas are equal to that of the circle OA.

Cor. 2. The area of a circle is equal to half the product of its circumference and radius.

Scho. A circle may be regarded as a regular polygon, having an infinite number of sides.

PROPOSITION VI. THEOREM.

The perimeters of two regular polygons having the same number of sides are to each other as their apothems, and their areas are to each other as the squares of their apothems.

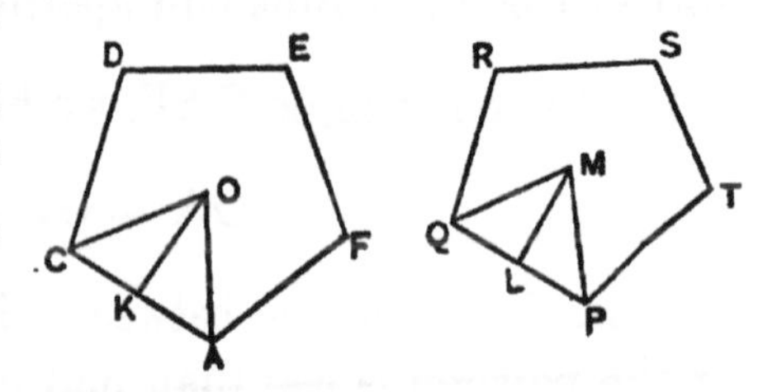

Let ACDEF and PQRST be two regular polygons having the same number of sides, and let OK and ML be their apothems.

Draw the radii OA, OC, MP, and MQ.

The isosceles triangles ACO and PQM are similar to each other, because they are corresponding triangles of similar polygons (P. 12, *Cor.* 1, B. 4); their altitudes are therefore homologous lines of those polygons. Hence, the perimeters of the polygons are to each other as their apothems, and their areas are to each other as the squares of their apothems (P. 1, *Cor.*), *which was to be proved.*

Cor. 1. The circumferences of two circles are to each other as their radii, or as their diameters, and their areas are to each other as the squares of their radii, or as the squares of their diameters.

Cor. 2. Because similar sectors and similar segments are like parts of the circles to which they belong, their arcs are to each other as their radii, and their areas are to each other as the squares of their radii.

PROPOSITION VII. PROBLEM.

To inscribe a square in a circle; also to circumscribe a square about that circle.

Let ACDE be the given circle.

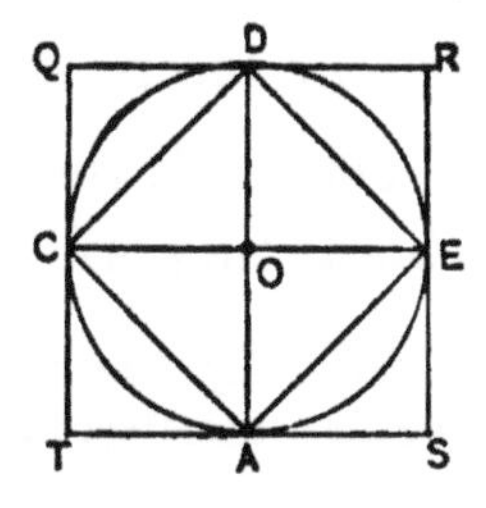

1°. Draw two diameters, AD and CE, perpendicular to each other; they will divide the circumference into four equal arcs. Draw the corresponding chords AC, CD, DE, and EA; they will form an inscribed square (P. 3).

2°. Draw tangents to the circumference at the points A, C, D, and E; they will form a circumscribed square, (P. 3, *Cor.* 2).

Scho. Having an inscribed and a circumscribed square, we may, by the method of continued duplication, construct regular inscribed and circumscribed polygons having 8, 16, 32, &c., sides.

Proposition VIII. Theorem.

The side of a regular inscribed hexagon is equal to the radius.

Let ACDEFG be a regular hexagon inscribed in the circle OA.

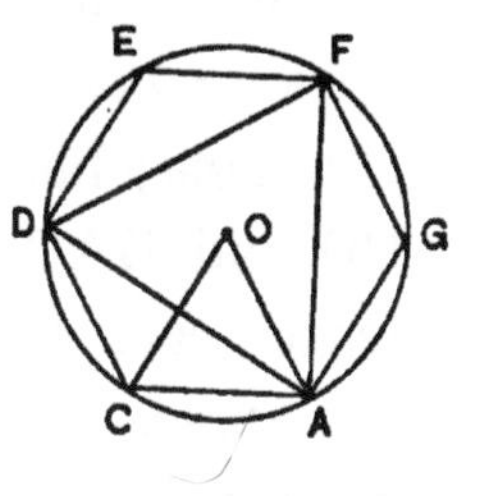

Draw the radii OA and OC.

The angle COA being an angle at the centre is equal to 4 right angles, *divided by* 6, that is, to $\frac{2}{3}$ of a right angle; hence, the sum of the angles ACO and OAC is equal to $\frac{4}{3}$ of a right angle, (P. 24, *Cor.* 1, B. 1); but ACO and OAC are equal because they lie opposite the equal sides OA and OC; hence, each is $\frac{2}{3}$ of a right angle. The triangle ACO is therefore equiangular and consequently equilateral, (P. 13, *Cor.*, B. 1); hence, the side AC is equal to the radius OA, *which was to be proved.*

Cor. 1. If the radius of a circle is applied to the circumference six times, as a chord, the resulting polygon will be a regular inscribed hexagon.

Cor. 2. If the alternate vertices A, D, and F, are joined by chords they will form a regular inscribed triangle.

Scho. By the method of continued duplication we may construct regular inscribed polygons of 12, 24, 48, &c., sides.

PROPOSITION IX. THEOREM.

The side of a regular inscribed decagon is equal to the greater segment obtained by dividing the radius in extreme and mean ratio.

Let AC be any side of a regular decagon inscribed in the circle OA.

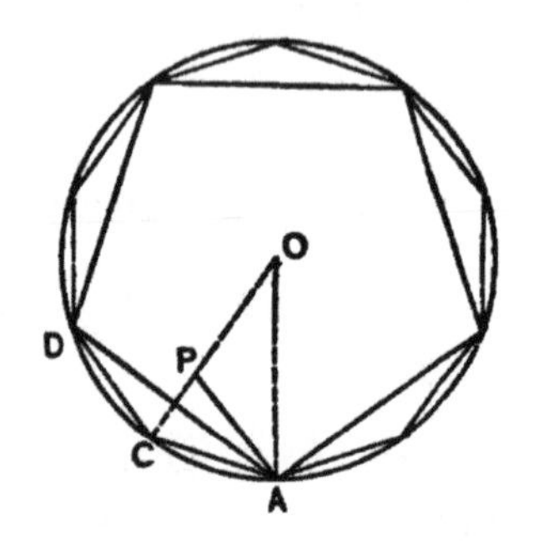

Draw OA and OC; also draw AP bisecting OAC.

The angle COA being at the centre of a regular decagon is equal to 4 right angles *divided by* 10, that is, to $\frac{2}{5}$ of a right angle; hence, each of the angles ACO and OAC is equal to $\frac{4}{5}$ of a right angle, and consequently, each of the angles OAP and PAC is equal to $\frac{2}{5}$ of a right angle. The triangle OAP is therefore isosceles, and consequently the side AP is equal to OP. Now, the triangles ACO and CPA have the angle C in common, and the angles COA and PAC equal; they are therefore similar: hence, the triangle CPA is isosceles, the side AC being equal to AP, and consequently to OP. Because ACO and CPA are similar, we have,

$$CO : CA :: CA : CP,$$

or, $$CO : OP :: OP : CP;$$

hence, OC is divided in extreme and mean ratio at P, (*Prob.* 5, *Scho.*, B. 3), and AC is equal to the greater segment, *which was to be proved.*

Cor. 1. If the radius of a circle is divided in extreme and mean ratio, and the greater segment applied ten times to the circumference as a chord, the resulting polygon will be a regular inscribed decagon.

Cor. 2. If the alternate vertices of the decagon are joined by chords, they will form a regular inscribed pentagon.

Scho. By the method of continued duplication, we may construct regular inscribed polygons of 20, 40, 80, &c., sides.

PROPOSITION X. PROBLEM.

The perimeters of a regular inscribed polygon and of a similar circumscribed polygon being given, to find the perimeters of regular inscribed and circumscribed polygons having double the number of sides.

Let AC be any side of the given inscribed polygon and DE the corresponding side of a similar circumscribed polygon, O the common centre of the polygons, and H the middle of the arc AHC.

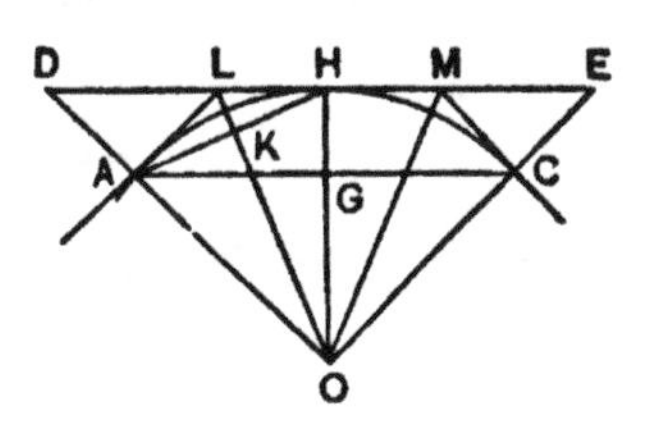

Draw AH, and at A and C draw the tangents AL and CM; then will AH be one side of the required inscribed polygon, and LM one side of the required circumscribed polygon.

Denote the perimeters of the given inscribed and circumscribed polygons, by p and P, and the perimeters of the required inscribed and circumscribed polygons by p' and P′.

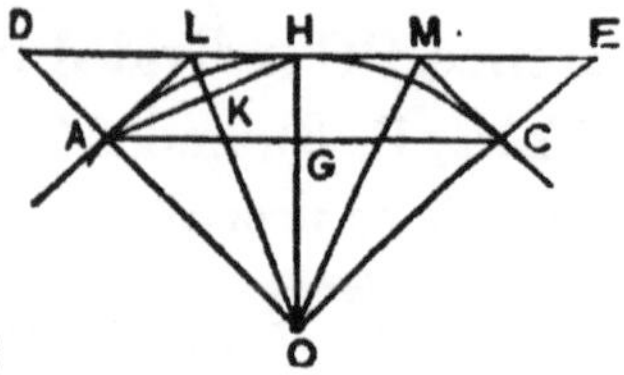

1°. The lines AG and DH are like parts of the perimeters p and P; hence,

$$p : \mathrm{P} :: \mathrm{AG} : \mathrm{DH}, \quad (1)$$

and from the similar triangles OAG and ODH, we have,

$$\mathrm{AG} : \mathrm{DH} :: \mathrm{OA}, \text{ or } \mathrm{OH} : \mathrm{OD}; \quad (2)$$

But, from P. 14, B. 3, we have,

$$\mathrm{OH} : \mathrm{OD} :: \mathrm{LH} : \mathrm{DL}; \quad (3)$$

hence, from (1), (2), and (3), we have,

$$p : \mathrm{P} :: \mathrm{LH} : \mathrm{DL};$$

whence, by composition,

$$p : p + \mathrm{P} :: \mathrm{LH} : \mathrm{LH} + \mathrm{DL}, \text{ or } \mathrm{DH},$$

or, $$2p : p + \mathrm{P} :: 2\mathrm{LH} : \mathrm{DH}. \quad (4)$$

Now, 2LH, or LM, and DH are like parts of P′ and P; hence,

$$2\mathrm{LH} : \mathrm{DH} :: \mathrm{P}' : \mathrm{P}; \quad (5)$$

from (4) and (5), we have,

$$2p : p + \mathrm{P} :: \mathrm{P}' : \mathrm{P},$$

or, $$\mathrm{P}' = \frac{2p\mathrm{P}}{p + \mathrm{P}}. \quad (a)$$

2°. Because HK and HL are like parts of p' and P′, we have,

$$p' : \mathrm{P}' :: \mathrm{HK} : \mathrm{HL}, \quad (6)$$

and from the similar triangles HKL and AGH, we have,

$$\mathrm{HK} : \mathrm{HL} :: \mathrm{AG} : \mathrm{AH}; \quad (7)$$

but, AG and AH are like parts of p and p'; hence,

$$AG : AH :: p : p'; \quad . \ . \ . \ . \ . \ (8)$$

from (6), (7) and (8), we have,

$$p' : P' :: p : p',$$

or,

$$p' = \sqrt{p \times P'} \quad . \ . \ . \ . \ . \ . \ . \ (b)$$

Scho. From formula (*a*) we can find the value of P', and then from formula (*b*) we can find the value of p'.

Proposition XI. Problem.

To find the numerical value of a circumference whose diameter is equal to 1.

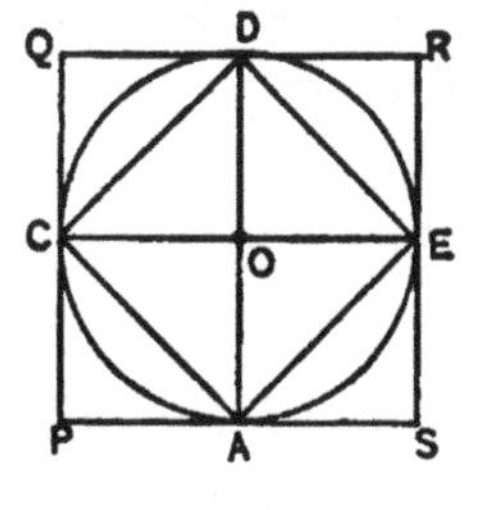

Let ACDE be a square inscribed in a circle OA, whose diameter AD is equal to 1; and let PQRS be a circumscribed square.

From P. 8, *Cor.* 2, B. 4, we have,

$$DE = DO\sqrt{2}, \quad \text{or} \quad DE = \tfrac{1}{2}\sqrt{2};$$

we also have each side of the circumscribed square equal to 1; hence, if we denote the perimeter of the inscribed square by p and that of the circumscribed square by P, we shall have

$$p = 2\sqrt{2} = 2.828427, \quad \text{and} \quad P = 4.$$

Substituting these values in formulas (*a*) and (*b*) of the preceding proposition and reducing, we find

$$P' = 3.313709, \quad \text{and} \quad p' = 3.061468,$$

in which p' and P' are the perimeters of regular inscribed and circumscribed octagons. Denoting these in turn by p and P

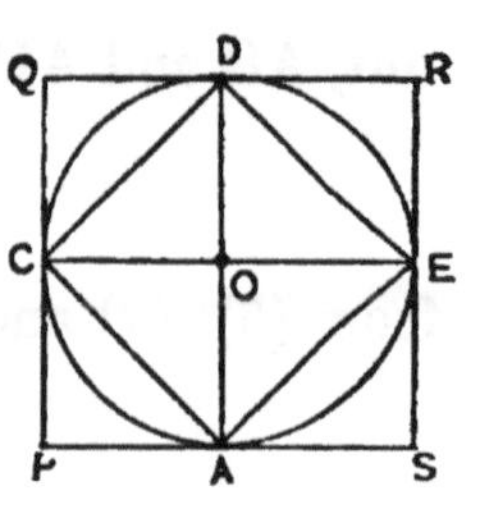

and substituting in formulas (*a*) and (*b*) and reducing, we have

$$P' = 3.182598 \quad \text{and} \quad p' = 3.121445,$$

in which p' and P' are the perimeters of regular inscribed and circumscribed polygons of 16 sides. By continuing this process of continued substitution and reduction, we can find in succession the perimeters of regular inscribed and circumscribed polygons of 32, 64, 128, &c., sides. If the operation is carried on till we come to the polygons of 2048 sides, we shall find

$$P' = 3.141592 \quad \text{and} \quad p' = 3.141588.$$

Now, these values of P' and p' differ from each other by less than *a hundredth thousandth* part of 1, and because the first is greater and the second less than the circumference, either of them differs from the circumference by less than they differ from each other. Hence, 3.14159 is an approximate value of the circumference, true to within less than the unit of the fifth decimal place. This number is usually denoted by the Greek letter π (called *pi*). By continuing the process above indicated the value of π may be found to any degree of accuracy. For most purposes of computation, it will be sufficiently accurate to assume

$$\pi = 3.1416.$$

Cor. 1. Because the circumferences of any two circles are to each other as their diameters, π is the numerical measure of *the ratio of the diameter of any circle to its circumference.*

Cor. 2. If we denote the circumference of a circle whose radius is r by c, we shall have

$$1 : \pi :: 2r : c, \quad \text{or} \quad c = 2\pi r,$$

that is, *the circumference is equal to its radius multiplied by* 2π.

Cor. 3. If we denote the area of a circle whose radius is r by a, we shall have, (P. 5, *Cor.* 2),

$$a = 2\pi r \times \tfrac{1}{2}r = \pi r^2,$$

that is, *the area of a circle is equal to the square of its radius multiplied by* π.

Cor. 4. The arc of a sector whose radius is r, and whose angle at the centre is $n°$, is equal to

$$2\pi r \times \frac{n°}{360°}, \text{ or to } \frac{\pi r n}{180}.$$

Cor. 5. The area of a sector whose radius is r and whose angle at the centre is $n°$, is equal to

$$\pi r^2 \times \frac{n°}{360°}, \text{ or to } \frac{\pi r^2 n}{360}.$$

BOOK VI.

PLANES AND ANGLES FORMED BY PLANES.

Definitions.

88. A **plane**, as already defined, is a surface such that a straight line through any two of its points lies wholly in the surface.

A plane is supposed to extend to an infinite distance in all directions; its position is indicated by two or more of its lines.

89. A line is **perpendicular** to a plane, when it is perpendicular to all the lines of the plane that pass through its **foot**; that is, through the point in which it meets the plane.

Conversely, the plane is then perpendicular to the line.

90. Two planes are **parallel** when they cannot meet, how far soever both may be extended.

91. A line is **parallel** to a plane, when it cannot meet the plane, how far soever both may be extended.

Conversely, the plane is then parallel to the line.

A line that is neither perpendicular, nor parallel, to a plane is said to be oblique to it.

PROPOSITION I. THEOREM.

One plane, and only one, can be passed through three points not in the same straight line.

Let A, C, and D be three points not in the same straight line.

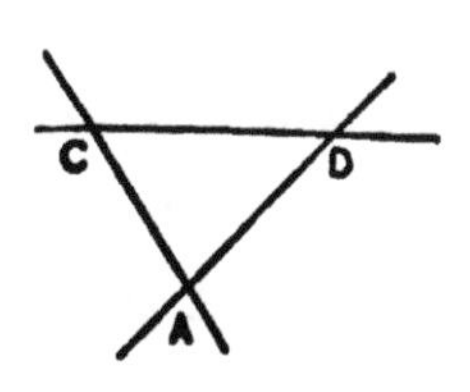

Draw a line through A and C.

Now, it is obvious that a plane can be passed through AC. If we suppose the plane to be revolved about AC, as an axis, it will ultimately reach the point D, and in this position it will pass through A, C, and D; but if it is further revolved about AC in either direction it will no longer contain D. Hence, one plane, and only one, can be passed through A, C, and D, *which was to be proved.*

Cor. 1. Three points not in a straight line determine, that is, fix the position of a plane.

Cor. 2. A straight line and a point not in that line determine the position of a plane.

Cor. 3. Two lines, AC and AD, that intersect, determine the position of a plane. For, a plane through one line and any point of the other will contain both lines and will be fixed in position.

Cor. 4. Two parallel lines determine the position of a plane. For by definition the two lines are in the same plane, and because this plane passes through one of the lines and a point in the other it is fixed in position.

PROPOSITION II. THEOREM.

If two planes intersect, their intersection is a straight line.

Let the planes KL and PQ intersect, and let A and C be two points of their intersection.

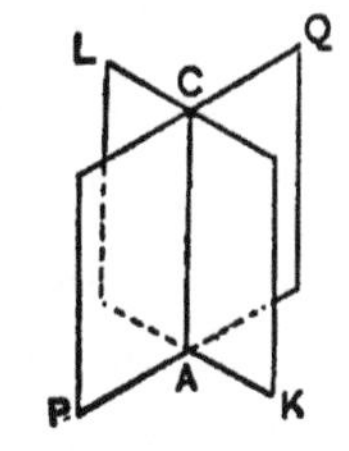

Draw the straight line AC.

From the definition of a plane, AC must lie wholly in both planes; furthermore no point outside of AC can lie in both, otherwise we should have two planes through a line and a point without it, which is impossible. Hence, the intersection is a straight line, *which was to be proved.*

Cor. A single straight line does not determine a plane.

PROPOSITION III. THEOREM.

If a straight line is perpendicular to each of two intersecting lines at their point of intersection, it is perpendicular to the plane of those lines.

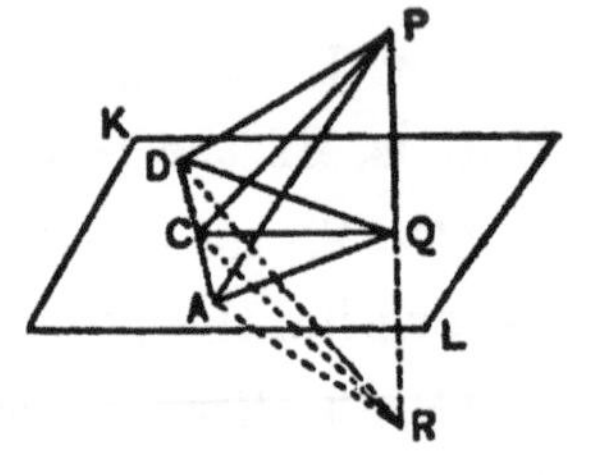

Let AQ and DQ be two lines in the plane KL, and let PQ be perpendicular to both; then is it perpendicular to the plane KL.

Draw any line, QC, through Q in the plane KL, and through any point C of this line, draw AD, cutting AQ and DQ at A and D; also draw PA, PC, and PD. Prolong PQ beyond KL, till QR is equal to PQ, and draw RA, RC, and RD.

Because AQ is perpendicular to PR at its middle point, AR is equal to AP (P. 16, B. 1), and for like reason DR is equal to DP; hence, the triangle ARD is equal to APD, and consequently CR is equal to CP. The line CQ has, therefore, two of its points, Q and C, equally distant from P and R; hence, PR is perpendicular to CQ. But, CQ is any line of KL passing through Q; hence, PQ is perpendicular to the plane KL, *which was to be proved.*

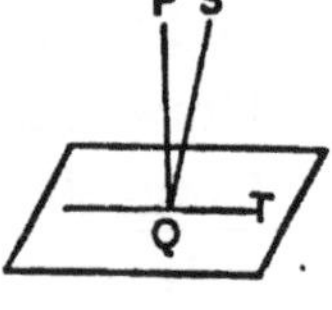

Cor. 1. Only one perpendicular can be drawn to a plane at a given point. For, if there could be two, as PQ and SQ, a plane could be passed through them cutting the given plane in a line, QT, and we should then have two perpendiculars to this line at the same point, which is impossible, (P. 2, B. 1).

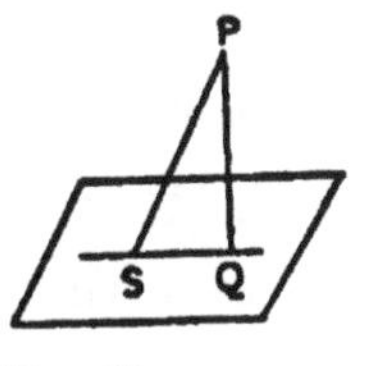

Cor. 2. Only one perpendicular can be drawn to a plane from a point without. For, if two could be drawn, as PQ and PS, we should have two perpendiculars to QS from the same point P, which is impossible, (P. 5, B. 1).

Proposition IV. Theorem.

If from a point without a plane a perpendicular is drawn to the plane, and also oblique lines meeting the plane at different points:

1°. *Any two oblique lines that meet the plane at equal distances from the foot of the perpendicular are equal;*

2°. *Of two oblique lines that meet the plane at unequal distances from the foot of the perpendicular, that which meets it at the greater distance is the greater;*

3°. *The perpendicular is less than any oblique line.*

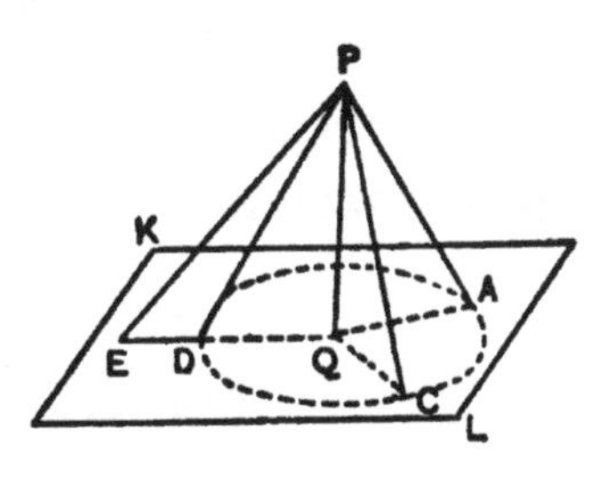

Let P be any point without the plane KL. Let PQ be perpendicular to KL; let PA and PC be oblique lines meeting KL at equal distances from Q; and let PA and PE be oblique lines meeting KL at unequal distances from Q.

1°. The lines PA and PC are equal. For, draw QA and QC: the right-angled triangles PQA and PQC have the side PQ common, and QA equal to QC; they are therefore equal in all their parts; hence, PA and PC are equal, *which was to be proved.*

2°. Draw QE and suppose it greater than QA; then is PE greater than PA. For, lay off QD equal to QA, and draw PD; then is PD equal to PA from what has just been proved. But, PE is greater than PD, (P. 15, B. 1); hence, PE is greater than PA, *which was to be proved.*

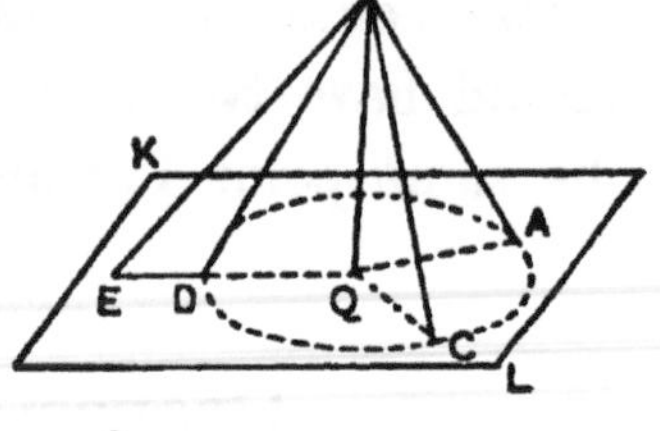

3°. The line PQ is less than any oblique line, as PA. For, in the triangle PQA, the angle A is less than the angle Q; hence, PQ is less than PA, *which was to be proved.*

Cor. 1. All the equal oblique lines that can be drawn from P to KL, meet that plane in a circumference whose centre is Q. Hence, to draw a line perpendicular to a given plane from a point without, find three points of the plane equally distant from the given point and then find the centre of a circumference passing through them; the line from the given point to this centre is the required perpendicular.

Cor. 2. The angle QEP is the **inclination** of PE to the plane KL. All equal lines from P to KL are equally inclined to that plane; and of two unequal lines, the less line has the greater inclination.

PROPOSITION V. THEOREM.

If from the foot of a perpendicular to a plane, a line is drawn perpendicular to any line of the plane, and if the point of intersection is joined with any point of the perpendicular, the last line is perpendicular to the line of the plane.

Let PQ be perpendicular to the plane KL, and let QD be

drawn from Q perpendicular to AC, a line of KL, and let D be joined with any point P on PQ; then is PD perpendicular to AC.

Lay off from D, equal distances DA and DC; draw QA, QC, and PA, PC.

Because QD is perpendicular to AC, and DA equal to DC, QA is equal to QC. Now, the right-angled triangles PQA and PQC, have PQ common, and QA equal to QC; they are therefore equal in all their parts, and consequently PA is equal to PC. Because P and D are each equally distant from A and C, PD is perpendicular to AC, *which was to be proved.*

Cor. 1. AC is perpendicular to the plane of PD and QD, (P. 3).

Cor. 2. The shortest distance from PQ to AC, is QD; for QD is shorter than any other line, PC, that can be drawn from PQ to AC.

Scho. The lines PQ and AC are said to be perpendicular to each other, though they do not intersect. The angle between any two lines in space is equal to an angle formed by drawing lines through any point respectively parallel to the given lines.

PROPOSITION VI. THEOREM.

If one of two parallel lines is perpendicular to a plane, the other one is also perpendicular to that plane.

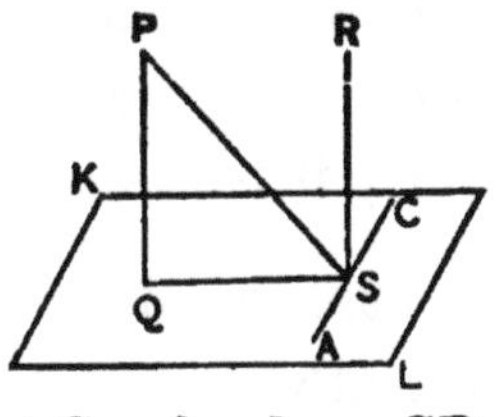

Let PQ and RS be parallel, and let PQ be perpendicular to the plane KL; then is RS perpendicular to KL.

Let QS be the intersection of the plane of the parallels with KL. Draw AC in the plane KL perpendicular to QS at S; also draw SP.

The line AC is perpendicular to the plane of PS and QS, and consequently to the plane of PQ and RS, (P. 5, *Cor.* 1); hence, ASR is a right angle. Because PQ is perpendicular to the plane KL, PQS is a right angle; and because PQ and RS are parallel, QSR is also a right angle, (P. 21, *Cor.* 2, B. 1). Hence, RS is perpendicular to SA and SQ at S, and consequently to their plane KL, *which was to be proved.*

Cor. 1. If a plane KL is perpendicular to each of two lines, PQ and RS, the lines are parallel. For, if not, a line could be drawn through Q parallel to RS, and this would be perpendicular to the plane KL and we should then have two lines through Q perpendicular to KL, which is impossible.

Cor. 2. If two lines, A and C, are parallel to a third line, D, not in their plane, they are parallel to each other. For if a plane is passed perpendicular to D it will also be perpendicular to both A and C.

PROPOSITION VII. THEOREM.

If a line lying without a plane is parallel to a line of the plane, it is parallel to that plane.

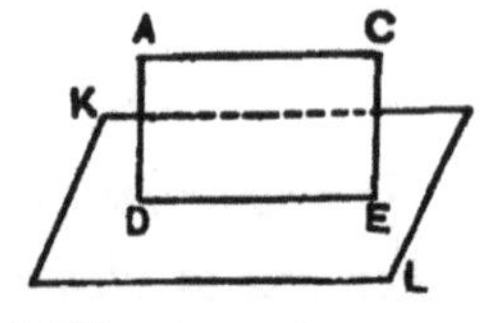

Let AC be parallel to the line DE of the plane KL; then is AC parallel to KL.

For, if AC could meet KL it would do so at some point of DE, because AC and DE are in the plane AE; but AC and DE cannot meet because they are parallel; hence, AC cannot meet the plane KL; it is therefore parallel to it, *which was to be proved.*

Cor. A plane through one of two parallels is parallel to the other.

PROPOSITION VIII. THEOREM.

If two planes are perpendicular to the same line they are parallel to each other.

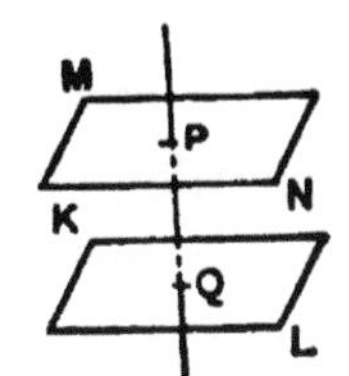

Let the planes KL and MN be perpendicular to PQ at Q and P; then are they parallel to each other.

For if the planes could meet at any point, the lines from that point to P and Q would both be perpendicular to PQ; but, this is impossible, (P. 5, B. 1); hence, the planes cannot meet; they are therefore parallel to each other, *which was to be proved.*

Cor. If the points P and Q coincide, the planes KL and MN also coincide; hence, only one plane can be perpendicular to the same line at the same point.

PROPOSITION IX. THEOREM.

If a plane intersect two parallel planes the lines of intersection are parallel.

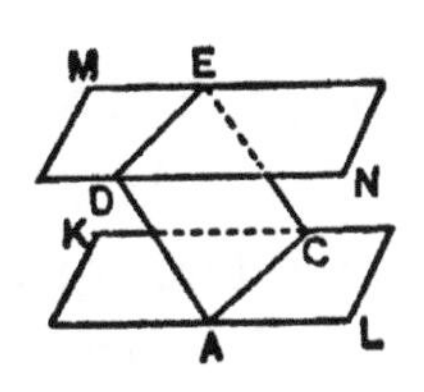

Let the plane CD intersect the parallel planes KL and MN in AC and DE; then are AC and DE parallel.

For, if they are not parallel they will meet, inasmuch as they lie in the plane CD, and if they meet, the planes KL and MN will meet; but, this is impossible; hence, AC and DE are parallel, *which was to be proved.*

PROPOSITION X. THEOREM.

If a line is perpendicular to one of two parallel planes it is also perpendicular to the other.

Let KL and MN be two parallel planes and let PQ be perpendicular to KL; then will it be perpendicular to MN.

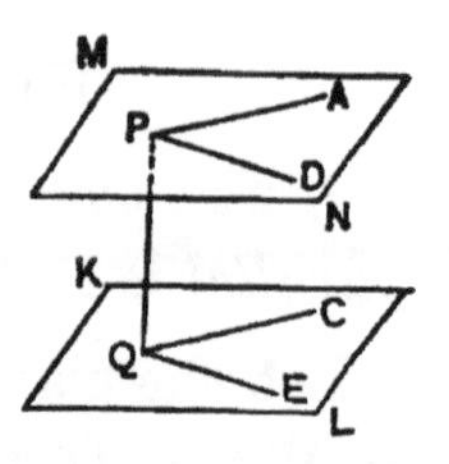

Pass a plane through PQ intersecting KL in QC and MN in PA. Now PQC is a right angle, and because QC and PA are parallel, (P. 9), QPA is also a right angle. Pass a second plane through PQ, intersecting KL in QE and MN in PD; for the same reason as before, QPD is a right angle. Hence, PQ is perpendicular to both PD and PA at P; it is therefore perpendicular to their plane MN, *which was to be proved.*

Cor. PQ is the shortest distance from P to KL.

PROPOSITION XI. THEOREM.

Parallel lines included between parallel planes are equal.

Let AC and DE be parallel lines between the parallel planes KL and MN; then are they equal.

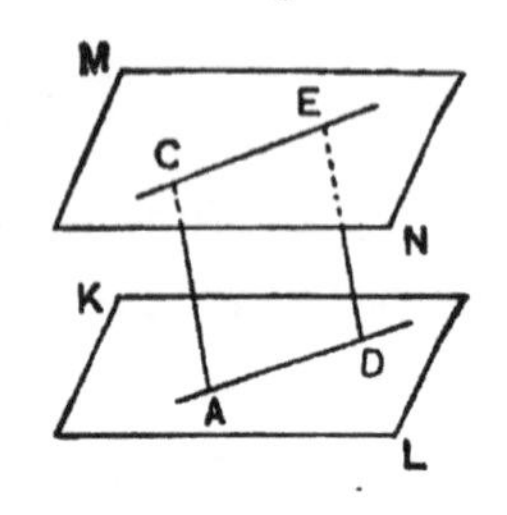

Draw AD and CE; these are the intersections of the plane AE with the planes KL and MN; they are therefore parallel; hence, ACED is a parallelogram; its opposite sides are therefore equal, (P. 26, B. 1); consequently AC and DE are equal, *which was to be proved.*

Cor. 1. If AC and DE are perpendicular to KL they are also perpendicular to MN, and because they are equal, it follows that parallel planes are everywhere equally distant.

Cor. 2. If lines are drawn from any two points of CE perpendicular to the plane KL they are equal; hence, a line that is parallel to a plane is everywhere equally distant from that plane.

PROPOSITION XII. THEOREM.

If the sides of two angles, not in the same plane, are parallel and lie in the same direction, the angles are equal, and their planes are parallel.

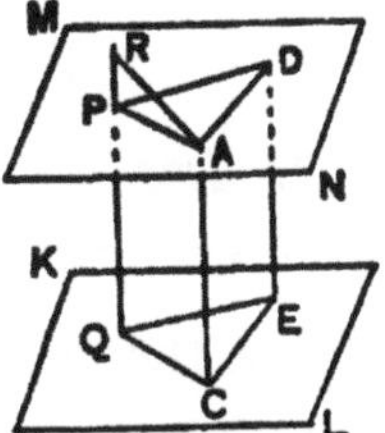

Let PA and PD be respectively parallel to QC and QE, and let them lie in the same direction; then is the angle EQC equal to DPA, and their planes, KL and MN, are parallel.

1°. Draw any line, AD cutting PA and PD at A and D; lay off QC equal to PA, QE equal to PD, and draw CE; also draw AC, DE, and PQ.

Because PA and QC are parallel and equal, APQC is a parallelogram, and consequently AC is parallel and equal to PQ; for a like reason, DE is also parallel and equal to PQ. Now, because AC and DE are parallel and equal to PQ, they are parallel and equal to each other; hence, AD and CE are parallel and equal. The sides of the triangle APD are therefore respectively equal to the sides of CQE; hence, these triangles are equal in all their parts; consequently, the angles EQC and DPA are equal, *which was to be proved.*

2°. The planes KL and MN are parallel. For if they are not, let a plane be passed through A parallel to KL and suppose it to cut one of the lines, as PQ, in some other point than P, say R; AR will then be parallel to CQ (P. 9), and we shall have two lines through A parallel to CQ, which is impossible. Hence, the plane MN is parallel to KL, *which was to be proved.*

Cor. 1. If three lines, AC, DE, and PQ, not in the same plane, are equal and parallel, the triangles APD and CQE, formed by joining their extremities are equal, and their planes are parallel.

Cor. 2. If two intersecting planes, PC and PE, are cut by two parallel planes, KL and MN, the angles APD and CQE formed by their intersections are equal.

PROPOSITION XIII. THEOREM.

If two lines are cut by three parallel planes, they are divided proportionally.

Let the lines AC and DE be cut by the parallel planes KL, MN, and PQ, the first at A, H, and C, and the second at D, F, and E; then is,

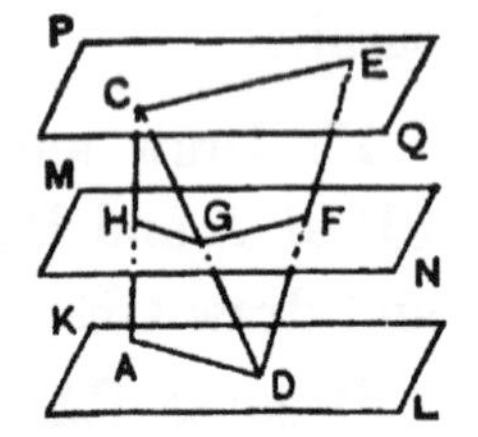

AH : HC :: DF : FE.

Draw DC, and let it pierce the plane MN at G. Draw AD, HG, GF, and CE.

The lines AD and HG are parallel, (P. 9); hence,

AH : HC :: DG : GC; . . . . . (1)

the lines GF and CE are also parallel; hence,

DF : FE :: DG : GC; . . . . . (2)

from (1) and (2), we have,

AH : HC :: DF : FE,

which was to be proved.

Cor. 1. If two lines are cut by any number of parallel planes, they are divided proportionally.

Cor. 2. If any number of lines are cut by any number of planes, they are divided proportionally.

Definitions.

92. A **diedral angle** is the inclination of two planes that meet. The planes are called **faces** of the angle, and the line in which they meet is called the **edge of the angle.**

A diedral angle is the same as a plane angle formed by two lines, one in each plane, and both perpendicular to the edge at the same point. Thus, if AC and DC are perpendicular to ML at C, the diedral angle of the planes LK and MN is equal to ACD. If this is a right angle, the planes are **perpendicular** to each other.

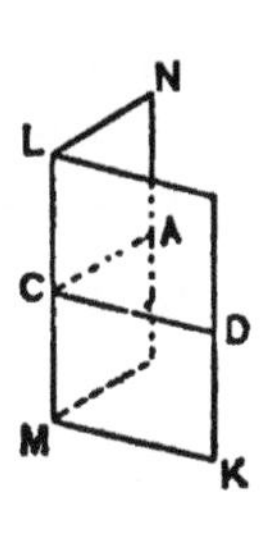

If two planes intersect, they form four diedral angles, equal two and two. These angles are called by the same names, and bear to each other the same relations, as the corresponding plane angles.

93. A **polyedral angle** is the angular space included within three or more planes meeting at a common point.

The planes are called **faces** of the polyedral angle, the lines in which the faces meet are called **edges**, and the point at which the edges meet is the **vertex.** The angle between two consecutive edges is called a **lateral angle**, and the angle between two consecutive faces is called an **interfacial angle.**

94. A **triedral angle** is a polyedral angle having three faces.

In the triedral angle whose vertex is O, the edges are OA, OC, and OD; the lateral angles are AOC, COD, and DOA; and the interfacial angles are the angles between the planes AOC, COD; COD, DOA; and DOA, AOC.

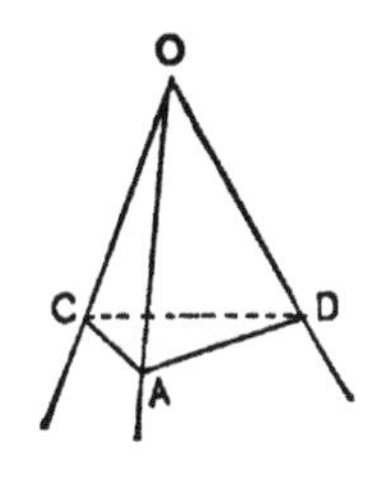

PROPOSITION XIV. THEOREM.

If a line is perpendicular to a plane, every plane through the line is perpendicular to that plane.

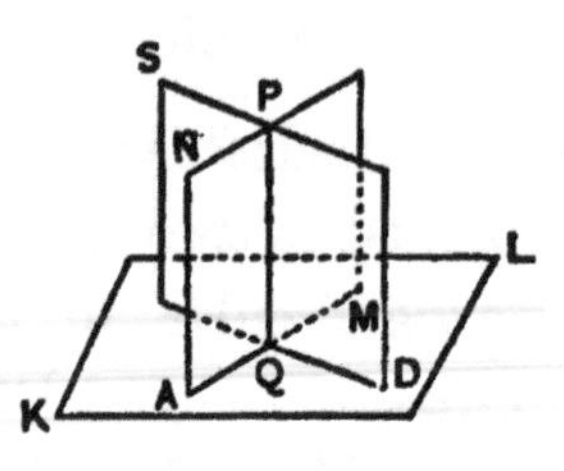

Let PQ be perpendicular to the plane KL, and let MN be any plane through PQ; then is MN perpendicular to KL.

Draw QD in KL perpendicular to AM, the line in which MN intersects KL.

Because PQ is perpendicular to the plane KL it is perpendicular to both AM and QD. But, PQ is in the plane MN, and QD is in the plane KL, and both are perpendicular to AM at Q; hence, the planes MN and KL are perpendicular to each other, (*Art.* 92), *which was to be proved.*

Cor. 1. Because QD is perpendicular to both QA and QP, it is perpendicular to their plane MN, that is, if two planes are perpendicular to each other, a line in one plane perpendicular to their intersection is perpendicular to the other plane.

Cor. 2. If three lines QD, QA, and QP, are perpendicular to each other at a common point, each is perpendicular to the plane of the other two.

Cor. 3. If two planes are perpendicular to a third plane, their intersection is perpendicular to that plane.

PROPOSITION XV. THEOREM.

The sum of any two lateral angles of a triedral angle is greater than the third lateral angle.

Let AOC, COD, and DOA be the lateral angles of the triedral angle O; then is the sum of any two, as AOC and COD, greater than the third, DOA.

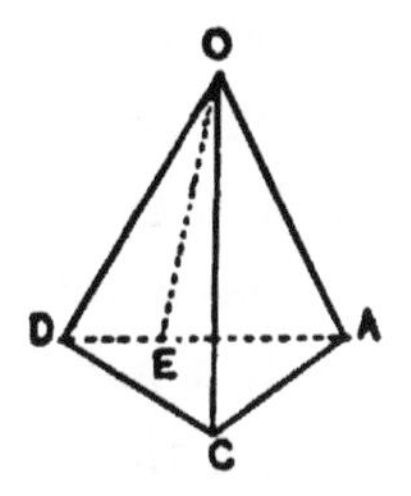

If DOA is less than, or equal to, either of the other lateral angles, the proposition needs no proof; let us then suppose DOA is greater than either of the others.

Draw any line AD, cutting OA and OD at A and D; then construct the angle AOE equal to AOC; also lay off OC equal to OE and draw AC and CD.

The triangles AOE and AOC have two sides and their included angle equal, each to each; they are therefore equal in all their parts; consequently AE is equal to AC.

In the triangle ACD we have

$$AC + CD > AD;$$

subtracting AC from the first member and its equal AE from the second member of this inequality, we have

$$CD > DE.$$

The triangles COD and DOE have DO common, CO equal to EO, and CD greater than DE; hence, (P. 9, B. 1),

$$COD > DOE;$$

adding AOC to the first member of this inequality and its equal AOE to the second member, we have

$$AOC + COD > DOE + AOE, \text{ or } AOD,$$

which was to be proved.

Proposition XVI. Theorem.

The sum of the lateral angles of a polyedral angle is less than four right angles.

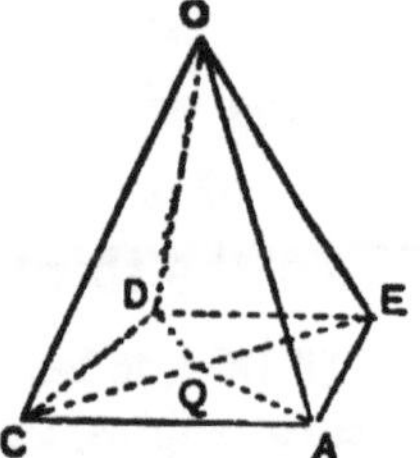

Let AOC, COD, DOE, and EOA be the lateral angles of a polyedral angle O; then is their sum less than four right angles.

Pass a plane cutting from the polyedral angle a polygon ACDE, and from a point within this polygon, as Q, draw lines to the vertices A, C, D, and E; these will divide the polygon into as many **basal triangles** as the polyedral angle has faces, and the sum of all the angles of the basal triangles will be equal to the sum of all the angles of the lateral triangles AOC, COD, DOE, and EOA. If we consider the triedral angle A, we have OAE + OAC greater than QAE + QAC (P. 15); in like manner we have at the triedral angle C, OCA + OCD greater than QCA + QCD; and so on for the other triedral angles. Hence, the sum of all the angles at the bases of the *lateral triangles* is greater than the sum of all the angles at the bases of the *basal triangles;* consequently the sum of all the angles at O is less than the sum of all the angles at Q. But, the sum of the angles at O is the sum of the lateral angles of the given polyedral angle, and the sum of the angles at Q is equal to four right angles; hence, the sum of the lateral angles of the polyedral angle is less than four right angles, *which was to be proved.*

Scho. The polyedral angle is supposed to be convex; that is, the faces are so arranged that any plane section of the angle is a convex, or salient, polygon.

PROPOSITION XVII. THEOREM.

If the lateral angles of two triedral angles are equal, each to each, the corresponding interfacial angles are also equal, each to each.

Let O and Q be two triedral angles in which the lateral angle

AOC is equal to EQF, COD to FQG, and DOA to GQE; then are the corresponding interfacial angles equal.

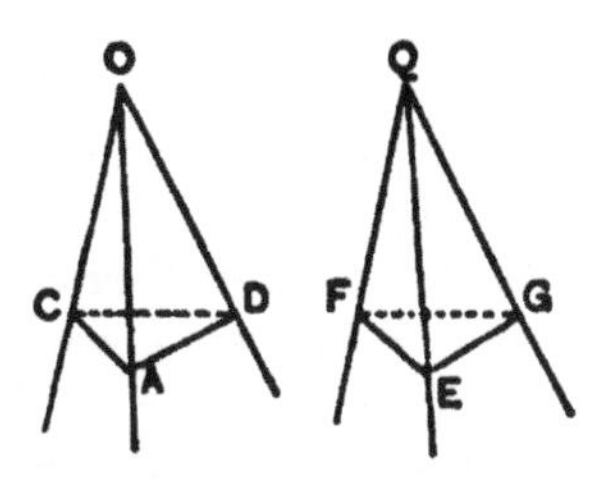

Take any point A, on OA, and from it draw AC and AD, the former in the plane AOC, the latter in AOD, and both perpendicular to OA; lay off QE equal to OA and from E draw EF and EG, the former in the plane EQF, the latter in EQG, and both perpendicular to QE; also draw CD and FG. The angle DAC is the interfacial angle whose edge is OA, and GEF is the interfacial angle whose edge is QE.

In the right-angled triangle, OAC and QEF, OA is equal to QE, and the angle COA to FQE; hence, these triangles are equal in all their parts; consequently AC is eqnal to EF, and OC to QF. In like manner it may be shown that AD is equal to EG, and OD to QG. Again, the triangles COD and FQG have two sides and their included angle equal, each to each; hence, these triangles are equal in all their parts; consequently, CD is equal to FG. Now, the triangles ACD and EFG have their sides equal, each to each; hence, the triangles are equal in all their parts; consequently, the angles DAC and GEF are equal, that is, the interfacial angles whose edges are OA and QE are equal. In like manner it may be shown that the interfacial angles whose edges are OC and QF are equal; also, that the interfacial angles whose edges are OD and QG are equal. Hence, the corresponding interfacial angles of the two triedral angles are equal, *which was to be proved.*

Scho. In the case considered, the faces of the triedral angles are similarly placed, and the triedral angles may be so placed as to coincide throughout. If the faces are not similarly placed, as shown in the diagram, the angles cannot be made to coincide throughout: in this case they are said to be **symmetrical**. Two triedral angles of this kind can be so placed that the lateral angles COD and FQG will coincide, the edge OA falling on one side and the edge QE on the other side of the common face. When so placed, for every point of OA on one side of the common plane there will be a corresponding point of QE on the other side, so situated that the line joining them is perpendicular to, and bisected by, that plane. The plane is then called a **plane of symmetry**.

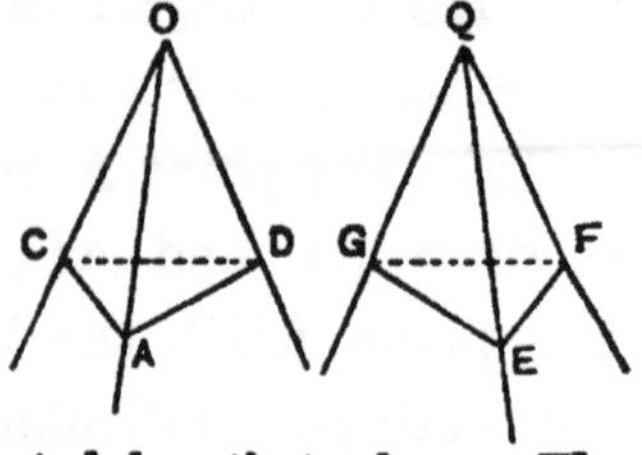

Definition.

95. Any two magnitudes are **symmetrical** with respect to a plane, when they are so related that for every point of one there is a corresponding point of the other, such that the line joining them is perpendicular to, and bisected by, the plane.

The student may acquire a rough idea of what is meant by symmetry if he will place his hands together, palm to palm. The hands when so placed fulfil the requirements of symmetry.

BOOK VII.

COMPARISON AND MEASURE OF POLYEDRONS.

Definitions.

96. A **polyedron** is a volume bounded by plane surfaces.

The polygons determined by the intersection of these planes are called **faces**; the lines in which the faces meet are called **edges**; and the points in which the edges meet are called **vertices**.

97. A **prism** is a polyedron, two of whose faces are parallel and equal polygons, the remaining faces being parallelograms; as ACDEF–P.

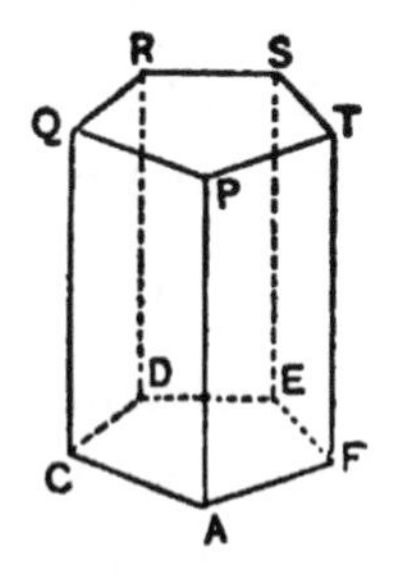

The parallel polygons are called **bases**: that on which the prism is supposed to stand is called the **lower base**, or simply the **base**; the other is called the **upper base**. The remaining faces are called **lateral faces**, and taken together they make up the **lateral surface**; the lines in which the lateral faces meet are called **lateral edges**.

98. The **altitude** of a prism is the perpendicular distance between its bases.

99. A **right prism** is a prism whose lateral edges are perpendicular to the planes of its bases: in this case the altitude is equal to any lateral edge.

Prisms are said to be *triangular*, *quadrangular*, *pentagonal*, &c., according to the number of sides of their bases.

100. A **right section** of a prism is a section made by a plane perpendicular to its lateral edges.

101. A **parallelopipedon** is a prism whose bases are parallelograms.

102. A **rectangular parallelopipedon** is a parallelopipedon whose bases and lateral faces are all rectangles; as AR.

A **cube** is a rectangular parallelopipedon whose faces are equal squares.

103. A **pyramid** is a polyedron, one of whose faces is any polygon, the others being triangles having a common vertex; as ACDE–V.

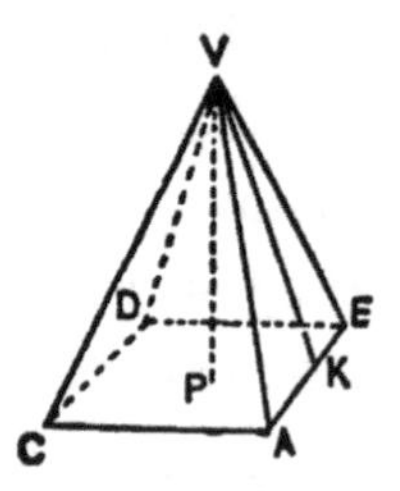

The polygon is the **base**, and the triangles are **lateral faces**. The lateral faces taken together make up the **lateral surface**; the lines in which the lateral faces meet are **lateral edges**; and the point at which the lateral edges meet is the **vertex**.

104. The **altitude** of a pyramid is the perpendicular distance from its vertex to the plane of its base; as VP.

105. A **right pyramid** is one whose base is a regular polygon, the line from the centre of this polygon to the vertex being perpendicular to the base.

It is obvious from P. 4. B. 6, that the lateral edges of a right pyramid are all equal; it is also plain that the lateral faces are all equal.

106. The **slant height** of a right pyramid is the altitude of any lateral face; as VK.

107. A **frustum of a pyramid** is that part of a pyramid included between the base and a secant plane parallel to the base.

The intersection of the secant plane with the pyramid is called the **upper base** of the frustum, and the distance between the upper and the lower base is the **altitude**. If the frustum is part of a right pyramid its **slant height** is the altitude of any lateral face.

108. Two polyedrons are **similar** when their faces are similar and similarly placed.

109. A **diagonal of a polyedron** is a line joining any two vertices not in the same plane.

Proposition I. Theorem.

A secant plane parallel to the base of a prism divides the lateral edges and the altitude proportionally, and the section which it cuts from the prism is equal to the base.

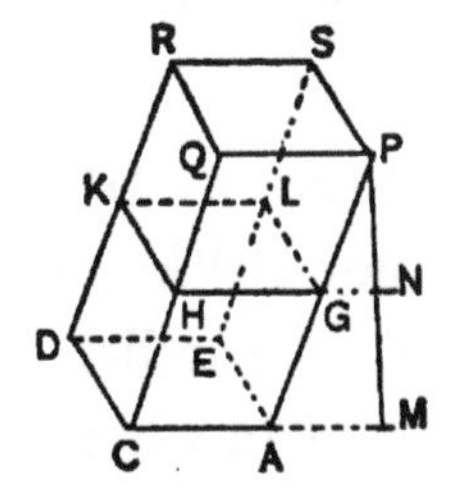

Let ACDE–P be a prism whose altitude is PM; let HL be a section made by a plane parallel to the base; and let N be the point in which this plane cuts PM.

1°. Because the planes CE, HL, and QS are parallel, they divide the lines AP, CQ, DR, ES, and MP proportionally (P. 13, *Cor.* 2, B. 6), *which was to be proved.*

2°. Because the planes CE and HL are parallel, the angles of GHKL are equal to the corresponding angles of ACDE, (P. 12, *Cor.* 2, B. 6), and because the corresponding sides of

GHKL and ACDE are parallels included between parallels, the polygons are equal to each other; hence, the section HL is equal to the base CE, *which was to be proved.*

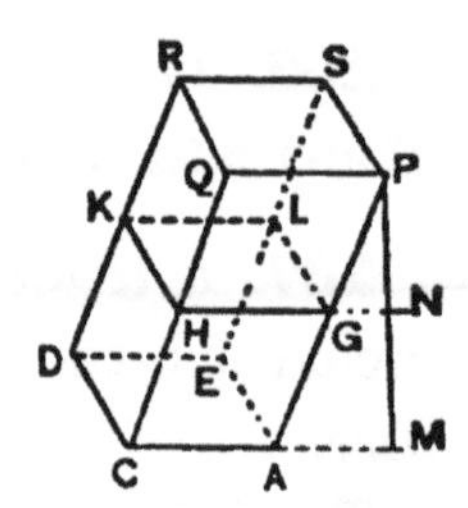

Cor. 1. If MN is equal to NP, AG is equal to GP, and the two segments of the prism are equal. For let the upper segment be placed on the lower one so that the section HL shall coincide with the base CE; then, because the lateral angles of the triedral angles G and A are equal, each to each, and similarly placed, GP will fall on AG, and P will coincide with G; for like reason Q will coincide with H, R with K, and S with L; hence, the segments will coincide throughout.

Cor. 2. If any number of planes are passed parallel to the base so as to divide MP into equal parts, they will also divide the prism into equal segments.

PROPOSITION II. THEOREM.

A secant plane parallel to the base of a pyramid divides the lateral edges and the altitude proportionally, and the section which it cuts from the pyramid is similar to the base.

Let ACDE–V be a pyramid whose altitude is VM; let QS be a section made by a plane parallel to the base, and let N be the point in which this plane cuts VM. Also let KL be a plane through V parallel to the base.

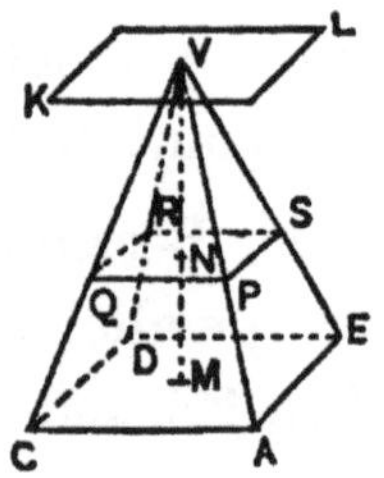

1°. Because the planes CE, QS, and KL are parallel, they divide the lines AV, CV, DV, EV, and MV proportionally, *which was to be proved.*

2°. Because the planes CE and QS are parallel, the angles

of the section QS are equal to the corresponding angles of the base. Again, because PQ is parallel to AC,

$$PQ : AC :: QV : CV, \quad . \quad . \quad . \quad . \quad (1)$$

and in like manner,

$$QR : CD :: QV : CV; \quad . \quad . \quad . \quad . \quad (2)$$

from (1) and (2), we have,

$$PQ : AC :: QR : CD.$$

In like manner it may be shown that the remaining sides of the section QS are proportional to the corresponding sides of the base CE. Hence, the section is similar to the base, *which was to be proved.*

Cor. 1. The section QS, is to the base CE, as the square of the distance of the section from the vertex, is to the square of the altitude of the pyramid. For, the section being similar to the base, we have, (P. 12, *Cor.* 2, B. 4),

$$PQRS : ACDE :: \overline{PQ}^2 : \overline{AC}^2. \quad . \quad . \quad . \quad (3)$$

But, $\qquad PQ : AC :: QV : CV :: NV : MV,$

or, $\qquad \overline{PQ}^2 : \overline{AC}^2 :: \overline{NV}^2 : \overline{MV}^2; \quad . \quad . \quad . \quad . \quad (4)$

from (3) and (4), we have,

$$PQRS : ACDE :: \overline{NV}^2 : \overline{MV}^2.$$

Cor. 2. If two pyramids having equal bases lying in the same plane, and equal altitudes, are cut by a plane parallel to the bases, the sections are equal.

PROPOSITION III. THEOREM.

The area of the lateral surface of a right prism is equal to the perimeter of its base multiplied by its altitude.

Let ACDEF–P be a right prism; then are all its lateral faces rectangles, whose bases are the sides of ACDEF, and whose altitudes are equal to the altitude of the prism.

The area of any lateral face, as AQ, is equal to its base, AC, multiplied by its altitude, AP; hence, the sum of the areas of all the lateral faces is equal to the sum of the sides of ACDEF, multiplied by the altitude of the prism AP, *which was to be proved.*

Proposition IV. Theorem.

The area of the lateral surface of a right pyramid is equal to the perimeter of its base multiplied by half its slant height.

Let ACDE–V be a right pyramid whose slant height is VK; then are all its lateral faces equal isosceles triangles, whose bases are the sides of ACDE, and whose altitudes are equal to VK.

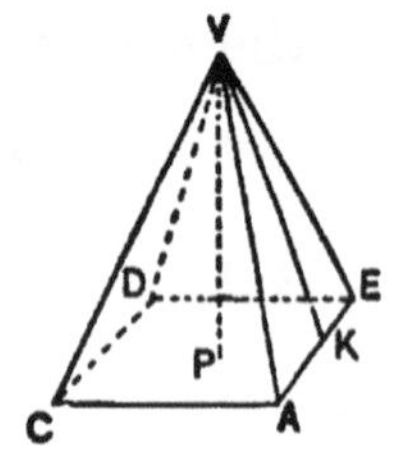

The area of any lateral face, as AEV, is equal to its base AE, multiplied by half its altitude KV; hence, the sum of the areas of all the lateral faces is equal to the sum of the sides of the base, multiplied by half the slant height, VK, *which was to be proved.*

Cor. The lateral surface of a frustum of a right pyramid is equal to half the sum of the perimeters of its bases, multiplied by its slant height. For any lateral face, as AQ, (see figure, P. 2), is equal to half the sum of its sides, AC and PQ, multiplied by the slant height; hence, the lateral surface is equal to half the sum of the perimeters of ACDE and PQRS, multiplied by the slant height.

PROPOSITION V. THEOREM.

The volume of any triangular prism is equal to that of a right prism, whose base is a right section of the given prism, and whose altitude is equal to the lateral edge of that prism.

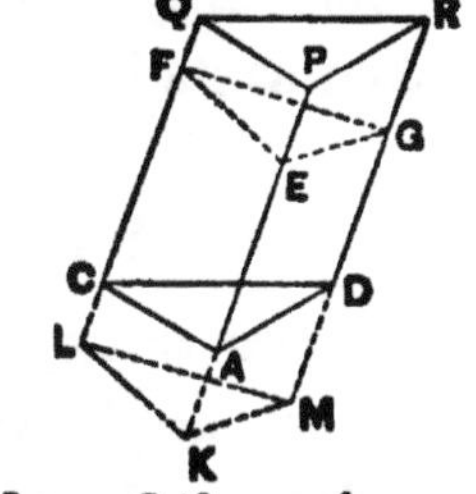

Let ACD–PQR be any triangular prism, and let EFG be one of its *right* sections. Prolong the lateral edges, making AK equal to PE, CL to QF, and DM to RG; draw KL, LM, and MK; then will EK be equal to PA, FL to QC, and GM to RD, and each will be equal to a lateral edge of the prism.

Because EK and FL are parallel and equal, EF is parallel and equal to KL; for like reason, FG is parallel and equal to LM, and GE to MK; hence, KLM–E is a right prism whose base is equal to EFG and whose altitude is equal to AP. Furthermore, the polyedron KLM–ACD is equal to the polyedron EFG–PQR, because all the parts of the one are equal to the corresponding parts of the other and are similarly placed. Now, if we take the first of these, from the polyedron KLM–PQR, there will remain the prism ACD–P, and if we take the second from the same polyedron, there will remain the prism KLM–E; hence, these prisms are equal, *which was to be proved.*

PROPOSITION VI. THEOREM.

The opposite lateral faces of a parallelopipedon are equal, each to each.

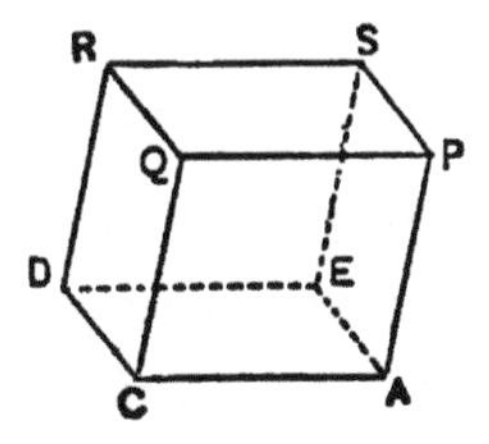

Let ACDE–P be a parallelopipedon whose bases are AD and PR; these bases are equal and parallel polygons by definition, and the lateral faces are parallelograms.

Because CA is equal and parallel to DE, and CQ to DR, the faces CP and DS are equal and parallel; in like manner it may be shown that the faces CR and AS are equal and parallel, *which was to be proved.*

Cor. Any two opposite faces of a parallelopipedon may be taken as its bases.

PROPOSITION VII. THEOREM.

If two triangular prisms have the faces that include a triedral angle in one equal to the faces that include a triedral angle in the other, each to each, and if these faces are similarly placed, the prisms are equal.

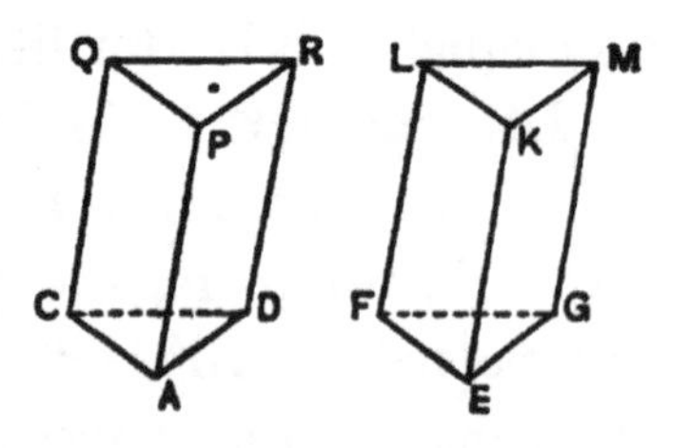

Let ACD–P and EFG–K be two triangular prisms having the face EFG equal to ACD, the face EL to AQ, and the face FM to CR.

Because the lateral angles about F are equal to those about C, and because they are similarly placed, the triedral angle F is equal to the triedral angle C, (P. 17, *Scho.*, B. 6). Let the prism EFG–K be placed on ACD–P so that the triedral angles F and C shall coincide; then will L fall on Q, M on R, and K on P, and consequently, the prisms will coincide throughout; they are therefore equal, *which was to be proved.*

Cor. If two right triangular prisms have equal bases and equal altitudes they are equal.

PROPOSITION VIII. THEOREM.

If a plane is passed through the diagonally opposite edges of a parallelopipedon, it divides the parallelopipedon into two equal triangular prisms.

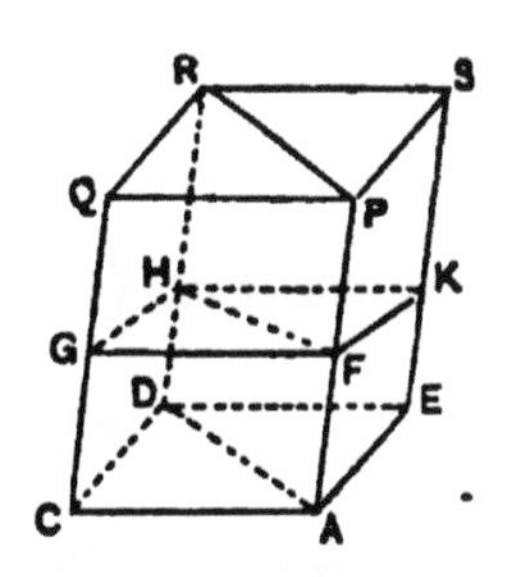

Let ACDE–P be a parallelopipedon, AR a plane through AP and DR, and let FGHK be one of its right sections; then are FGH and HKF right sections of the prisms ACD–P and DEA–P.

The side FG is parallel and equal to KH, the side GH is parallel and equal to FK, and the side HF is common; hence, the triangles FGH and HKF are equal. Now, the triangular prism ACD–P is equal to a right prism whose base is FGH and whose altitude is AP, and the triangular prism DEA–P is equal to a right prism whose base is HKF and whose altitude is AP. But, right prisms whose bases are FGH and HKF, and whose altitudes are AP, are equal (P. 7, *Cor.*); hence, the prisms ACD–P and DEA–P are equal, ***which was to be proved.***

PROPOSITION IX. THEOREM.

Two rectangular parallelopipedons having equal bases, are to each other as their altitudes.

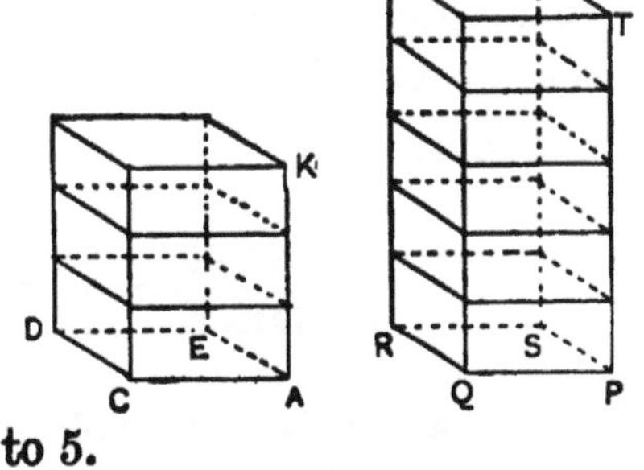

Let ACDE–K and PQRS–T be two rectangular parallelopipedons whose equal bases AD and PR lie in the same plane PD.

1°. Let their altitudes AK and PT be to each other as two whole numbers, say as 3 is to 5.

Divide AK into 3, and PT into 5 equal parts, and through the points of division pass planes parallel to the plane PD; these will divide the first parallelopipedon into 3 segments and the second prism into 5 segments, all equal to each other, (P. 1, *Cor.* 2); hence,

$$\text{ACDE–K} : \text{PQRS–T} :: 3 : 5; \quad . \quad . \quad . \quad (1)$$

but,
$$\text{AK} : \text{PT} :: 3 : 5; \quad . \quad . \quad . \quad . \quad . \quad (2)$$

hence, from (1) and (2), we have

$$\text{ACDE–K} : \text{PQRS–T} :: \text{AK} : \text{PT},$$

which was to be proved.

2°. Let the altitudes AK and PT be incommensurable. Let AK be divided into m equal parts, and let one of these be applied to PT and suppose that it is contained in it n times with a remainder less than that part; then will

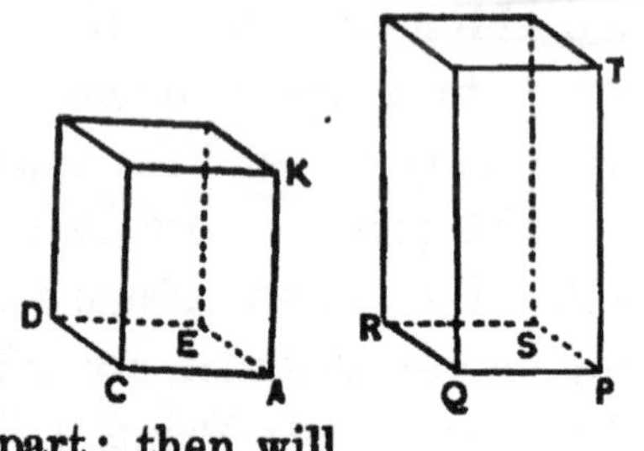

$$\frac{\text{PT}}{\text{AK}} = \frac{n}{m} \textit{ to within less than } \frac{1}{m}; \quad . \quad . \quad . \quad (3)$$

through the points of division of AK and PT pass planes parallel to PD; these planes will divide ACDE–K into m equal segments, and PQRS–T into n segments, each equal to those of ACDE–K, with a remainder less than one of these segments; hence,

$$\frac{\text{PQRS–T}}{\text{ACDE–K}} = \frac{n}{m} \textit{ to within less than } \frac{1}{m}. \quad . \quad . \quad (4)$$

Because m is any whole number, the ratios expressed by the first members of (4) and (3) are equal; hence,

$$\text{ACDE–K} : \text{PQRS–T} :: \text{AK} : \text{PT},$$

which was to be proved.

PROPOSITION X. THEOREM.

Two rectangular parallelopipedons, having equal altitudes, are to each other as their bases.

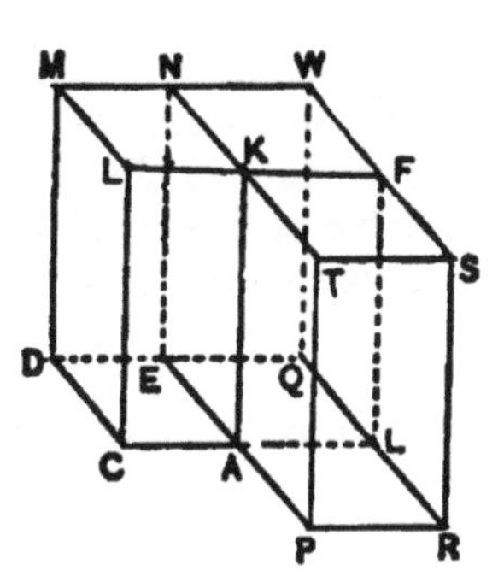

Let the rectangular parallelopipedons ACDE–K and PEQR–S have equal altitudes and let them be placed as shown in the diagram.

Prolong the plane CK till it meets the face RW in the line LF.

The parallelopipedons AM and AW have the same base AN; they are therefore to each other as their altitudes, that is,

$$ACDE\text{–}K : LAEQ\text{–}F :: ED : EQ; \quad . \quad . \quad (1)$$

the parallelopipedons AW and PW have the same base EW; they are therefore to each other as their altitudes, that is,

$$LAEQ\text{–}F : RPEQ\text{–}S :: EA : EP; \quad . \quad . \quad (2)$$

multiplying (1) by (2), term by term, and dividing the first couplet of the resulting proportion by LAEQ–F, we have,

$$ACDE\text{–}K : RPEQ\text{–}S :: ED \times EA : EQ \times EP.$$

But, $ED \times EA$ is equal to the base of the first parallelopipedon and $EQ \times EP$ is equal to the base of the second parallelopipedon; hence, these parallelopipedons are to each other as their bases, *which was to be proved.*

Definitions and Explanations.

110. The unit of volume is a cube, each of whose edges is equal to the linear unit.

111. The **measure of a volume** is the number of times it contains the assumed unit.

If two volumes have the same measure, they are said to be equal; if they are capable of superposition, they are said to be equal in all their parts.

The product of three lines is the product of the numbers that express the lengths of the lines. The product of a surface by a line is the product of the number of superficial units in the surface by the number of linear units in the line. In either case, the unit of the product is the unit of volume.

The *length*, the *breadth*, and the *height*, of a magnitude are called its **dimensions**.

Proposition XI. Theorem.

The volume of a rectangular parallelopipedon is equal to the product of its base and altitude, that is, to the product of its three dimensions.

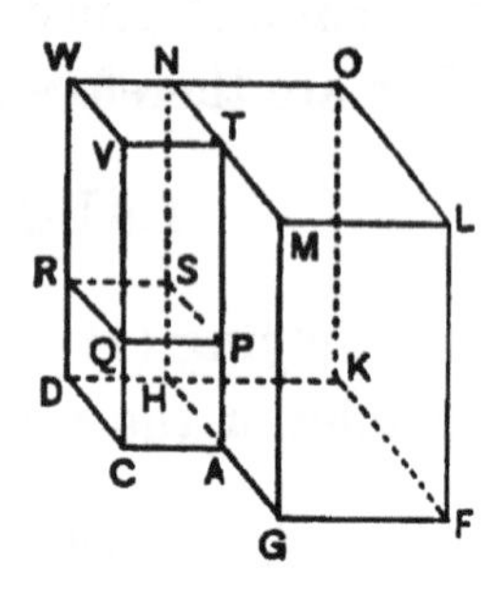

Let FN be a rectangular parallelopipedon and let AR be a cube, each of whose edges is a linear unit, and let them be placed as shown in the diagram.

Extend the faces AQ, CR, and DS till they meet the plane of the upper base of FN.

Now, the parallelopipedons FN and AW have the same altitude HN; they are therefore proportional to their bases; hence,

$$FGHK\text{-}L : ACDH\text{-}T :: FGHK : ACDH. \quad . \; . \; (1)$$

Again, the parallelopipedons AW and AR have the same base AD; they are therefore to each other as their altitudes; hence,

$$ACDH\text{-}T : ACDH\text{-}P :: HN : HS; \quad . \; . \; (2)$$

multiplying proportions (1) and (2), term by term, and dividing both terms of the first couplet of the resulting proportion by ACDH–T, we have,

$$\text{FGHK–L} : \text{ACDH–P} :: \text{FGHK} \times \text{HN} : \text{ACDH} \times \text{HS},$$

or, passing to numerical values,

$$\text{FGHK–L} : 1 :: \text{FGHK} \times \text{HN} : 1,$$

whence, $\text{FGHK–L} = \text{FGHK} \times \text{HN} = \text{HG} \times \text{HK} \times \text{HN}$,

which was to be proved.

PROPOSITION XII. THEOREM.

If two parallelopipedons have the same lower base, and if their upper bases lie between the same parallels, they are equal.

Let the parallelopipedons ACDE–K and ACDE–P, have the same lower base and let their upper bases lie between the parallels QK and RN.

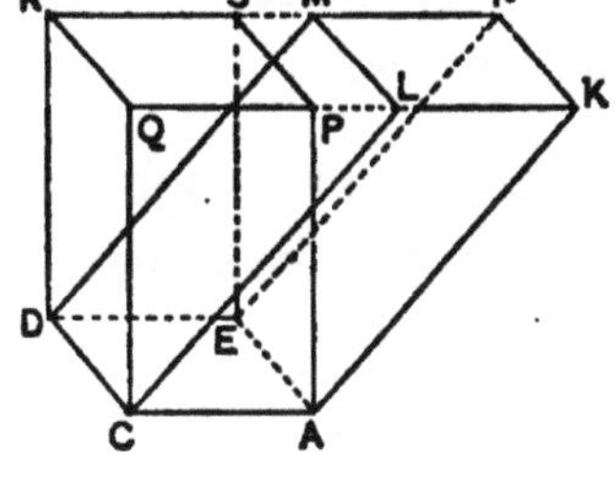

Now the base KM may fall entirely without the base PR, or it may partially coincide with that base; in either case, PK will be equal to QL, and SN to RM.

The triangular prisms KAP–S and LCQ–R, have the face KAP equal to LCQ, the face AESP to CDRQ, and the face KPSN to LQRM; they are therefore equal, (P. 7). Now if we take the first of these prisms from the polyedron ACDE – KQRN, there will remain the parallelopipedon ACDE–P, and we take the second prism from the same polyedron, there will remain the parallelopipedon ACDE–K; hence, the two parallelopipedons are equal, *which was to be proved.*

PROPOSITION XIII. THEOREM.

The volume of any parallelopipedon is equal to that of a right prism having the same base and the same altitude.

Let ACDE–K be any parallelopipedon and let ACDE–P be a right prism having the same base and the same altitude; then will their upper bases lie in the same plane.

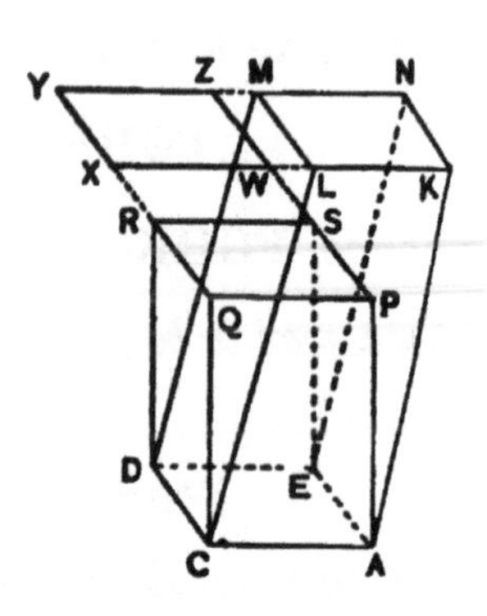

Prolong QR, PS, KL, and NM till, by their intersection, they form the parallelogram WXYZ; this parallelogram will be equal and parallel to the base ACDE.

Now, conceive a parallelopipedon to be formed whose lower base is ACDE and whose upper base is WXYZ. This parallelopipedon is equal to ACDE–P and also to ACDE–K, (P. 12); hence, ACDE–K is equal to ACDE–P, *which was to be proved.*

Cor. 1. The right prism ACDE–P is equal to a rectangular parallelopipedon having the same dimensions. For, through AP and CQ pass planes perpendicular to the face ER; these planes will determine a rectangular parallelopipedon ACFO–P, having the same dimensions as the prism ACDE–P, and equal to it (P. 12).

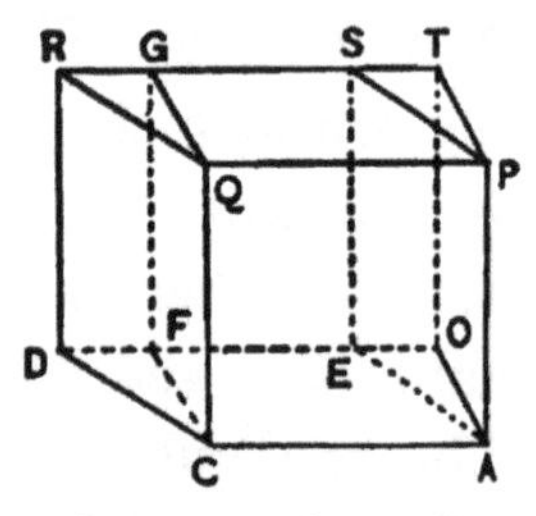

Cor. 2. The volume of any parallelopipedon is equal to that of a rectangular parallelopipedon having the same dimensions.

Cor. 3. The volume of any parallelopipedon is equal to

the product of its base and altitude, or to the product of its three dimensions.

Cor. 4. Any two parallelopipedons are to each other as the products of their three dimensions. If they have one dimension in common they are to each other as the products of the other two. If they have two dimensions in common they are to each other as their remaining dimensions.

Cor. 5. The volume of any triangular prism is equal to the product of its base and altitude (P. 8).

Proposition XIV. Theorem.

The volume of any prism is equal to the product of its base and altitude.

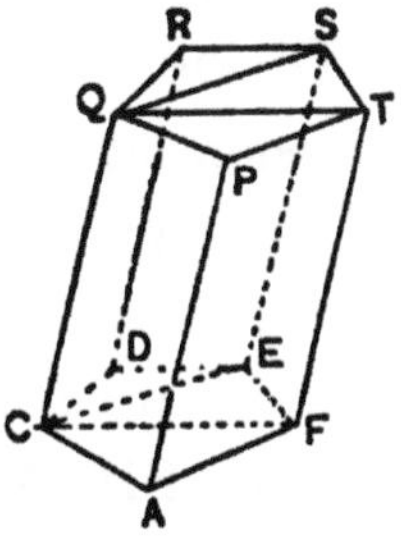

Let ACDEF-P be any prism. Through any lateral edge, as CQ, pass planes CT and CS, dividing the prism into triangular prisms; these will all have the same altitude as the given prism.

Now, each triangular prism is equal to the product of its base and altitude; hence, their sum is equal to the sum of their bases multiplied by their common altitude, that is, the volume of the given prism is equal to the product of its base and altitude, *which was to be proved.*

Cor. 1. Any two prisms are to each other as the products of their bases and altitudes.

Cor. 2. Two similar prisms are to each other as the cubes of their homologous edges. For, their bases are to each other as the squares of any two homologous edges, and their altitudes are to each other as those edges; hence, the prisms are to each other as the cubes of any two homologous edges.

PROPOSITION XV. THEOREM.

Triangular pyramids having equal bases and equal altitudes are equal.

Let ACD–V and *acd–v* be triangular pyramids having their bases equal and lying in the same plane; and let CZ be their common altitude; then, if the pyramids are not equal let the first, ACD–V, be the greater.

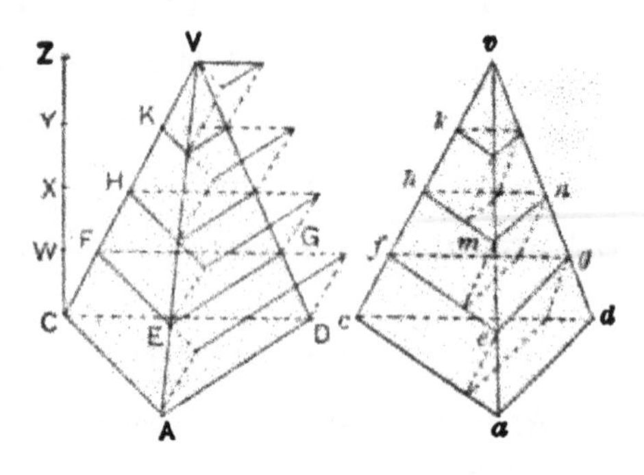

Divide CZ in *m* equal parts and through the points of division let planes be passed parallel to ACD; then will each of these planes intersect the two pyramids in equal triangles, (P 2, *Cor.* 2). On ACD, EFG, &c., as bases, construct exterior prisms having their lateral edges parallel and equal to CF, FH, &c.; also on the triangles *efg*, *mhn*, &c, as bases construct interior prisms having their lateral edges parallel and equal to *cf*, *fh*, &c. Now the sum of the exterior prisms is greater than the first pyramid, and the sum of the interior prisms is less than the second pyramid; hence, the difference between these sums is greater than the difference between the pyramids. But, the second external prism is equal to the first internal prism, because they have equal bases and equal altitudes; the third external prism is equal to the second internal prism; and so on: hence, the difference between the sum of the external and the sum of the internal prisms is equal to the first external prism. Consequently, the difference between the pyramids is less than that prism, that is, less than a prism whose base is ACD and whose altitude is one of the equal parts of CZ. But m is any number; if we suppose it increase indefinitely and finally to become greater than any assignable number, each part of CZ will become less than any assignable number, and the prism ACD–F will also become less than any assignable quantity, that is, the difference between the pyramids will be less than any assignable quantity; they are therefore equal to each other, *which was to be proved.*

PROPOSITION XVI. THEOREM.

Any triangular prism can be divided into three equal triangular pyramids.

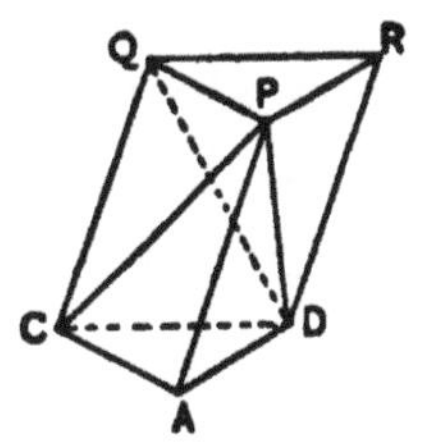

Let ACD–P be any triangular prism. Pass a plane through D, C, and P; also pass a plane through D, Q, and P; these planes divide the given prism into three pyramids ACD–P, CDQ–P, and RQD–P.

The pyramids CDQ–P and RDQ–P have their bases equal because they are halves of the parallelogram DCQR, and their altitudes equal because their bases are in the same plane and their vertices at the same point; hence, they are equal (P. 15). But the pyramid RQD–P may be considered as having the base PQR and the vertex D. Now the pyramids PQR–D and ACD–P are equal, because they have equal bases, PQR and ACD, and equal altitudes each being equal to the altitude of the given prism. Hence, the three pyramids into which the prism is divided are equal, *which was to be proved.*

Cor. 1. A triangular pyramid is one-third of a prism having the same base and the same altitude.

Cor. 2. The volume of any triangular pyramid is equal to one-third the product of its base and altitude.

PROPOSITION XVII. THEOREM.

The volume of any pyramid is equal to one-third the product of its base and altitude.

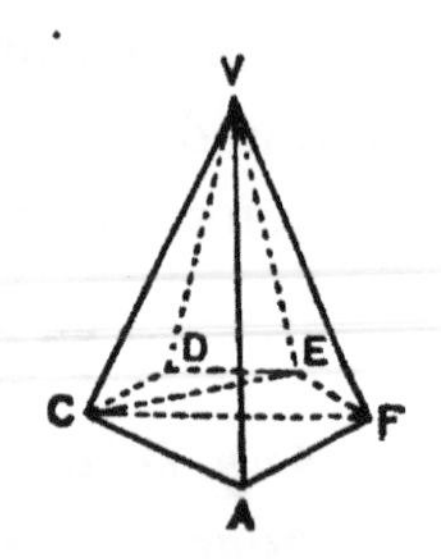

Let ACDEF–V be any pyramid. Through any lateral edge, as CV, pass planes, CVE and CVF, dividing it into triangular pyramids; these will have the same altitude as the given pyramid. Now, each triangular pyramid is equal to one-third the product of its base and altitude; hence, their sum is equal to the sum of their bases multiplied by one-third their common altitude, that is, the volume of the given pyramid is equal to one-third the product of its base and altitude, *which was to be proved.*

Cor. 1. Any two pyramids are to each other as the product of their bases and altitudes.

Cor. 2. Two similar pyramids are to each other as the cubes of their homologous edges. For, their bases are to each other as the squares of any two homologous edges, and their altitudes are to each other as those edges; hence, the pyramids are to each other as the cubes of any two homologous edges.

PROPOSITION XVIII. THEOREM.

A frustum of a triangular pyramid is equal to the sum of three pyramids whose common altitude is the altitude of the frustum and whose bases are the lower base of the frustum, the upper base of the frustum, and a mean proportional between these bases.

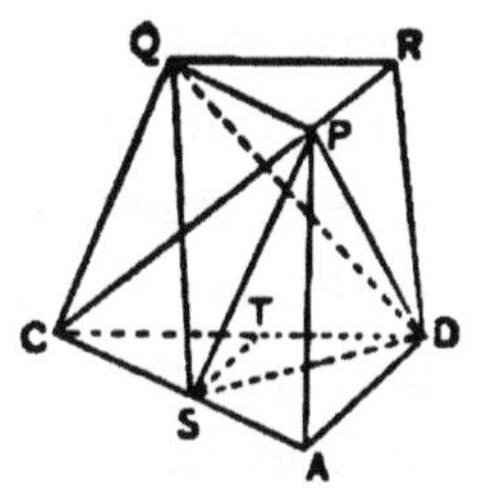

Let ACD–PQR be a frustum of a triangular pyramid.

First, pass a plane through D, C, and P; it will cut off a pyramid whose base is the lower base of the frustum, and whose altitude is that of the frustum. Then, pass a plane through D, P, and Q; it will cut off a pyramid whose base is the upper base of the frustum, and whose altitude is that of the frustum, and there will remain the pyramid whose base is DQC and whose vertex is P.

Draw PS parallel to QC; also draw SD and SQ; PS will be parallel to the plane DQC, (P. 7, B. 6), and the points S and P will be equally distant from that plane, (P. 11, *Cor.* 2, B. 6); hence, the pyramids DQC–P and DQC–S have the same base and the same altitude; they are therefore equal. But the last named pyramid may be regarded as having the base SCD, and the vertex Q. If we draw ST parallel to AD, or PR, the triangle SCT will be equal to PQR, and SCD will be a mean proportional between ACD, and SCT, or PQR, (P. 11, *Cor.* 2, B. 4); hence, the pyramid SCD–Q has for its base, a mean proportional between the two bases of the frustum, and for its altitude, the altitude of the frustum. The frustum is therefore equal to three pyramids whose common altitude is that of the frustum, and whose bases are the lower base of the frustum, the upper base of the frustum, and a mean proportional between these bases, *which was to be proved.*

Cor. 1. The volume of a triangular frustum is equal to the sum of its bases *plus* a mean proportional between them, multiplied by one-third of its altitude.

Cor. 2. The volume of a frustum of any pyramid is equal to the sum of its bases, *plus* a mean proportional between them, multiplied by one-third of its altitude. For, the planes that divide the entire pyramid into triangular pyramids will divide the frustum into triangular frustums having the same altitude as the given frustum. Now, if a plane parallel to the base cuts the given frustum in a section which is a mean proportional between its bases, it will cut each triangular frustum in a section which is a mean proportional between its bases; that is, the sum of the mean proportionals of the bases of the several triangular frustums is a mean proportional between the bases of the given frustum. The volume of the given frustum is the sum of the volumes of the triangular frustums; hence, it is equal to the sum of their upper bases *plus* the sum of their lower bases *plus* the sum of the mean proportionals of their bases multiplied by one-third their common altitude; that is, the volume of any frustum is equal to the sum of its bases *plus* a mean proportional between them, multiplied by one-third of its altitude.

General Scholium. Let the volume of a right prism be denoted by V, its lateral surface by L, the area of its base by B, the perimeter of its base by P, and its altitude by A; then from P. 3 and P. 14 we have,

$$L = PA, \quad \ldots \ldots \quad (a)$$

$$V = BA. \quad \ldots \ldots \quad (b)$$

Let the volume of a right pyramid be denoted by V, its lateral surface by L, the area of its base by B, the perimeter of its base by P, its altitude by A, and its slant height by S; then from P. 4 and P. 17, we have,

$$L = \tfrac{1}{2}PS, \quad \ldots \ldots \quad (c)$$

$$V = \tfrac{1}{3}BA. \quad \ldots \ldots \quad (d)$$

Let the volume of a right pyramidal frustum be denoted by V, its lateral surface by L, the areas of its bases by B and *b*, the corresponding perimeters by P and *p*, its altitude by A, and its slant height by S; then from P. 4, *Cor.*, and P. 18, *Cor.*, we have,

$$L = \tfrac{1}{2}S(P + p), \quad \ldots \ldots \quad (e)$$

$$V = \tfrac{1}{3}A(B + b + \sqrt{Bb}). \quad \ldots \ldots \quad (f)$$

BOOK VIII.

MEASURE OF THE CYLINDER, CONE, AND SPHERE.

Definitions.

112. A **cylinder** is a volume that may be generated by a rectangle revolving about one of its sides.

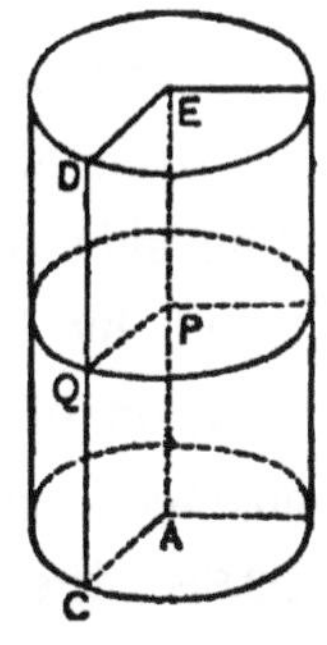

If the rectangle AD is revolved about the side AE it will generate a cylinder. The side AE is the **axis**; the surface generated by CD is the **lateral surface**; any position of CD is an **element** of the lateral surface; the circles generated by AC and ED are **bases**; and the distance between the bases is the **altitude**.

Any line of the rectangle, as PQ, parallel to AC, will generate a circle equal and parallel to the base; hence, every section parallel to the base is a circle. Any section through the axis is a rectangle double the generating rectangle.

In Elementary Geometry the term *cylinder* is used in a restricted sense; whereas in the Higher Geometry it is applied to any volume bounded in part by plane surfaces and in part by a curved surface whose elements are parallel straight lines. The term *cone*, in Elementary Geometry, is used in a similarly restricted sense.

113. An **inscribed prism** is one whose bases are inscribed in those of the cylinder; as ACDEFG–P. Its edges are elements of the lateral surface, and if the bases are regular polygons the line, KL, joining their centres, is the **axis** of the prism.

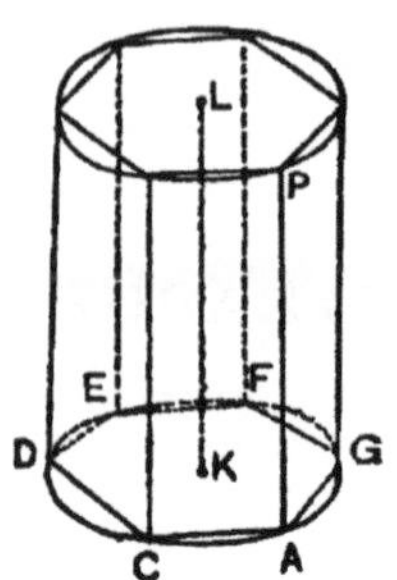

114. A **cone** is a volume that may be generated by a right-angled triangle revolving about a side adjacent to the right angle.

If the triangle ACV, right-angled at A, is revolved about AV, it will generate a cone. The side AV is the **axis**; the surface generated by CV is the **lateral surface**; any position of CV is an **element** of the lateral surface; the circle generated by AC is the **base**; the distance, VA, from the vertex to the base is the **altitude**; and the distance, VC, from the vertex to the circumference of the base is the **slant height.** Any line of the triangle, as PQ, parallel to AC, will generate a circle parallel to the base; hence, every section parallel to the base is a circle. Any section through the axis is an isosceles triangle, double the generating triangle.

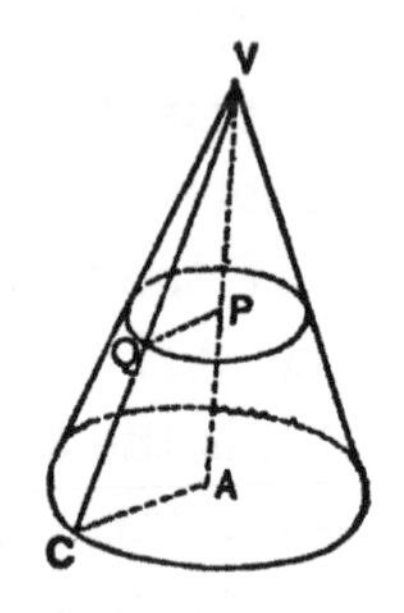

115. An **inscribed pyramid** is one whose vertex is the vertex of the cone and whose base is inscribed in that of the cone; as ACDEFG–V. Its edges are elements of the lateral surface, and if its base is a regular polygon the line from its centre to the vertex is the axis of the pyramid.

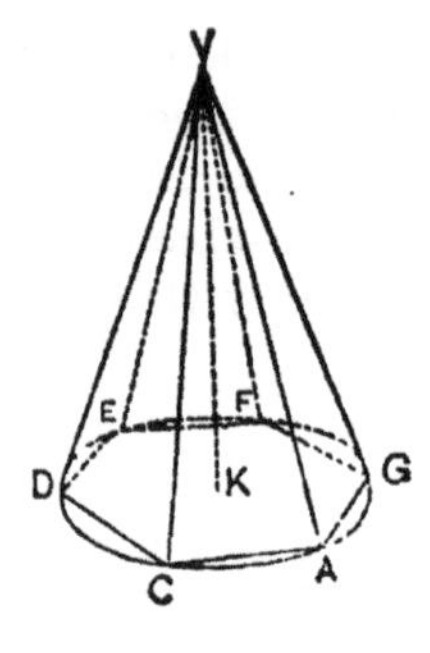

116. A **frustum of a cone**, or a conic frustum, is that

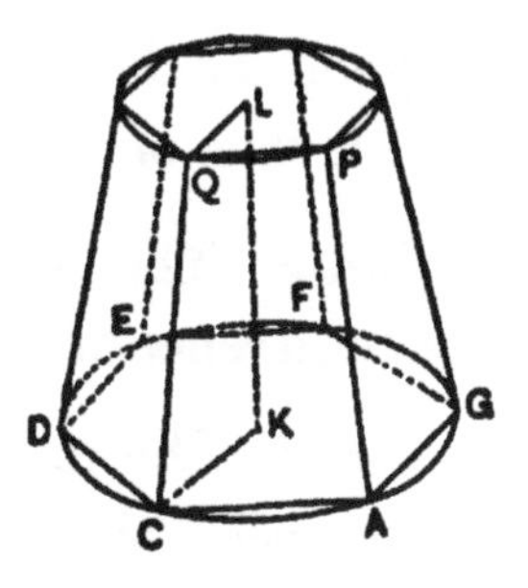

part of a cone included between the base and a section parallel to the base.

If the bi-rectangular trapezoid LKCQ is revolved about the side LK, adjacent to the two right angles, it will generate a conic frustum, whose bases are the circles KC and LQ, whose altitude is LK, and whose slant height is QC.

117. An **inscribed frustum of a pyramid** is one whose bases are inscribed in those of the conic frustum; as ACDEFG–P. If its bases are regular polygons, the line joining their centres is the axis of the conic frustum.

118. Similar cylinders are those that may be generated by the revolution of similar rectangles, and **similar cones** are those that may be generated by the revolution of similar triangles about homologous sides.

Proposition I. Theorem.

The limit of an inscribed prism, whose base is a regular polygon, is the circumscribing cylinder.

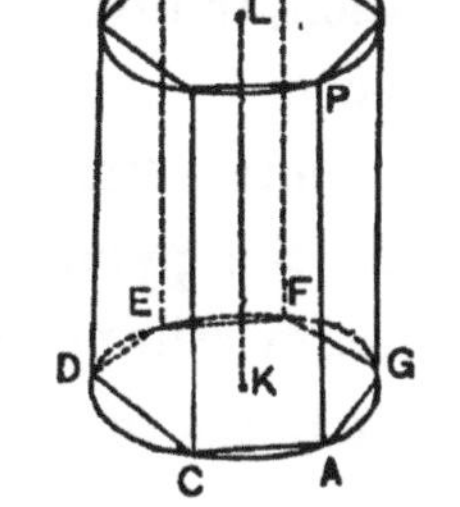

Let ACDEFG–P be an inscribed prism, whose base is a regular polygon.

If the number of sides of the base is increased by the method of continued duplication, the base of the prism will approach that of the cylinder; when the number of sides becomes infinite the two bases will coincide, and the prism will coincide with the cylinder; hence, the circumscribing cylinder is the limit of the inscribed prism, *which was to be proved.*

Cor. 1. A cylinder is a right prism, whose base is a regular polygon having an infinite number of sides. Hence, whatever is true of the prism is true of the cylinder, (Art. 66).

Cor. 2. The lateral surface of a cylinder is equal to the circumference of its base multiplied by its altitude, (P. 3, B. 7).

Cor. 3. The volume of a cylinder is equal to the product of its base and altitude, (P. 14, B. 7). Hence, any two cylinders are to each other as the products of their bases and altitudes.

Cor. 4. Similar cylinders are to each other as the cubes of their altitudes, or as the cube of any two homologous lines, (P. 14, *Cor.* 2, B. 7).

Proposition II. Theorem.

The limit of an inscribed pyramid, whose base is a regular polygon, is the circumscribing cone.

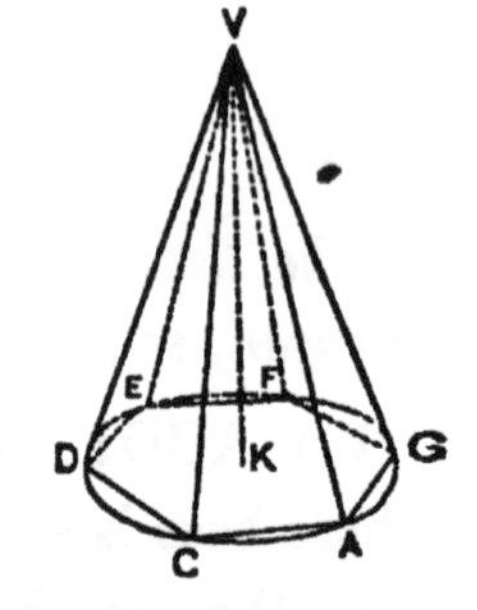

Let ACDEFG–V be an inscribed pyramid whose base is a regular polygon.

If the number of sides of the base is increased by the method of continued duplication, the base of the pyramid will approach that of the cone; when the number of sides becomes infinite, the two bases will coincide, and the pyramid will coincide with the cone; hence, the circumscribing cone is the limit of the inscribed pyramid, *which was to be proved.*

Cor. 1. A cone is a right pyramid, whose base is a regular polygon having an infinite number of sides.

Cor. 2. A secant plane parallel to the base of a cone divides its slant height and its altitude proportionally, (P. 2, B. 7).

Cor. 3. Any two sections of a cone parallel to its base are to each other as the squares of their distances from the vertex, (P. 2, *Cor.* 1, B. 7).

Cor. 4. The lateral surface of a cone is equal to the circumference of its base multiplied by half its slant height, (P. 4, B. 7).

Cor. 5. The volume of a cone is equal to one-third the product of its base and altitude, (P. 17, B. 7).

Cor. 6. Similar cones are to each other as the cubes of their altitudes, or as the cubes of any two homologous lines, (P. 17, *Cor.* 2, B. 7).

PROPOSITION III. THEOREM.

The limit of an inscribed frustum of a pyramid, whose bases are regular polygons, is the circumscribing conic frustum.

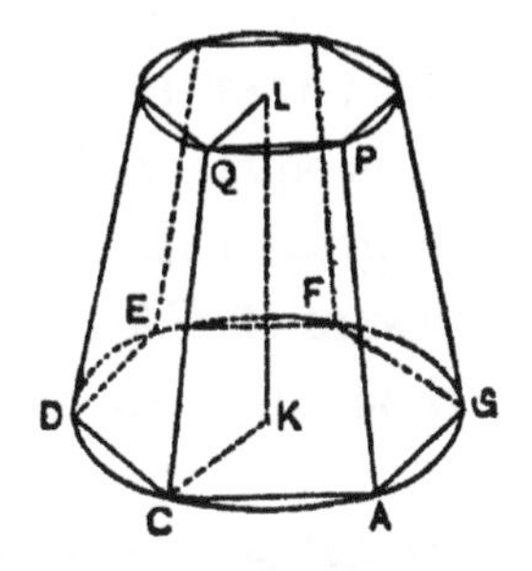

Let ACDEFG–P be an inscribed frustum whose bases are regular polygons. If the number of sides of each base is increased by the method of continued duplication, the bases of the pyramidal frustum will approach those of the conic frustum; when the number of sides becomes infinite the bases of the two frustums will coincide, and the pyramidal frustum will coincide with the conic frustum; hence, the circumscribing conic frustum is the limit of the inscribed pyramidal frustum, *which was to be proved.*

Cor. 1. A frustum of a cone is a frustum of a right pyramid, whose bases are regular polygons having an infinite number of sides.

Cor. 2. The lateral surface of a conic frustum is equal to half the sum of the circumferences of its bases multiplied by its slant height, (P. 4, *Cor.*, B. 7).

Cor. 3. The volume of a conic frustum is equal to the sum of its bases *plus* a mean proportional between the bases, multiplied by one-third of the altitude, (P. 18, *Cor.* 2, B. 7).

PROPOSITION IV. THEOREM.

If a straight line is revolved about an axis wholly exterior to it, the surface generated is equal to the revolving line, into the circumference generated by its middle point.

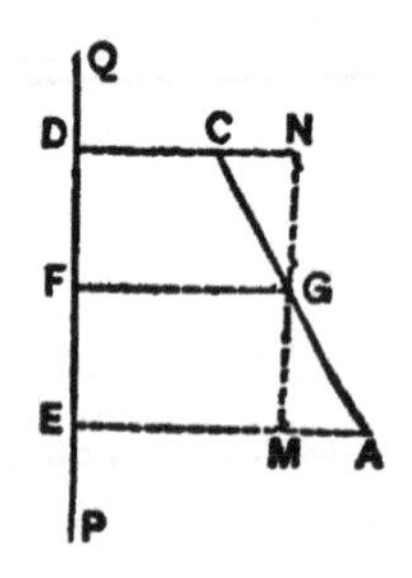

Let AC be any line, PQ an axis exterior to it, and let G be the middle point of AC.

Through A, G, and C, draw AE, GF, and ND, perpendicular to PQ, and through G draw MN parallel to PQ. The right-angled triangles AMG and CNG, have the hypothenuse AG equal to GC, and the angle AGM to CGN; they are therefore equal in all their parts; hence, MA is equal to CN, that is, FG is as much less than EA as it is greater than DC; the line FG is therefore equal to half the sum of EA and DC, and consequently the circumference FG is equal to half the sum of the circumference EA and DC.

If AC is revolved about PQ it generates the lateral surface of a conic frustum whose slant height is AC and whose bases are the circumferences EA and DC; but this surface is equal to AC multiplied by half the sum of the circumferences EA and DC, that is, it is equal to AC, multiplied by the circumference FG, *which was to be proved.*

Scho. The proposition holds true when AC meets PQ at D, in which case AC generates the lateral surface of a cone; it also holds true when AC is parallel to PQ, in which case AC generates the lateral surface of a cylinder.

Definitions.

119. A **sphere** is a volume bounded by a surface all of whose points are equally distant from a point within, called **the centre.**

Any line from the centre to a point of the surface is a **radius**; any line through the centre and limited by the surface is a **diameter.** All radii of the same sphere are equal, and all diameters of the same sphere are equal.

A sphere may be generated by a semicircle revolving about its diameter, as an axis. In this revolution every point of the semi-circumference will generate the circumference of a circle whose plane is perpendicular to the axis, and whose radius is the distance of that point from the axis.

120. A plane is **tangent** to the surface of a sphere when it touches it in a single point. This point is called **the point of contact.**

121. A **zone** is a portion of the surface included between two parallel planes. The distance between the planes is the **altitude** of the zone, and the lines in which the planes cut the surface are the **bases** of the zone.

If one of the limiting planes is tangent to the surface, the zone has but one base.

122. A **spherical segment** is a portion of a sphere included between two parallel planes. The distance between the planes is the **altitude** of the segment, and the sections which the planes cut from the sphere are the **bases** of the segment.

If one of the limiting planes is tangent to the surface of the sphere, the segment has but one base.

123. A **spherical sector** is a volume that may be generated by any sector of a semicircle when the semicircle is revolved about its diameter.

The spherical sector is usually bounded by a zone and two conic surfaces; if the axis coincides with one of the extreme radii of the circular sector, the spherical sector is bounded by a zone and one conical surface. One of the conic surfaces may reduce to a plane.

PROPOSITION V. THEOREM.

Any plane section of a sphere is a circle.

Let the sphere OQ be cut by any plane ACD; let OQ be a radius of the sphere perpendicular to the plane ACD, and piercing it at P.

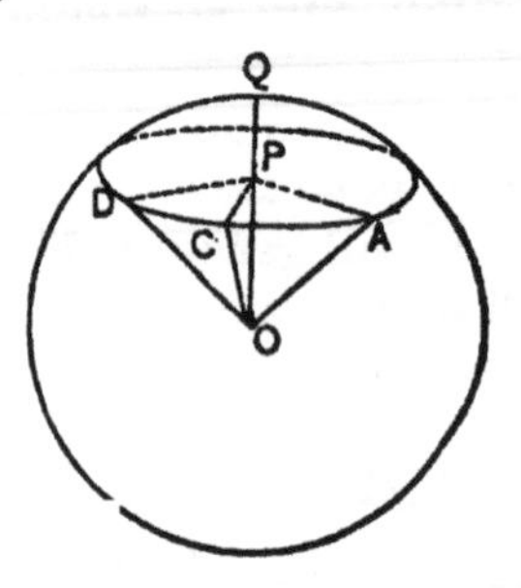

Let A, C, and D, be any three points of the intersection of the plane with the surface of the sphere, and draw OA, OC, and OD.

Because OA, OC, and OD are radii of the sphere they are equal; consequently the points A, C, and D, are in the circumference of a circle whose centre is P, (P. 4, *Cor.*, B. 6). But A, C, and D, are any points in the line of intersection; hence, the intersection is a circle, *which was to be proved.*

Cor. 1. If the cutting plane passes through the centre of the sphere, the radius of the section is equal to the radius of the sphere; such a section is called **a great circle.** If the cutting plane does not pass through the centre of the sphere, the radius of the section is less than that of the sphere; such a section is called **a small circle.**

Cor. 2. The radius of the sphere that passes through the centre of a small circle is perpendicular to the plane of that circle; *conversely,* a line through the centre of a small circle, and perpendicular to its plane, passes through the centre of the sphere.

Cor. 3. The plane of a great circle bisects the sphere. For the two parts into which it divides the sphere may be placed so as to coincide; otherwise there would be points on the surface of the sphere unequally distant from its centre.

Cor. 4. One great circle can always be passed through any two points on the surface of a sphere, and but one. For, one plane can always be passed through the given points and the centre of the sphere, and but one; this plane cuts the sphere in a great circle.

If the points are at the extremities of a diameter, an infinite number of great circles can pass through them.

PROPOSITION VI. THEOREM.

If a plane is perpendicular to any radius of a sphere at its extremity, it is tangent to the surface of the sphere; conversely, if a plane is tangent to the surface of a sphere, it is perpendicular to the radius drawn to the point of contact.

1°. Let OP be any radius of the sphere, and let KL be a plane perpendicular to OP at P; let OQ be a line drawn from O to any point of KL, except P.

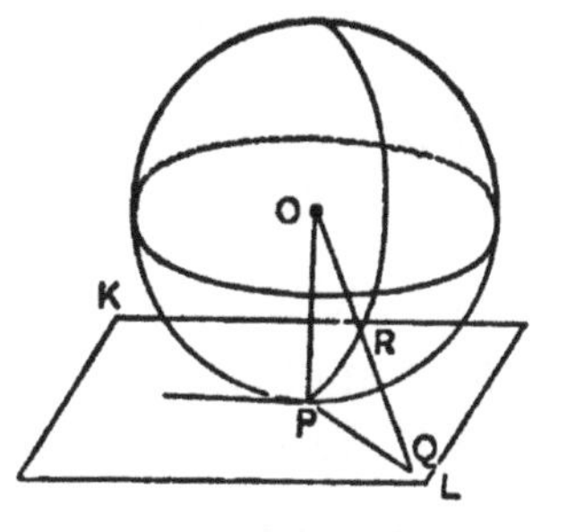

Because OP is perpendicular, and OQ oblique to KL, OQ is greater than OP; consequently, Q lies without the sphere; hence, every point of the plane KL, except P, lies without the sphere; consequently, KL is tangent to the surface at P, *which was to be proved.*

2°. Let the plane KL be tangent to the sphere at P, and let Q be any point of KL, except P.

Because KL is tangent to the surface at P, Q lies without the sphere, that is, OQ is greater than OP; hence OP is the shortest line from O to KL; it is therefore perpendicular to KL, *which was to be proved.*

Definitions.

124. If an arc is divided into any number of equal parts, the corresponding chords form a **regular broken line.** If the arc is a semi-circumference the broken line is a **regular semi-perimeter,** and the area included between it and the diameter is a **regular semi-polygon.** The parts of the broken line are called **sides.**

125. If lines are drawn from the extremities of any side perpendicular to the diameter, the part of the diameter intercepted between them is called the **projection** of the side. Thus KL and LM are projections of CD and DE.

Proposition VII. Theorem.

The surface of a sphere is equal to the circumference of a great circle multiplied by its diameter.

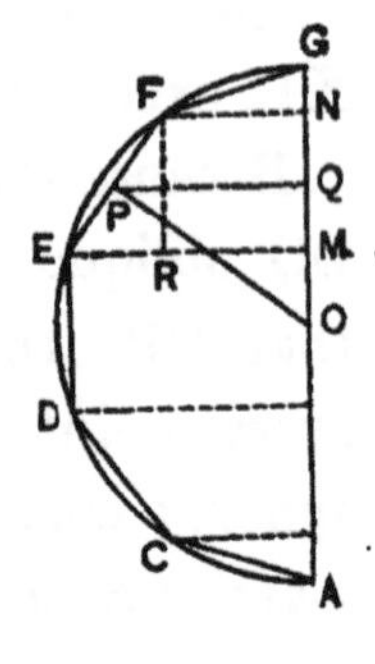

Let ACDEFG be a regular semi-perimeter whose apothem is OP. Through P, the middle of EF, draw PQ perpendicular to AG; draw EM and FN perpendicular to AG; also, draw FR parallel to AG.

If the semi-perimeter is revolved about AG as an axis, any side as EF will generate the surface of a conic frustum whose measure is

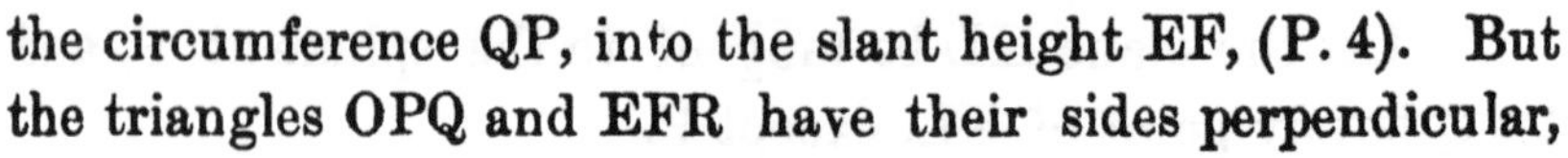

the circumference QP, into the slant height EF, (P. 4). But the triangles OPQ and EFR have their sides perpendicular, each to each; they are therefore mutually equiangular and consequently similar; hence,

$$OP : QP :: EF : RF, \text{ or } MN;$$

whence, $$QP \times EF = OP \times MN,$$

or, $$\textit{circum}\ QP \times EF = \textit{circum}\ OP \times MN,$$

that is, the surface generated by any side is equal to a circumference whose radius is the apothem, multiplied by the projection of that side on the axis. Hence, the surface generated by the entire semi-perimeter is equal to the circumference whose radius is the apothem, multiplied by the sum of the projections of all its sides, that is, by AG.

If the number of sides of the semi-polygon is made infinite by the method of continued duplication, the semi-perimeter will become the semi-circumference whose diameter is AG, the apothem will become the radius of the semi-circumference, and the surface generated will become the surface of a sphere; hence, the surface of a sphere is equal to the circumference of a great circle multiplied by its diameter, *which was to be proved.*

Cor. 1. If we denote the radius of a sphere by R, the circumference of a great circle will be equal to $2\pi R$, (P., B. 5), and consequently, the surface of the sphere will be equal to $2\pi R \times 2R$, or to $4\pi R^2$, that is, to *four great circles.*

Cor. 2. The surfaces of any two spheres are to each other as the squares of their radii.

Cor. 3. The area of any zone is equal to the circumference of a great circle multiplied by its altitude. For, the generating arc may be regarded as a regular broken line, having an infinite number of sides, and whose projection is equal to the altitude of the zone.

Cor. 4. Any two zones of the same sphere are to each other as their altitudes.

Cor. 5. The lateral surface of a circumscribed cylinder is equal to the circumference of a great circle multiplied by the diameter of the sphere; it is therefore equal to the surface of the sphere. If we include the bases, the surface of a circumscribed cylinder is to the surface of the sphere as 6 is to 4, or as 3 is to 2.

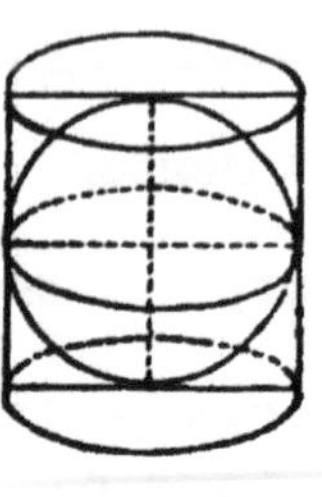

PROPOSITION VIII. THEOREM.

The volume of a sphere is equal to its surface multiplied by one-third of its radius.

Let ACDEFGH be a regular semi-polygon whose centre is O, and let the radii OC, OD, &c., be drawn, dividing it into equal isosceles triangles; furthermore, let us suppose that the semi-polygon has an even number of sides, in which case every side is oblique to AH.

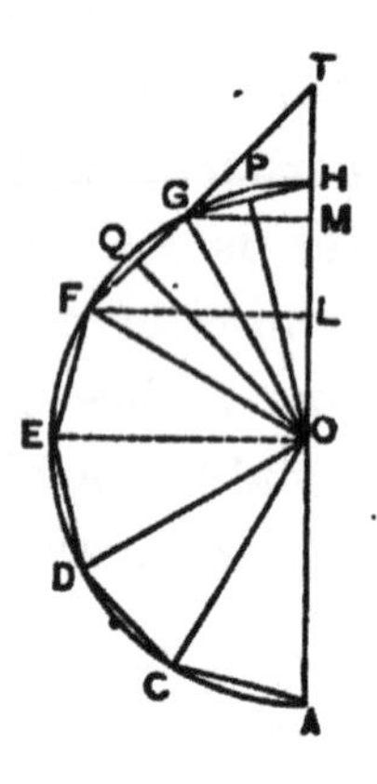

Draw OP perpendicular to GH, and OQ perpendicular to FG; draw FL and GM perpendicular to AH; and prolong FG till it meets AH at T.

If the regular semi-polygon is revolved about AT as an axis, it will generate a volume equal to the sum of the volumes generated by the triangles OHG, OGF, &c. Now, the volume generated by the triangle OGH is made up of two cones whose vertices are O, and H, and whose common base is the circle generated by MG; denoting this volume by the expression *vol* OGH, we have,

$$vol\ \mathrm{OGH} = \pi\overline{\mathrm{MG}}^2 \times \tfrac{1}{3}\overline{\mathrm{OH}} = \tfrac{1}{3}\pi\overline{\mathrm{MG}} \times \overline{\mathrm{MG}} \times \overline{\mathrm{OH}}. \; . \; . \; (1)$$

But, $\overline{MG} \times \overline{OH} = \overline{GH} \times \overline{OP}$, because each is equal to twice the triangle OGH; hence,

$$vol\ OGH = \tfrac{1}{3}\pi\overline{MG} \times \overline{GH} \times \overline{OP}. \quad . \quad . \quad . \quad . \quad (2)$$

The surface generated by $\overline{GH}$ is equal $2\pi\overline{MG} \times \tfrac{1}{2}\overline{GH}$, (P. 2, *Cor*. 4), or to $\pi\overline{MG} \times \overline{GH}$; denoting this surface by the expression *surf* GH, we have, from (2),

$$vol\ OGH = surf\ GH \times \tfrac{1}{3}OP. \quad . \quad . \quad . \quad . \quad . \quad (3)$$

Again, we have, in like manner,

$$vol\ OFT = surf\ FT \times \tfrac{1}{3}OQ, \quad . \quad . \quad . \quad . \quad . \quad (4)$$

and,

$$vol\ OGT = surf\ GT \times \tfrac{1}{3}OQ; \quad . \quad . \quad . \quad . \quad . \quad (5)$$

subtracting (5) from (4), member from member, remembering that OQ is equal to OP, we have,

$$vol\ OFG = surf\ FG \times \tfrac{1}{3}OP.$$

In like manner it may be shown that the volume generated by any of the isosceles triangles is equal to the surface generated by its base, multiplied by one-third of the apothem. Hence, the volume generated by the semi-polygon is equal to the surface generated by the semi-perimeter, multiplied by one-third of the apothem.

If now we suppose the number of sides of the semi-polygon to be made infinite by the method of continued duplication, the semi-polygon will become a semicircle, its apothem will become the radius, and the volume generated will be the volume of a sphere; hence, the volume of a sphere is equal to its surface multiplied by one-third of its radius, *which was to be proved.*

Cor. 1. If we denote the radius of the sphere by R, its

surface is equal to $4\pi R^2$; consequently its volume is equal to $4\pi R^2 \times \frac{1}{3}R$, or to $\frac{4}{3}\pi R^3$.

Cor. 2. Any two spheres are to each other as the cubes of their radii.

Cor. 3. The volume of a spherical sector is equal to the zone which forms its base, multiplied by one-third of its radius. For, the arc of the sector which generates the zone may be regarded as a regular broken line having an infinite number of sides; hence, the volume generated is equal to the zone multiplied by one-third of the radius of the sphere.

General Scholium. Let the parts of a cylinder, of a cone, and of a conic frustum, be denoted by the same symbols as in the general scholium, B. 7; also let the radii of their lower bases be denoted by R; and let the radius of the upper base of the frustum be denoted by r: then, we have, from P. 1, *Cors.* 2 and 3; P. 2, *Cors.* 4 and 5; P. 3, *Cors.* 2 and 3,

1°. For the cylinder;

$$L = PA = 2\pi RA \quad \dots\dots\dots \quad (A)$$

$$V = BA = \pi R^2 A \quad \dots\dots\dots \quad (B)$$

2°. For the cone;

$$L = \tfrac{1}{2}PS = \pi RS \quad \dots\dots\dots \quad (C)$$

$$V = \tfrac{1}{3}BA = \tfrac{1}{3}\pi R^2 A \quad \dots\dots\dots \quad (D)$$

3°. For the conic frustum;

$$L = \tfrac{1}{2}S(P + p) = \pi S(R + r) \quad \dots\dots \quad (E)$$

$$V = \tfrac{1}{3}A(B + b + \sqrt{Bb}) = \tfrac{1}{3}\pi A(R^2 + r^2 + Rr) \quad \dots \quad (F)$$

Let the volume of a sphere be denoted by V, its surface by S, and its radius by R; then, from P. 7, and P. 8, we have,

$$S = 4\pi R^2 \quad \dots\dots\dots \quad (G)$$

$$V = \tfrac{4}{3}\pi R^3 \quad \dots\dots\dots \quad (H)$$

Let the surface of a zone on a sphere whose radius is R be denoted by Z, and let its altitude be denoted by A; then, from P. 7, *Cor.* 3, we have,

$$Z = \pi R^2 A \quad \dots\dots\dots \quad (I)$$

BOOK IX.

SPHERICAL GEOMETRY.

Definitions.

126. A **spherical angle** is an angle on the surface of a sphere included between the arcs of two great circles.

The arcs are called **sides**, and the point at which they meet is called the **vertex** of the angle.

The angle between two curves meeting at a point is the same as the angle between the tangents to the curves at their common point.

127. A **spherical polygon** is a part of the surface of a sphere, bounded by arcs of three or more great circles.

The bounding arcs are **sides**, the points at which the sides meet are **vertices**, and the arcs of great circles joining two vertices not consecutive are **diagonals** of the polygon.

128. A **spherical triangle** is a spherical polygon of three sides.

Spherical triangles are classified in the same manner as plane triangles.

129. A **spherical pyramid** is a portion of a sphere bounded by a spherical polygon, and by sectors of circles whose arcs are the sides of the polygon.

The spherical polygon is the **base** of the pyramid, and the centre of the sphere is its **vertex**.

130. A **lune** is a portion of the surface of a sphere, bounded by semi-circumferences of two great circles.

The semi-circumferences are **sides**, and the angle between them is the **angle of the lune**. The volume bounded by the lune and by the planes of its sides is called a **spherical wedge.**

131. The **axis of a circle** is a line perpendicular to the plane of the circle through its centre.

132. The **poles** of any circle of a sphere are the points in which the axis of the circle meets the surface of the sphere.

In what follows each side of a spherical polygon is supposed to be less than a semi-circumference; the distance between any two points on the surface is supposed to be measured on the arc of a great circle passing through them; all the polygons treated of are supposed to be *salient* and to be situated on the same sphere or on equal spheres. Furthermore, all the arcs referred to are supposed to be arcs of great circles unless the contrary is indicated, either by words, or by the nature of the case.

PROPOSITION I. THEOREM.

Either pole of a circle is equally distant from every point of its circumference.

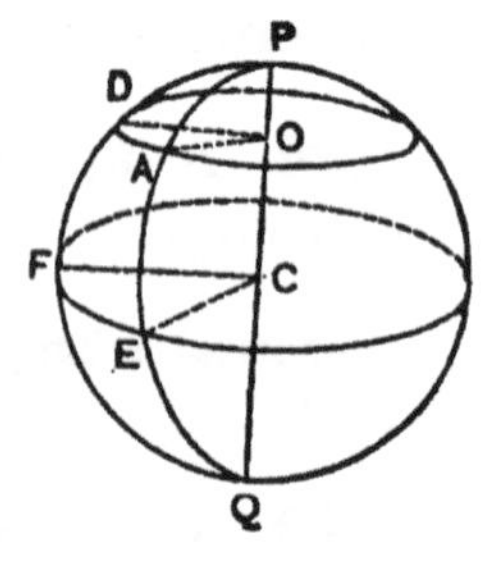

Let the circle OA be a circle of the sphere whose centre is C, and let PQ be its axis: this axis passes through C, and the points P and Q in which it pierces the surface are the poles of the circle OA. Let PAQ be a semicircle whose diameter is PQ.

If this semicircle is revolved about PQ, it will generate the sphere, and the point A will generate the circumference of the circle OA; hence, the distance of every point of the circumference OA, from P, is the arc PA, and the distance of every point from Q is the arc QA, *which was to be proved.*

Cor. 1. The middle point, E, of the arc PEQ generates the circumference of a great circle CE; hence, every point in the circumference of a great circle is at a quadrant's distance from each of its poles.

Cor. 2. If two points of the surface of a sphere, not the extremities of a diameter, are each at a quadrant's distance from a third point, the latter point is the pole of a great circle passing through two former points.

Scho, The arc PA is called the spherical radius of the circle OA: when the pole of an arc and its spherical radius are known, the arc can be drawn on the surface of a sphere with the same facility that a circle can be drawn on a plane surface.

Proposition II. Theorem.

A spherical angle is measured by the arc of a great circle described from the vertex as a pole, and limited by the sides of the angle.

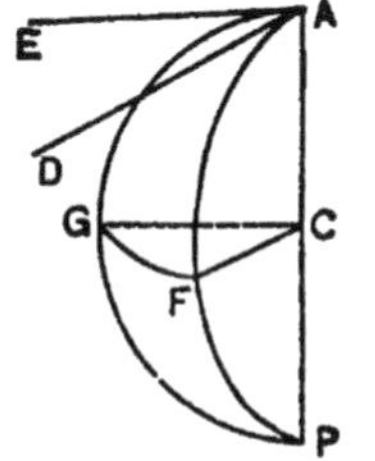

Let FAG be a spherical angle and let AD and AE be tangent to its sides at A; also, let FG be an arc of the great circle whose pole is A.

Draw the radii CF, CG, and CA.

Because DA and EA are tangent to FA and GA, at A, they are perpendicular to CA at that point; hence, the angle DAE is the measure of the diedral angle whose faces are FCA and GCA. Because FA and GA are quadrants, the radii FC and GC are perpendicular to the radius CA at C; hence, the angle FCG is also the measure of the diedral angle whose faces are FCA and GCA. The angle DAE is therefore equal to FCG; but, the angle FCG is measured by the arc FG; hence, DAE, or FAG, is measured by the arc FG, *which was to be proved.*

Cor. The arc FG measures the angle of the lune whose sides are PFA and PGA.

Scho. If arcs of two great circles intersect, as AC and DE, they form four angles about V; of these angles, the opposite ones are equal, and any two adjacent ones are supplementary.

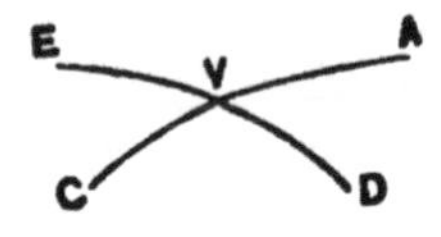

Proposition III. Theorem.

Any side of a spherical triangle is less than the sum of the other two.

Let ACD be a spherical triangle, on a sphere whose centre is O.

Draw the radii OA, OC, and OD.

The planes AOC, COD, and DOA form a triedral angle O, whose lateral angles are measured by the sides of ACD; but any one of these angles is less than the sum of the other two, (P. 15, B. 6); hence, any side of ACD is less than the sum of the other two, *which was to be proved.*

Cor. The sum of the lateral angles of the polyedral angle O is less than 4 right angles, (P. 16, B. 6); hence, the sum of the sides of a spherical triangle is less than 4 quadrants.

Proposition IV. Theorem.

If the vertices of one spherical triangle are poles of the sides of a second spherical triangle, then are the vertices of the second triangle poles of the sides of the first triangle.

Let the vertices of the spherical triangle ACD be the poles of the sides of the spherical triangle EFG.

Draw the arcs DE and CE.

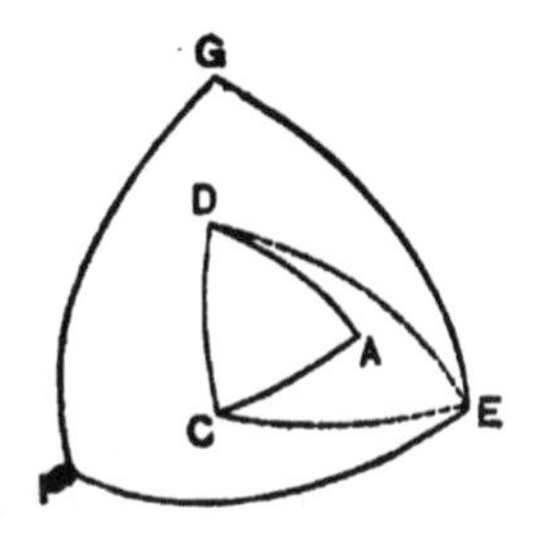

Because D is the pole of FE, the arc DE is a quadrant, and because C is the pole of EG, the arc CE is a quadrant; hence E is the pole of CD (P. 1, *Cor.* 2); in like manner it may be shown that F is the pole of DA, and G the pole of CA, *which was to be proved.*

Scho. 1. The triangles ACD and EFG are called polar triangles, each being the polar triangle of the other.

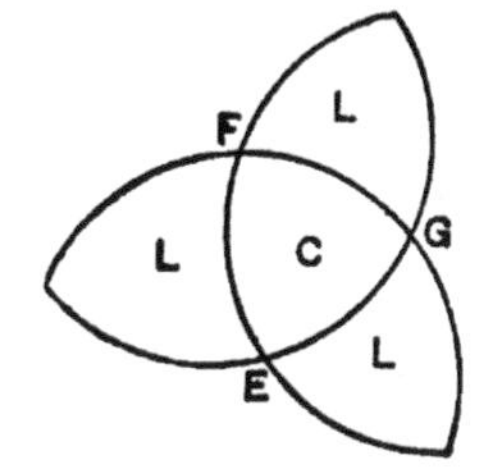

Scho. 2. If all the sides of EFG are prolonged in both directions they will determine a group of four triangles, a central one, C, and three lateral ones, L. In like manner if the sides of ACD are prolonged they will determine a similar group. The term *polar triangles* is applied only to the central ones of each group.

Proposition V. Theorem.

Any angle of a spherical triangle is the supplement of the side lying opposite to it in the polar triangle.

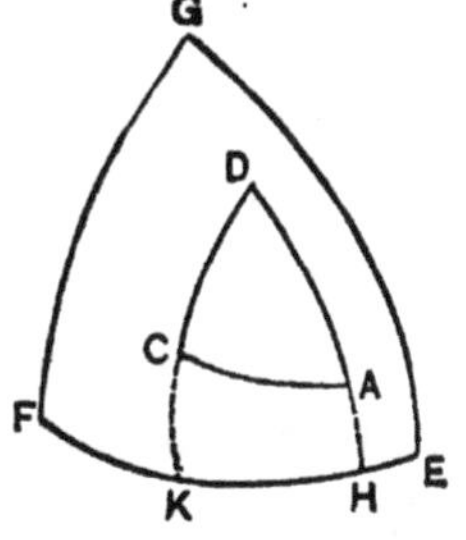

Let ACD be any spherical triangle, and let EFG be its polar triangle.

Prolong DA and DC till they meet EF, at H and K.

Because DH and DK are quadrants, HK is the measure of the angle D, (P. 2). Because E is the pole of DK, EK is a quadrant, and because F is the pole of DH, FH is a quadrant; consequently EK + FH, or EF + HK, is equal to a semi-circumference, that is, the arc HK is the supplement of FE; hence, D is the supplement of FE; in like manner it may be shown that A is the supplement of FG, and C the supplement of GE, *which was to be proved.*

Cor. Any side of either triangle is the supplement of the opposite angle of its polar triangle.

Definition.

133. Two spherical triangles are said to be **symmetrical** when all their parts are equal, each to each, but placed in a reverse order in the two triangles.

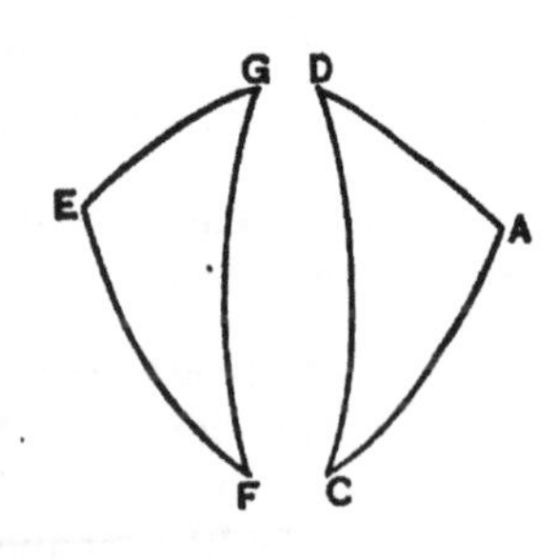

Thus, ACD and EFG are symmetrical if the side AC is equal to EF, CD to FG, DA to GE, and the angle A to E, C to F, D to G.

Symmetrical triangles correspond to symmetrical triedral angles at the centre, and also to symmetrical spherical pyramids whose vertices are at the centre.

Proposition VI. Problem.

To construct a spherical triangle that shall be symmetrical with respect to a given spherical triangle.

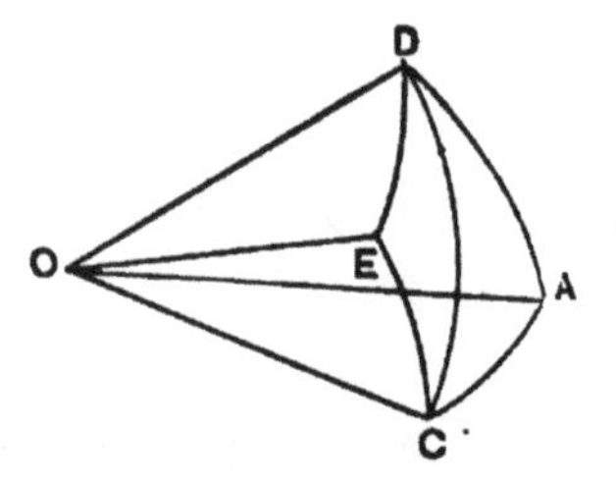

Let ACD be a spherical triangle lying on a sphere whose centre is O.

With D as a pole and DA as a spherical radius, describe an arc; with C as a pole and CA as a spherical radius, describe a second arc cutting the first at E. Draw the arcs EC and ED; also draw the radii OA, OC, OD, and OE.

The side EC is equal to AC, ED is equal to AD, and DC is common; hence, the sides of ECD and ACD are equal each to each, but taken in a reverse order. The lateral angles EOC, EOD, and COD, are respectively equal to AOC, AOD, and COD; hence the corresponding interfacial angles are equal, (P. 17, B. 6), that is, the angle E is equal to A, the angle ECD to ACD, and the angle EDC to ADC; hence, the angles of the triangles ECD and ACD are equal, each to each, but taken in a reverse order. The triangle ECD is therefore symmetrical with respect to ACD, *which was to be proved.*

Cor. If DA is equal to DC, DE is also equal to DC, and the symmetrical triangles are both isosceles. In this case if the triangle ECD is placed on CDA so that the angle EDC shall coincide with its equal CDA, the arc ED will coincide with CD, the arc CD with AD, and the two triangles will coincide throughout.

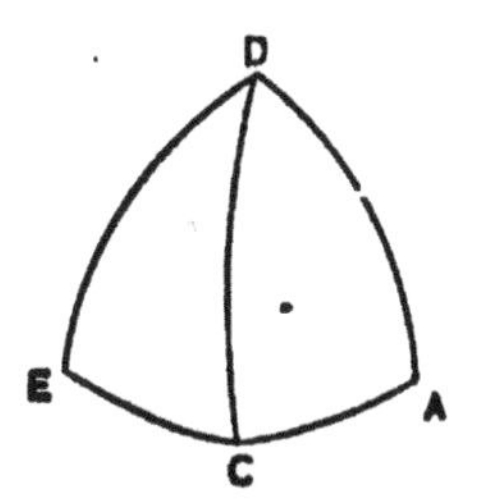

Scho. Symmetrical triangles cannot be made to coincide unless they are isosceles.

PROPOSITION VII. THEOREM.

If two spherical triangles have two sides and the included angle of one, equal to two sides and the included angle of the other, each to each, their remaining parts are equal, each to each.

Let FGH and ACD be two spherical triangles in which FG is equal to AC, GH to CD, and the angle FGH to ACD; also, let ACD and ECD be symmetrical.

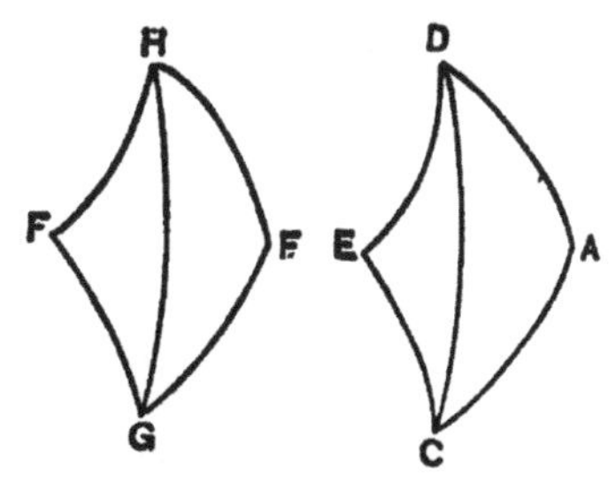

The triangle FGH can be placed on ACD, or else on its symmetrical triangle ECD so as to coincide with it, as may be shown by a course of reasoning similar to that employed in demonstrating P. 8, B. 1; hence, the side HF is equal to DA, the angle F to A, and the angle FHG to ADC, *which was to be proved.*

PROPOSITION VIII. THEOREM.

If two spherical triangles have two angles and the included side of one, equal to two angles and the included side of the other, each to each, their remaining parts are equal, each to each.

Let FGH and ACD be two spherical triangles, in which the angle FGH is equal to ACD, the angle FHG to ADC, and the side GH to CD; also, let ECD be symmetrical with ACD.

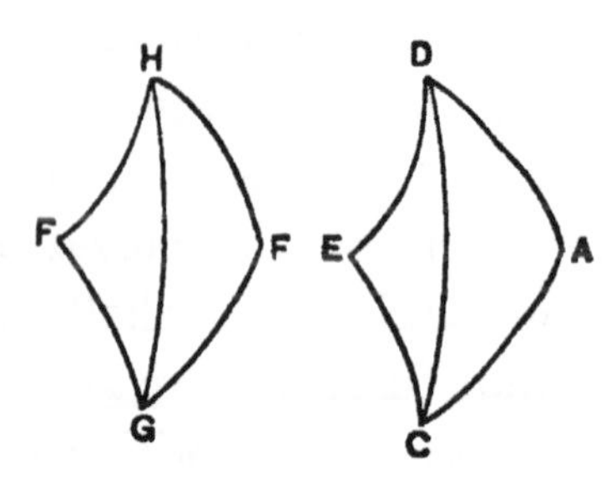

The triangle FGH can be placed on ACD, or else on its symmetrical triangle, so that they will coincide throughout, as may be shown by a course of reasoning similar to that employed in demonstrating P. 10, B. 1; hence, the side FG is equal to AC, the side HF to DA, and the angle F to E, or to A, *which was to be proved.*

NOTE.—Let the student demonstrate this and the preceding theorem in detail, following the methods indicated.

PROPOSITION IX. THEOREM.

If two spherical triangles have their sides equal, each to each, their angles are equal, each to each.

Let FGH and ACD be two spherical triangles, in which FG is equal to AC, GH to CD, and HF to DA.

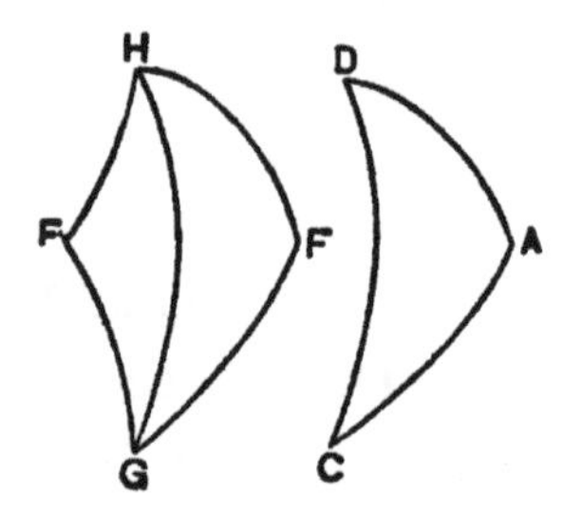

If radii are drawn from the vertices of each triangle to the centre of the sphere on which it lies, they will be the edges of two triedral angles whose lateral angles are equal, each to each; hence their interfacial angles are equal, each to each (P. 17, B. 6). But these are the angles of the given triangles; hence the angles of the given triangles are equal, each to each, *which was to be proved.*

Cor. If two spherical triangles have their angles equal, each to each, the sides of their polar triangles will be equal, each to each (P. 5), and consequently the sides of the given triangles are equal, each to each.

PROPOSITION X. THEOREM.

If two sides of a spherical triangle are equal, their opposite angles are equal; conversely, if two angles are equal, their opposite sides are equal.

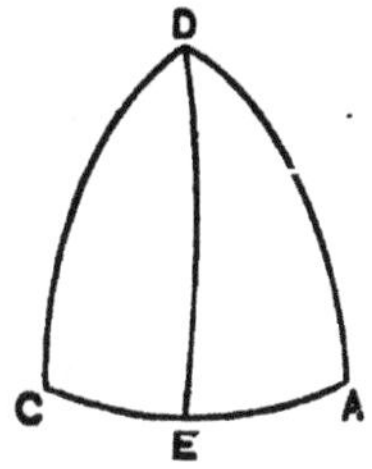

1°. Let ACD be a spherical triangle in which DA is equal to DC, and let DE be an arc drawn from D to the middle point of AC.

The triangles CED and AED have DE common, EC equal to EA, and CD to AD; their angles are therefore equal, each to each, (P. 9); hence, the angle C is equal to A, *which was to be proved.*

2°. Let ACD be a triangle in which the angles C and A are equal, and let its polar triangle be denoted by P.

The sides of P, lying opposite C and A, are equal, (P. 5, *Cor.*); consequently, their opposite angles are equal from what has just been proved; hence, the sides of the given triangle lying opposite to these angles are equal, *which was to be proved.*

Cor. 1. If all the sides of a spherical triangle are equal, all its angles are equal, and the reverse.

Cor. 2. The arc drawn from the vertex of an isosceles triangle to the middle of its base, is perpendicular to the base, and bisects the angle at the vertex.

PROPOSITION XI. THEOREM.

In any spherical triangle, the greater side lies opposite to the greater angle; and conversely, the greater angle lies opposite to the greater side.

1°. Let ACD be a spherical triangle, in which the angle D is greater than C.

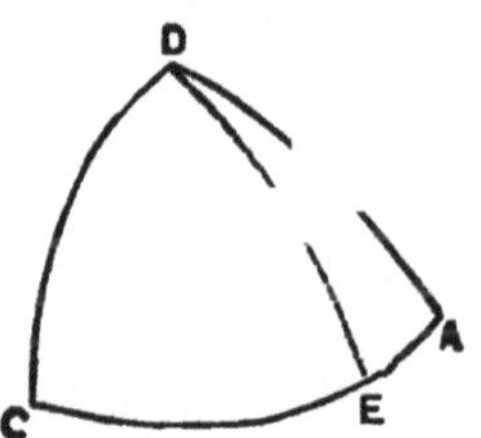

Draw the arc DE, making the angle CDE equal to C; then is DE equal to CE, (P. 10).

The sum DE and EA is greater than DA, (P. 3); but, DE + EA is equal to CE + EA, or to CA; hence, CA is greater than DA, *which was to be proved.*

2°. Let CA be greater than DA; then is the angle D greater than C. For, if D is equal to C, we have CA *equal* to DA, or if D is less than C, CA is *less* than DA, both of which are contrary to the hypothesis; hence, D must be greater than C, *which was to be proved.*

PROPOSITION XII. THEOREM.

The sum of the angles of a spherical triangle is less than six right angles and greater than two right angles.

Let ACD be any spherical triangle, and let EFG be its polar triangle.

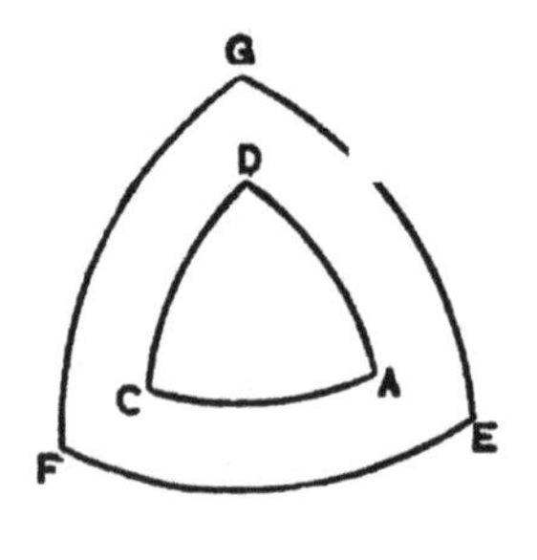

Because each angle of ACD is the supplement of the opposite side of EFG, the sum of the angles of ACD is equal to 6 right angles *minus* the sum of the sides of EFG; hence, this sum is less than 6 right angles. Again, because the sum of the sides of EFG is less than 4 right angles, (P. 3, *Cor.*), the sum of the angles of ACD is greater than 2 right angles, *which was to be proved.*

Cor. All the angles of a spherical triangle may be right angles, or all may be obtuse.

Scho. 1. If DA and DC are quadrants, A and C are right angles and the triangle is bi-rectangular.

Scho. 2. If in addition, AC is a quadrant, the angle D is a right angle and the triangle is tri-rectangular.

Scho. 3. If three planes are placed through the centre of a sphere, perpendicular to each other, they divide its surface into 8 equal parts, each of which is a *tri-rectangular triangle.* Hence, a tri-rectangular triangle is one-eighth of the surface of the sphere, or it is equal to one-half of a great circle. The same planes divide the sphere into 8 equal spherical pyramids, each of which is equal to $\frac{1}{8}$ of $\frac{4}{3}\pi R^3$, or to $\frac{1}{6}\pi R^3$, (Gen. Scho., B. 8). The volume of a tri-rectangular spherical pyramid is equal to the tri-rectangular triangle which forms its base multiplied by $\frac{1}{3}$ of the radius.

PROPOSITION XIII. THEOREM.

The area of a lune is to the surface of the sphere as the angle of the lune is to four right angles.

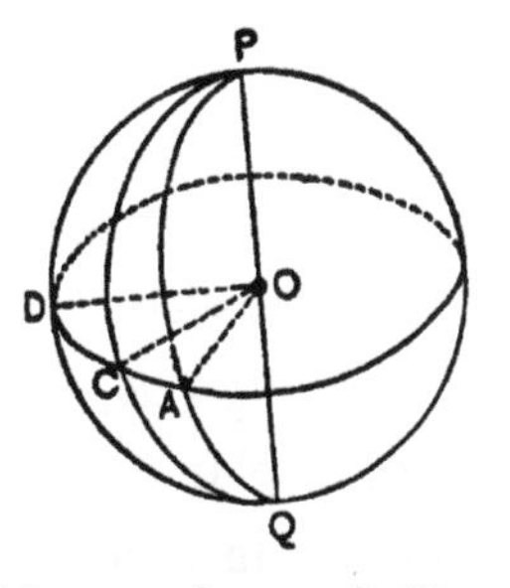

Let PAQ be a semicircumference whose centre is O, and whose diameter is PQ; and let PA be a quadrant.

If the semicircumference is revolved about PQ, as an axis, through an angle AOC it will generate a lune whose angle is APC, or AOC, and if it is revolved through four right angles it will generate the surface of the sphere. Now, it is obvious that the surface generated whilst revolving through any angle is proportional to that angle; hence, the surface of the lune, is to the surface of the sphere, as the angle of the lune, is to four right angles, *which was to be proved.*

Cor. 1. If we denote the area of the lune by L, its angle, in terms of a right angle as a unit, by A, the area of the tri-rectangular triangle by T, remembering that the surface of the sphere is equal to 8 tri-rectangular triangles, we shall have,

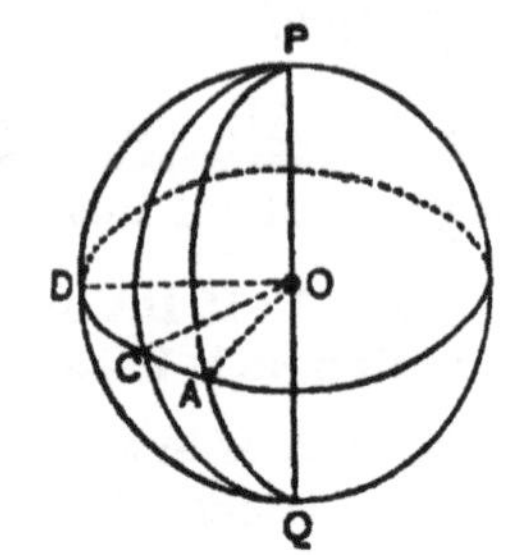

$$L : 8T :: A : 4,$$

whence, $$L = 2A \times T,$$

that is, the area of a lune is equal to twice the angle of the lune, multiplied by the area of the tri-rectangular triangle.

Cor. 2. Whilst the semi-circumference generates the lune the semicircle generates the corresponding wedge, and whilst the semi-circumference generates the surface of the sphere, the semicircle generates the volume of the sphere; hence, the volume of a spherical wedge, is to the volume of the sphere, as the angle of the wedge, is to four right angles.

PROPOSITION XIV. THEOREM.

The areas of two symmetrical spherical triangles are equal.

Let ACD and EFG be two symmetrical spherical triangles.

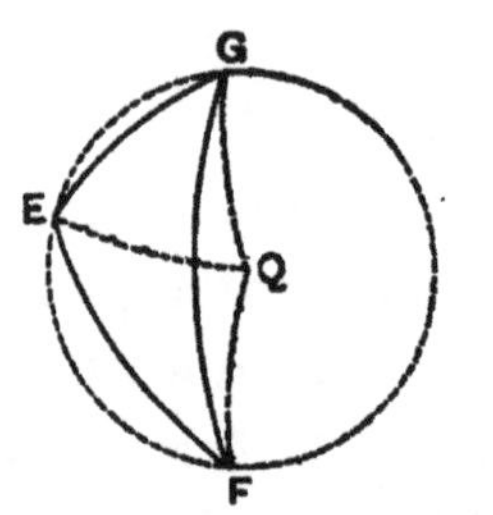

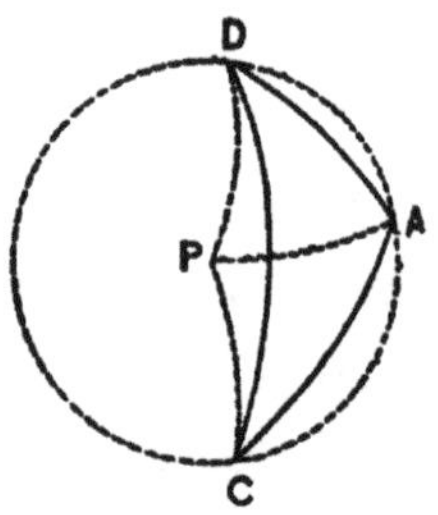

Through A, C, and D, pass a plane cutting the sphere in a small circle whose pole is P. Draw the arcs PA, PC, and PD; these arcs will be equal (P. 1). From F as a pole, with CP as a spherical radius, describe an arc, and from G as a pole, with the same arc as a spherical radius, describe a second arc cutting the first at Q; then draw the arcs QE, QF, and QG.

Because QF and QG are equal to PC and PD the triangles QFG and PCD are isosceles and symmetrical, (P. 6, *Cor.*); they are therefore equal in all their parts; hence, the angle FGQ is equal to CDP. Now, the points Q and P may both lie without the triangles EFG and ACD, or they may both lie within those triangles; in either case, the angle EGQ is equal to ADP, and consequently, all the parts of the triangle EGQ are equal to the corresponding parts of ADP, (P. 7); hence, QE is equal to PA; the triangles EGQ and ADP are therefore isosceles and equal, as are also the triangles EFQ and ACP.

If Q and P lie without the triangles EFG and ACD, EFG is equal to the sum of EGQ and EFQ, diminished by FGQ; and ACD is equal to the sum of ADP and ACP, diminished by CDP; if Q and P lie within EFG and ACD, EFG is equal to the sum of EFQ, EGQ, and FGQ; and ACD is equal to the sum of ACP, ADP, and CDP. Hence, in either case the areas of EFG and ACD are equal, *which was to be proved.*

PROPOSITION XV. THEOREM.

If the semi-circumferences of two great circles intersect on the surface of a hemisphere, the sum of the opposite triangles which they form is equal to a lune whose angle is equal to the angle formed by the semi-circumferences.

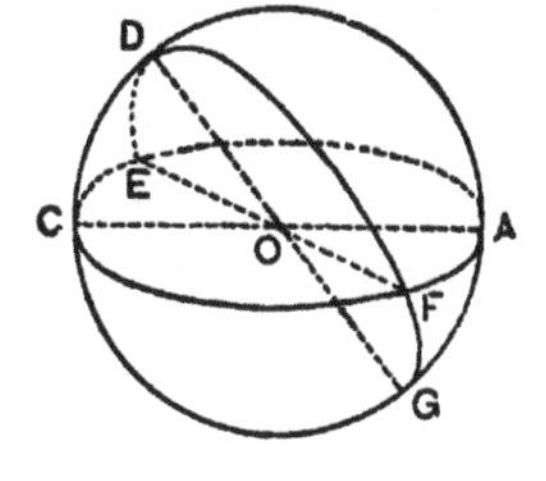

Let ADC and FDE be two semi-circumferences intersecting at D, on the surface of the hemisphere whose great circle is AFCE.

Prolong the arcs DA and DF till they meet at G, forming the lune GFDA.

Because EDF and DFG are semi-circumferences, ED is

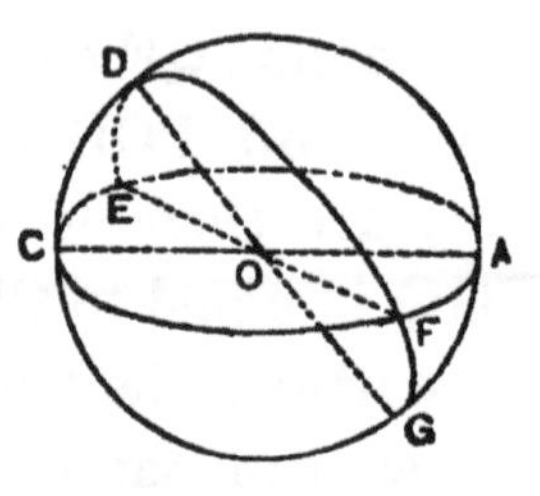

equal to FG; for like reason, CD is equal to AG; and because CEA and EAF are semi-circumferences, CE is equal to AF; hence, the triangles CED and AFG are mutually equilateral, and consequently, mutually equiangular, (P. 9); they are therefore symmetrical and equal in area. But the sum of the triangles FDA and AGF is equal to the lune whose angle is ADF; hence, the sum of the triangles FDA and CDE is also equal to that lune, *which was to be proved.*

Definitions.

134. The **spherical excess of a spherical triangle** is equal to the excess of the sum of its angles above two right angles.

If the right angle is taken as a unit, and if the angles of the triangle are denoted by A, C, and D, and the spherical excess by E, we have

$$E = A + C + D - 2.$$

If the angles are expressed in degrees, the spherical excess is equal to the sum of the angles of the triangle *minus* 180°, divided by 90°. Thus, if A = 85°, B = 105°, and C = 50°, we have

$$E = \frac{85° + 105° + 50° - 180°}{90°} = \frac{60°}{90°} = \frac{2}{3}.$$

135. The **spherical excess of a spherical polygon** is equal to the excess of the sum of its angles over the product of two right angles by the number of sides of the polygon *less* two; for if the polygon is divided into triangles by diagonals drawn from any vertex, the spherical excess of the polygon is equal to the sum of the numbers expressing the spherical excess of each triangle.

PROPOSITION XVI. THEOREM.

The area of a spherical triangle is equal to its spherical excess, multiplied by the area of a tri-rectangular triangle.

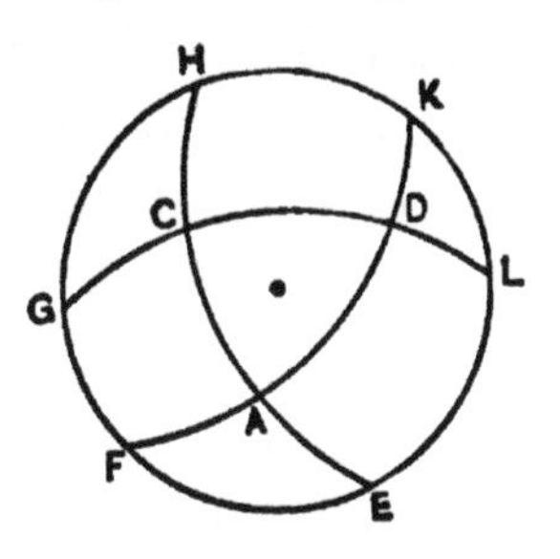

Let ACD be a spherical triangle lying on the surface of a hemisphere whose great circle is EGL.

Prolong the sides of the triangle till they meet the circumference EGL. Denote the angles of the triangle, expressed in terms of a right angle, by A, D, and C, and denote the area of a tri-rectangular triangle by T.

By the preceding proposition, the sum of the triangles AHK and AEF is equal to a lune whose angle is A, that is, to $2A \times T$, (P. 15); in like manner, the sum of the triangles CLE and CGH is equal to $2C \times T$, and the sum of the triangles DGF and DLK is equal to $2D \times T$. But the sum of the six triangles named, is equal to the area of the hemisphere *plus* twice the area of the triangle ACD. Because the hemisphere is equal to 4T, we have

$$2 \times \textit{area } ACD = 2A \times T + 2C \times T + 2D \times T - 4T;$$

whence, by reducing and factoring,

$$\text{area } ACD = (A + C + D - 2)\,T,$$

which was to be proved.

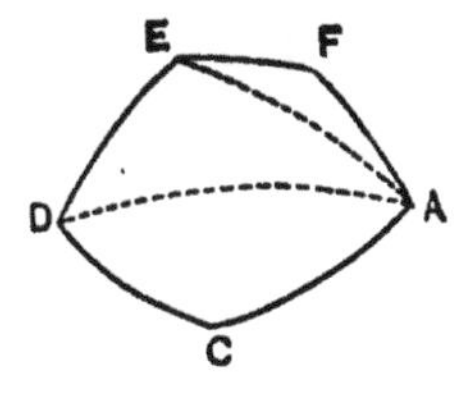

Cor. The area of a spherical polygon ACDEF is equal to its spherical excess, multiplied by the area of a tri-rectangular triangle. For if we divide the polygon into triangles by diagonals drawn from the vertex A, the area of the polygon will be equal to the sum of

Cor. Any side of either triangle is the supplement of the opposite angle of its polar triangle.

Definition.

133. Two spherical triangles are said to be **symmetrical** when all their parts are equal, each to each, but placed in a reverse order in the two triangles.

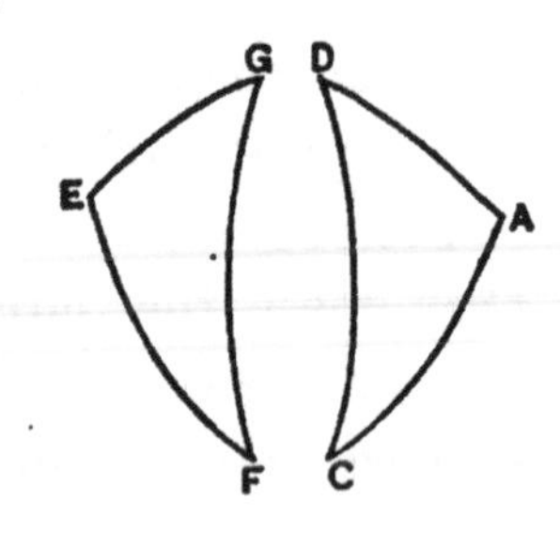

Thus, ACD and EFG are symmetrical if the side AC is equal to EF, CD to FG, DA to GE, and the angle A to E, C to F, D to G.

Symmetrical triangles correspond to symmetrical triedral angles at the centre, and also to symmetrical spherical pyramids whose vertices are at the centre.

Proposition VI. Problem.

To construct a spherical triangle that shall be symmetrical with respect to a given spherical triangle.

Let ACD be a spherical triangle lying on a sphere whose centre is O.

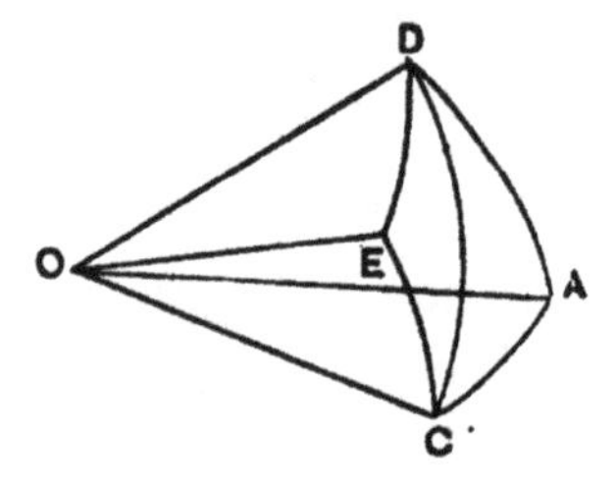

With D as a pole and DA as a spherical radius, describe an arc; with C as a pole and CA as a spherical radius, describe a second arc cutting the first at E. Draw the arcs EC and ED; also draw the radii OA, OC, OD, and OE.

The side EC is equal to AC, ED is equal to AD, and DC is common; hence, the sides of ECD and ACD are equal each to each, but taken in a reverse order. The lateral angles EOC, EOD, and COD, are respectively equal to AOC, AOD, and COD; hence the corresponding interfacial angles are equal, (P. 17, B. 6), that is, the angle E is equal to A, the angle ECD to ACD, and the angle EDC to ADC; hence, the angles of the triangles ECD and ACD are equal, each to each, but taken in a reverse order. The triangle ECD is therefore symmetrical with respect to ACD, *which was to be proved.*

Cor. If DA is equal to DC, DE is also equal to DC, and the symmetrical triangles are both isosceles. In this case if the triangle ECD is placed on CDA so that the angle EDC shall coincide with its equal CDA, the arc ED will coincide with CD, the arc CD with AD, and the two triangles will coincide throughout.

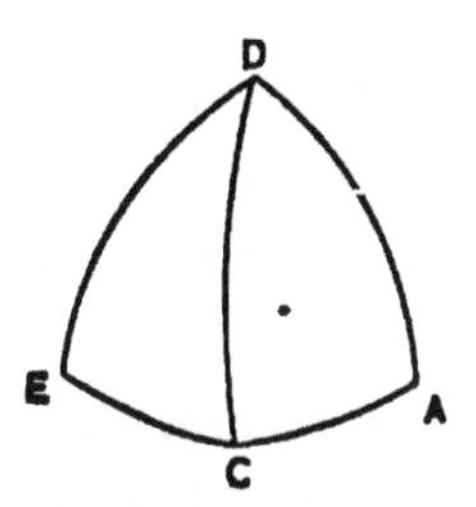

Scho. Symmetrical triangles cannot be made to coincide unless they are isosceles.

Proposition VII. Theorem.

If two spherical triangles have two sides and the included angle of one, equal to two sides and the included angle of the other, each to each, their remaining parts are equal, each to each.

Let FGH and ACD be two spherical triangles in which FG is equal to AC, GH to CD, and the angle FGH to ACD; also, let ACD and ECD be symmetrical.

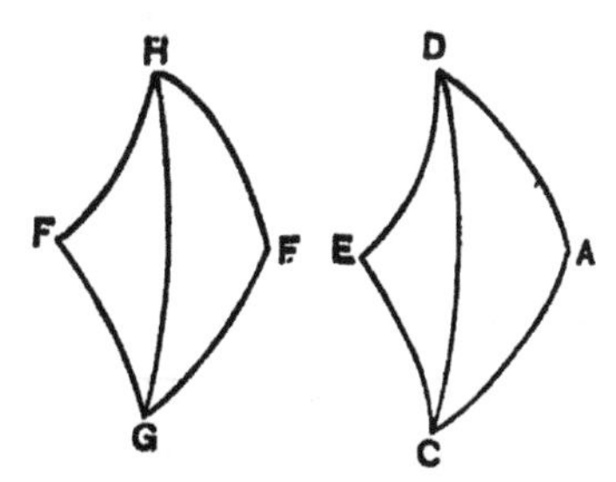

The triangle FGH can be placed on ACD, or else on its symmetrical triangle ECD so as to coincide with it, as may be shown by a course of reasoning similar to that employed in demonstrating P. 8, B. 1; hence, the side HF is equal to DA, the angle F to A, and the angle FHG to ADC, *which was to be proved.*

It will be shown that every conic section is a **parabola**, **an ellipse**, or **an hyperbola**.

I. THE PARABOLA.

Definitions.

3. A **parabola** is a plane curve that may be generated by a point, moving so that it shall always be equally distant from a fixed line and from a given point.

The fixed line is called the **directrix**; the given point is the **focus**, the line through the focus perpendicular to the directrix is the **axis**; and the point in which the axis meets the curve is the **vertex** of the parabola.

Proposition I. Problem.

To construct a parabola when its directrix and focus are given.

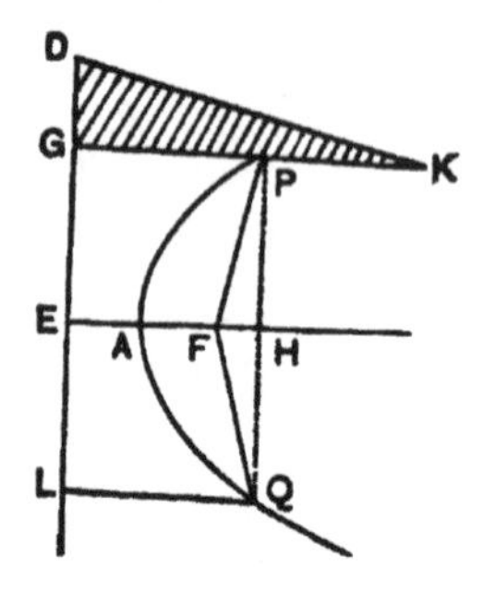

Let LD be the directrix, and F the focus. Through F, draw a line EH, perpendicular to ED; it will be the axis.

Let KGD be a triangular ruler, right-angled at G. Take a string equal in length to KG; fasten one end at K and the other at F; then press a pencil against the string, keeping its point in contact with the ruler, and move the ruler so that the edge GD shall coincide with LD; the pencil will trace out an arc of a parabola. For, in every position we shall have GP equal to FP.

Cor. 1. The point A, midway between E and F, is a point of the curve, and because it lies on EH, it is the vertex of the parabola.

Cor. 2. The curve is symmetrical with respect to EH. For, draw PH perpendicular to EH and prolong it till it meets the curve at Q; join F with Q, and draw QL perpendicular to LD. Because LQ and GP are parallels between parallels, they are equal; but LQ is equal to FQ and GP to FP, from the definition of the curve; hence, FQ, and FP are equal to each other, and consequently HQ and HP are equal, (P. 15, *Cor.* 1, B. 1), that is, for every point on one side of EH there is a corresponding symmetrical point on the other side.

PROPOSITION II. THEOREM.

If a point lies without a parabola, its distance from the directrix is less than its distance from the focus; if it lies within the curve, its distance from the directrix is greater than its distance from the focus.

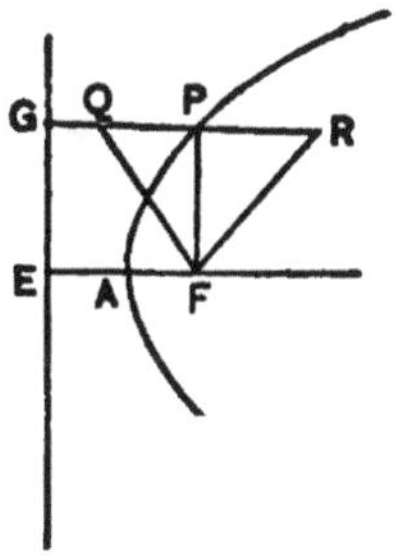

1°. Let Q be a point lying without a parabola whose directrix is EG, and whose focus is F.

Draw QG perpendicular to EG and prolong it till it meets the curve at P; draw QF and PF.

In the triangle FQP, we have,

$$FP < FQ + QP;$$

or, since FP is equal to GP,

$$GP < FQ + QP.$$

If we subtract QP from both members of the last inequality, remembering that GP–QP is equal to GQ, we have,

$$GQ < FQ,$$

which was to be proved.

2°. Let R be a point lying within the curve.

Draw RG perpendicular to EG, cutting the curve at P: also draw RF and PF.

In the triangle FPR, we have,

$$FP + PR > FR;$$

But FP is equal to GP, and consequently FP + PR is equal to GR; hence,

$$GR > FR,$$

which was to be proved.

Cor. If the distance of a point from EG is less than its distance from F, the point lies without the curve. For, if it were on the curve its distance from EG would be equal to its distance from F, or if it were within the curve its distance from EG would be greater than its distance from F.

PROPOSITION III. THEOREM.

The bisectrix of the angle formed by two lines drawn from any point of a parabola, one to the focus, and the other perpendicular to the directrix, is tangent to the curve at that point.

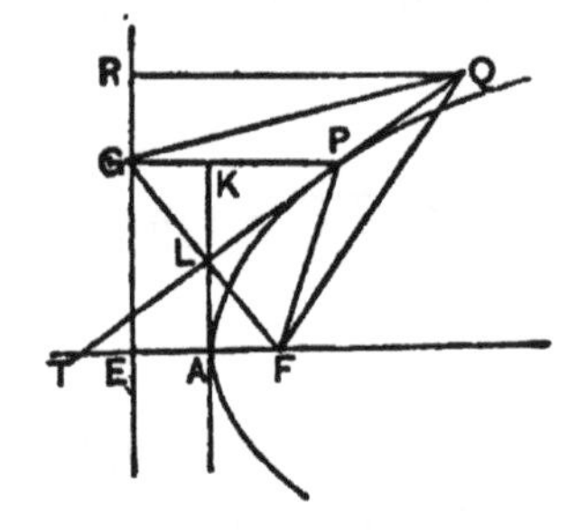

Let P be any point of a parabola whose directrix is EG and whose focus is F. Suppose PG to be perpendicular to EG, and let PT bisect the angle GPF.

Take any point, Q, of PT, except P, and draw QR perpendicular to EG; also draw QG, QF, and FG.

Because GP is equal to FP, and the angle GPT to TPF, the line PT is perpendicular to FG at its middle point, L; hence, GQ is equal to FQ, (P. 16, B. 1). But, RQ is less than GQ, (P. 14, B. 1); hence, RQ is less than FQ; consequently, Q lies without the curve, (P. 2, *Cor.*). Therefore, every point of PT, except P, lies without the parabola; hence, PT is tangent to the curve at P, *which was to be proved.*

Cor. 1. The line AK, perpendicular to EF at A is tangent to the parabola at A. It is called the **vertical tangent.**

Cor. 2. The angles FPT and GPT are equal by hypothesis, and FTP and GPT are equal because they are alternate angles; consequently, the triangle FTP is isosceles; hence, FT is equal to FP, that is, the focus is equally distant from the foot of the tangent, and from the point of contact.

Cor. 3. Because EA and AF are equal, and because AK is parallel to EG, AK passes through L; hence, the perpendicular from the focus to any tangent, meets it at a point of the vertical tangent.

Definitions.

4. The **ordinate** of a point is the perpendicular distance from the axis to the point.

5. The **parameter** of a parabola is the double ordinate through the focus.

6. A **normal** is the perpendicular to a tangent at the point of contact.

7. A **subtangent** is the distance from the foot of the tangent to the foot of the ordinate of the point of contact.

8. A **subnormal** is the distance from the foot of the ordinate of the point of contact to the foot of the normal.

PROPOSITION IV. THEOREM.

The parameter of a parabola is equal to twice the distance from the directrix to the focus.

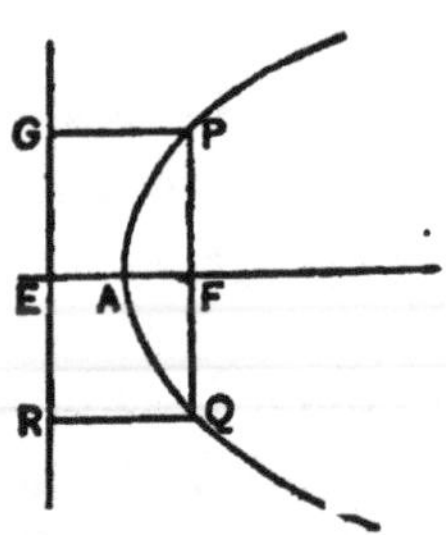

Let QAP be a parabola whose directrix is RG and whose focus is F; let QFP be perpendicular to EF, and let QR and PG be perpendicular to RG; then is QP, or 2FP, the parameter of the curve, (*Art.* 5).

From the definition of the parabola we have FP equal to GP; hence, 2FP is equal to 2GP, or to 2EF, *which was to be proved.*

Cor. The distance from the vertex to the focus is equal to one-fourth of the parameter.

PROPOSITION V. THEOREM.

The subtangent is bisected at the vertex.

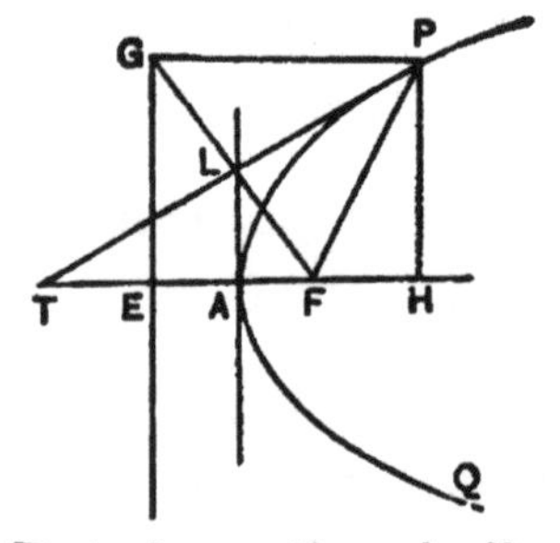

Let QAP be any parabola; P, any point of the curve; PT, the corresponding tangent; HP, the ordinate of P; and AL, the vertical tangent; then is TH the subtangent.

Draw FL perpendicular to PT; then, because FT is equal to FP, (P. 3, *Cor.* 2), TL is equal to LP. But from the similar triangles TAL and THP, we have,

$$TA : AH :: TL : LP;$$

hence, TA is equal to AH, *which was to be proved.*

Cor. 1. The distance AH is called the **abscissa** of P; the subtangent is, therefore, equal to twice the abscissa of the point of contact.

Cor. 2. The ordinate HP is equal to 2AL.

PROPOSITION VI. THEOREM.

The subnormal is equal to one-half the parameter.

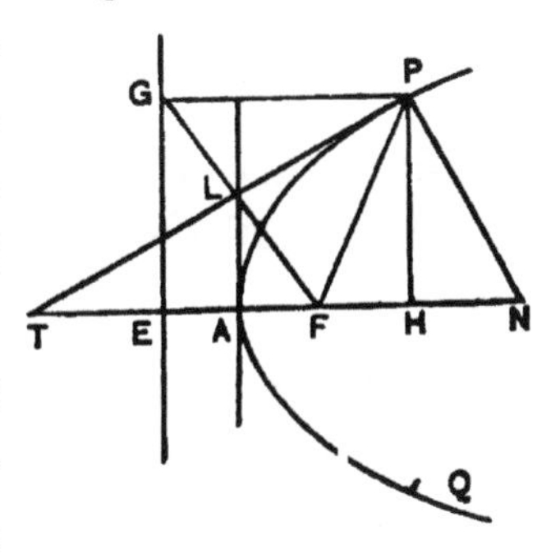

Let QAP be a parabola; P, any point of the curve; PT, the corresponding tangent; PN, the corresponding normal; HP, the ordinate of P; and AL, the vertical tangent.

Because the sides of the triangles NHP and FAL are parallel, each to each, the triangles are similar; hence,

$$HN : HP :: AF : AL;$$

but, HP is equal to 2AL, (P. 5, *Cor.* 2); consequently NH is equal to 2AF, or to half the parameter, (P. 4, *Cor.*), *which was to be proved.*

PROPOSITION VII. PROBLEM

To draw a tangent to a parabola at a given point of the curve.

Let QAP be a parabola whose directrix is EG, and whose focus is F; also let EN be the axis, P any point of the curve, and HP its ordinate.

FIRST CONSTRUCTION. — Draw PF; lay off FT equal to FP, and draw PT; then will PT be the required tangent, (P. 3, *Cor.* 2).

SECOND CONSTRUCTION.—Lay off AT equal to HA and draw PT; then will PT be the required tangent, (P. 5).

THIRD CONSTRUCTION.—Lay off HN equal to EF and draw PN; PN will be the normal at P, (P. 6).

Draw PT perpendicular to PN; then, from the definition, (*Art.* 6), PT will be the required tangent.

PROPOSITION VIII. THEOREM.

The square of the ordinate of any point is equal to the parameter of the curve, multiplied by the abscissa of the point.

Let QAP be a parabola; P, any point of the curve; HP, the ordinate of P; AH, its abscissa; PT, the corresponding tangent; and PN, the normal.

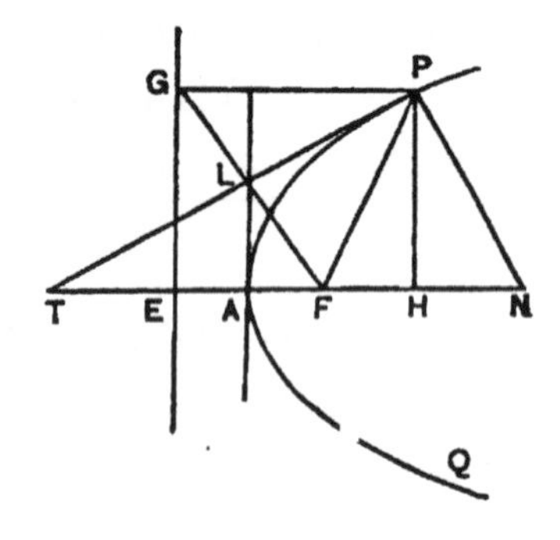

The triangle NTP is right-angled at P, and HP is perpendicular to NT; hence, (P. 18, B. 3).

$$\overline{HP}^2 = TH \times HN; \quad . \quad . \quad . \quad . \quad . \quad (1)$$

but, TH is equal to 2AH; and HN is equal to one-half of the parameter, or denoting the parameter by $2p$, HN is equal to p. Substituting these in (1), we have,

$$\overline{HP}^2 = 2p \times AH, \quad . \quad . \quad . \quad . \quad . \quad (2)$$

which was to be proved.

Scho. The preceding demonstration rests ultimately on the definition of a parabola; hence, the property demonstrated is characteristic of that curve, that is, any curve which possesses that property must be a parabola.

PROPOSITION IX. THEOREM.

If a conic surface is cut by a plane whose inclination to the plane of the base is equal to that of an element of the surface, the section is a parabola.

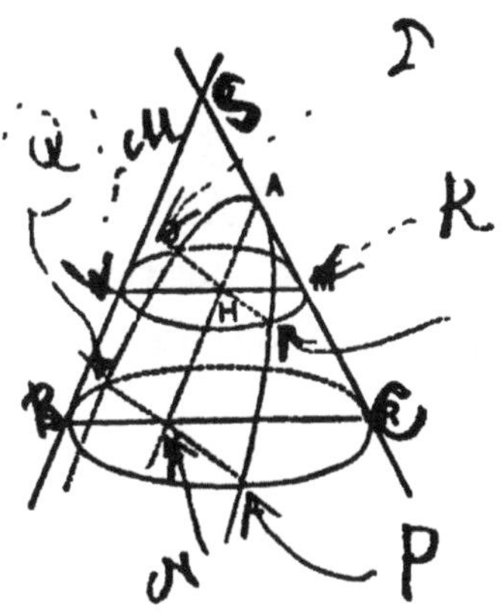

Let SVR be any meridian plane of the conic surface whose vertex is V, and whose base is RLT; and let MPQ be a section parallel to the base. Let TQAPL be a section made by a plane perpendicular to SVR, having its inclination RKA equal to the angle RSV, and let this plane intersect the plane SVR in AK, the circle MPQ in QP, and the circle RST in TL.

The plane SVR divides the conic surface into two symmetrical parts, and because it is perpendicular to the planes TAL, MPQ, and RLT, it divides the corresponding curves symmetrically; hence, AK is the axis of TQAPL, MN is a diameter of MPQ, and RS is a diameter of RLT. Furthermore, HP is perpendicular to the plane SVR, (P. 14, *Cor.* 3, B. 6), and consequently to AK and MN; and for like reason, KL is perpendicular to AK and RS.

From P. 18, *Cor.*, B. 3, we have,

$$\overline{HP}^2 = NH \times HM, \text{ and } \overline{KL}^2 = SK \times KR;$$

but, SK and NH are equal, because they are parallels between parallels; hence, the second members of these equations are equimultiples of HM and KR; consequently, we have,

$$\overline{HP}^2 : \overline{KL}^2 :: HM : KR. \quad . \quad . \quad . \quad (1)$$

From the similar triangles AHM and AKR, we have,

$$HM : KR :: AH : AK. \quad . \quad . \quad . \quad . \quad (2)$$

Hence, from (1) and (2), we have,

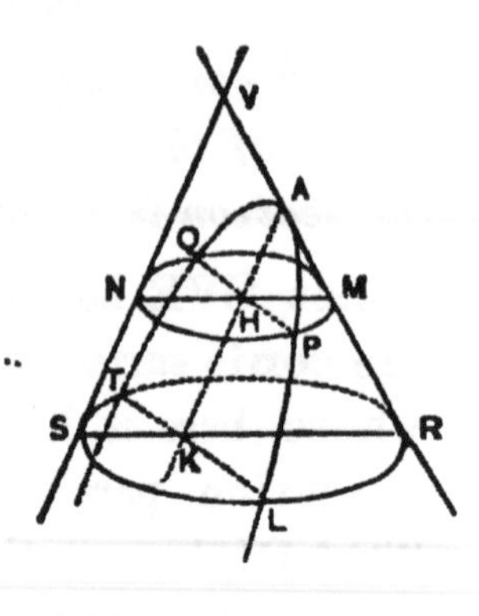

$$\overline{HP}^2 : \overline{KL}^2 :: AH : AK,$$

or, $$\overline{HP}^2 = \frac{\overline{KL}^2}{AK} \times AH \ldots (3)$$

But $\frac{\overline{KL}^2}{AK}$ is a fixed quantity, for a given position of the plane TAL; placing this equal to $2p$, we have,

$$\overline{HP}^2 = 2p \times AH.$$

Hence, from P. 8, *Scho.*, the section TQAPL is a parabola, *which was to be proved.*

II. THE ELLIPSE.

Definitions.

9. An **ellipse** is a plane curve that may be generated by a point, moving so that the sum of its distances from two fixed points is equal to a given line.

The moving point is called the **generatrix**; the fixed points are **foci**; the straight line through the foci, and limited by the curve, is the **transverse axis**; the point midway between the foci is the **centre**; the perpendicular to the transverse axis at the centre, and limited by the curve, is the **conjugate axis**; any line through the centre, and limited by the curve, is a **diameter**.

10. The **ordinate** of any point is its perpendicular distance from the transverse axis.

11. **Focal lines** of a point are lines from the foci to that point.

12. The **excentricity** is the distance from the centre to either focus, *divided* by half the transverse axis.

13. The **parameter of the curve** is the double ordinate through either focus.

Proposition X. Problem.

To construct an ellipse when the foci and one position of the generatrix are given.

Let F and F′ be the foci, and P, any position of the generatrix; let PF and F′P be focal lines drawn to P.

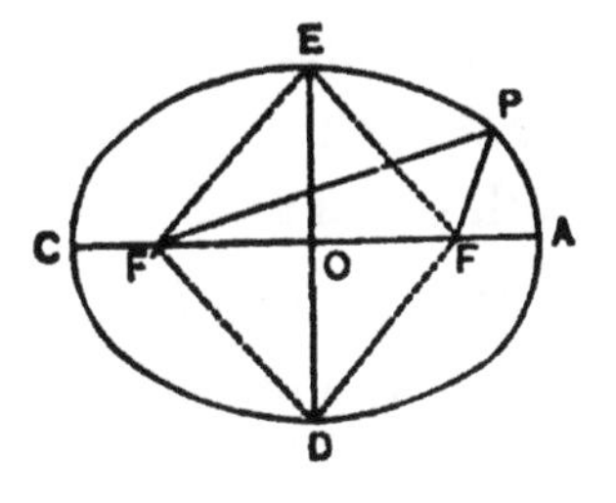

Take a string equal in length to the sum of FP and F′P, and fasten one end at F and the other end at F′; then press a pencil against the string so as to make it tense, and move the pencil around F and F′; the point of the pencil will trace out the required curve; for, in every position, the sum of its distances from the foci is equal to the length of the string.

Draw AC through F and F′, and it will be the transverse axis; through O, the middle of FF′, draw DE, perpendicular to AC, and it will be the conjugate axis.

Cor. 1. When the generatrix is at A, the sum of its focal lines is equal to 2AF + FF′, and when it is at C, this sum is equal to 2CF′ + FF′; but, these sums are equal; hence, CF′ is equal to AF, and consequently CO is equal to AO. Again, because 2AF + FF′ is equal to AF + FF′ + F′C, or to AC, the sum of the focal lines to any point of the curve is equal to the transverse axis.

Cor. 2. When the generatrix is at E, or at D, its focal

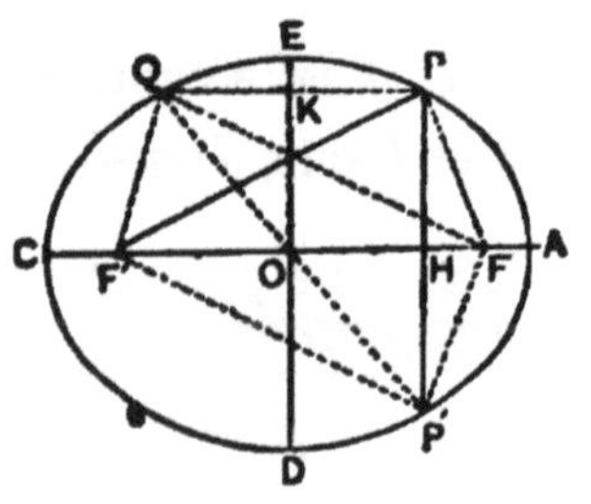

dicular to, and is bisected by, the transverse axis. Again, let the part DPE be revolved about DE through a half revolution; P will fall on the perpendicular PQ, making KQ equal to KP, F will fall on F', the line FP will fall on F'Q, and the angle OFP on the angle OF'Q; hence, the triangles FF'P and FF'Q are equal in all their parts, and consequently, Q is a point of the curve, that is, for every point P, there is a second point Q, such that the line joining them is perpendicular to, and is bisected by, the conjugate axis. Hence, the curve is symmetrical with respect to both of its axes, *which was to be proved.*

Cor. 1. The figure P'FQF' is a parallelogram, (P. 27, B. 1); hence, the angle FQF' is equal to FP'F', FQP' to QP'F', and F'QP' to FP'Q.

Cor. 2. The diagonals of P'FQF' bisect each other, (P. 29, B. 1); but, P'Q is a diameter; hence, every diameter is bisected at the centre.

PROPOSITION XIII. THEOREM.

The bisectrix of the angle formed by one of the focal lines of a point of the curve, and the prolongation of the other focal line, is tangent to the curve at that point.

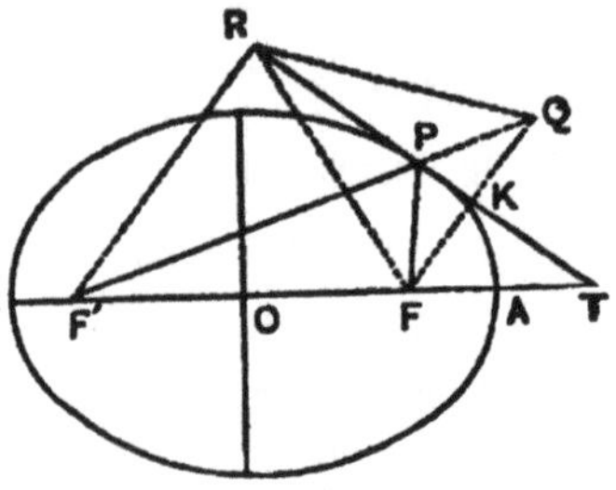

Let P be any point of an ellipse whose transverse axis is AC, and let FP and F'P be the corresponding focal lines; let PT bisect the angle between FP and the prolongation of F'P.

On the prolongation of F'P, lay off PQ equal to PF and

draw QF. Let R be any point of PT, except P, and draw RF, RF′, and RQ.

Because PF is equal to PQ, and the angle FPK to QPK, PT is perpendicular to FQ, at its middle point, K; consequently, RQ is equal to RF. Now, in the triangle F′QR, we have,

$$F'R + RQ > F'Q;$$

but, RQ is equal to RF, and F′Q is equal to the sum of F′P and FP, or to AC; consequently,

$$F'R + FR > AC;$$

hence, R lies without the curve, (P. 11, *Cor.*), that is, every point of PT, except P, lies without the curve; PT is therefore tangent to the ellipse at P, *which was to be proved.*

Cor. 1. The angles FPT and F′PR are equal; for each is equal to QPK; either of these angles is half the supplement of FPF′.

Cor. 2. Perpendiculars to AC, at A and C, are tangents to the curve, at A and C.

PROPOSITION XIV. PROBLEM.

To draw a tangent to an ellipse at a given point.

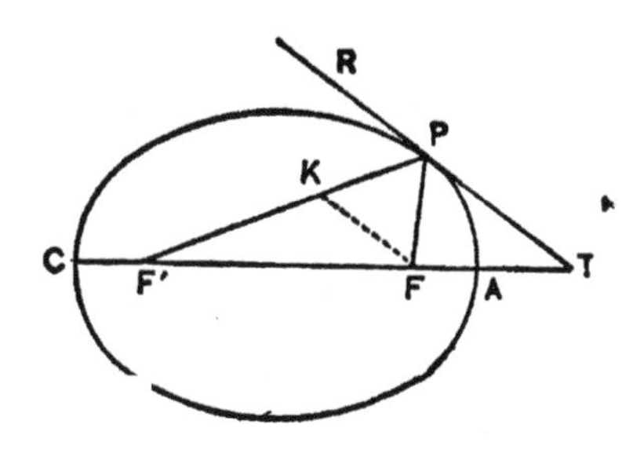

Let P be any point of the ellipse whose foci are F and F′; let FP and F′P be the corresponding focal lines, and suppose F′P to be the greater. Lay off PK equal to PF and draw FK, forming the isosceles triangle

Now, the triangles OFK and OF′K′ have the side OF equal to OF′, the side OK to OK′, and the angle FOK to F′OK′; hence, F′K′ is equal to FK. But, RK′ and AC are chords of the circle OA; hence, RF′ × F′K′, or RF′ × FK, is equal to AF′ × F′C, or to $\overline{OE}^2$, (P. 10, *Cor.* 3), that is, *the product of the perpendiculars from the foci to any tangent is equal to the square of the semi-conjugate axis.*

Cor. 2. Because PT is perpendicular to FQ at its middle point, it bisects the angle F′TQ; hence, (P. 14, B. 3),

$$F'T : QT :: F'P : PQ,$$

but, QT is equal to FT, and PQ to PF; consequently,

$$F'T : FT :: F'P : FP.$$

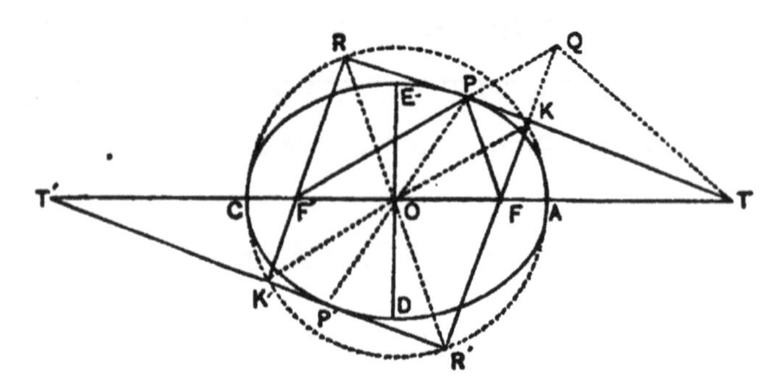

Hence, by composition and division

$$F'T+FT : F'T-FT :: F'P+FP : F'P-FP,$$

or, $\quad 2OT : 2OF :: 2OA : F'P-FP. \quad . \quad . \quad . \quad . \quad . \quad (1)$

Scho. The circle OA is said to circumscribe the ellipse.

Proposition XVII. Theorem.

The semi-transverse axis of an ellipse is a mean proportional between the distance from the centre to the foot of the tangent, and the distance from the centre to the foot of the ordinate of the point of contact.

Let PT be tangent to the ellipse whose centre is O; let HP be the ordinate of the point of contact, and let F′P and FP, be the focal lines of that point.

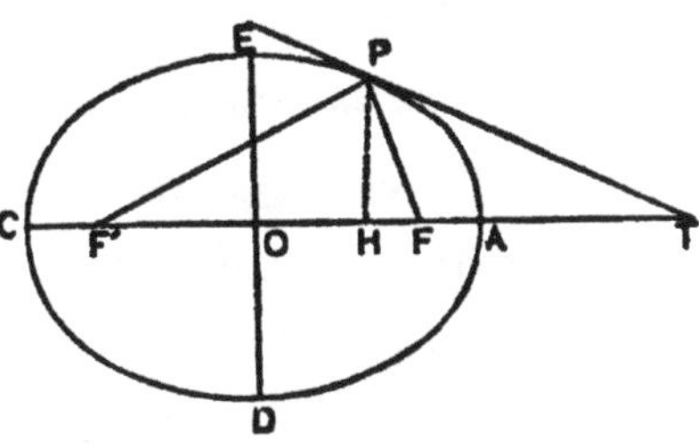

From the right-angled triangles F′HP and FHP, we have,

$$\overline{F'P}^2 = \overline{F'H}^2 + \overline{HP}^2, \quad . \quad . \quad . \quad . \quad . \quad (1)$$

and

$$\overline{FP}^2 = \overline{FH}^2 + \overline{HP}^2. \quad . \quad . \quad . \quad . \quad . \quad (2)$$

Subtracting (2) from (1), member from member, we have,

$$\overline{F'P}^2 - \overline{FP}^2 = \overline{F'H}^2 - \overline{FH}^2,$$

whence, by factoring the two members,

$$(F'P + FP)\,(F'P - FP) = (F'H + FH)\,(F'H - FH);$$

but, $F'P + FP$ is equal to $2OA$, $F'H + FH$ to $2OF$, and $F'H - FH$ to $(F'O + OH) - (FO - OH)$, or to $2OH$; hence, the last equation becomes

$$2OA \times (F'P - FP) = 2OF \times 2OH,$$

or,

$$2OA : 2OH :: 2OF : F'P - FP. \quad . \quad . \quad . \quad (3)$$

From proportion (1), P. 16, *Cor.* 2, we have, by alternation,

$$2OT : 2OA :: 2OF : F'P - FP; \quad . \quad . \quad . \quad (4)$$

combining (4) and (3), and dividing each term by 2, we have

$$OT : OA :: OA : OH, \quad . \quad . \quad . \quad . \quad . \quad (5)$$

which was to be proved.

PROPOSITION XVIII. THEOREM.

The subtangent in an ellipse is equal to the corresponding subtangent in the circumscribing circle.

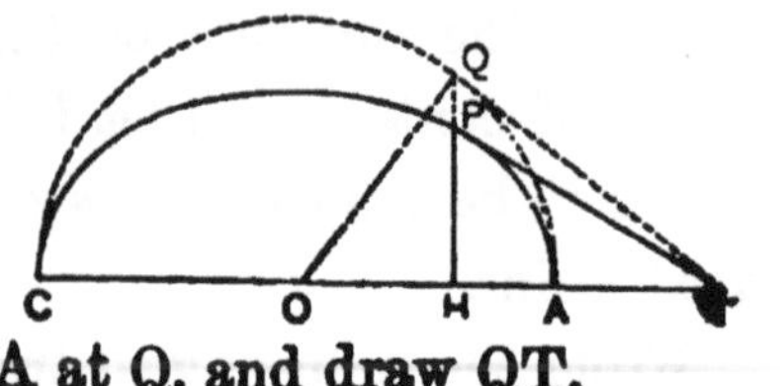

Let PT be tangent to the ellipse whose semi-transverse axis is OA, at any point, P; let the ordinate HP be prolonged till it meets the circumference OA at Q, and draw QT.

From P. 17, we have,

$$OT : OA :: OA : OH,$$

or,

$$OT : OQ :: OQ : OH.$$

Now, the triangles OQT and OHQ, have the angle QOT common, and from what has just been shown the including sides of this angle are proportional; hence, the triangles are similar; the angle OQT is therefore equal to OHP, that is, it is a right angle; hence, QT is tangent to the circumference OA at Q. The line HT is therefore a common subtangent, corresponding to the two tangents PT and QT, *which was to be proved.*

Scho. This principle is used in drawing a tangent to an ellipse at a given point as follows: let P be a given point on the ellipse; draw its ordinate, and prolong it to meet the circumscribing circumference at Q; draw QT perpendicular to the radius OQ, and from the point T, in which it meets the prolongation of the transverse axis, draw TP; this will be the required tangent.

PROPOSITION XIX. THEOREM.

The square of any ordinate of an ellipse is equal to the square of the ratio of the semi-axes, multiplied by the product of the segments into which the ordinate divides the transverse axis.

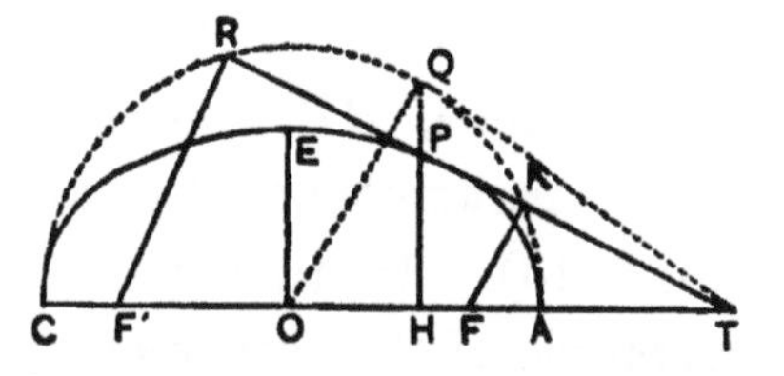

Let P be any point of the ellipse whose foci are F and F′. Let PT be tangent to the curve at P, and let FK and F′R be perpendicular to it.

Prolong the ordinate HP to meet the circumference OA, at Q, and draw QT; QT will be perpendicular to the radius OQ. The right-angled triangles THP, TKF, and TRF′, have the angle at T in common; they are therefore similar; hence,

$$HP : FK :: HT : KT, \quad . \quad . \quad . \quad . \quad . \quad (1)$$

and,

$$HP : F'R :: HT : RT; \quad . \quad . \quad . \quad . \quad . \quad (2)$$

multiplying (1) by (2), term by term, we have,

$$\overline{HP}^2 : FK \times F'R :: \overline{HT}^2 : KT \times RT. \quad . \quad . \quad (3)$$

But, (P. 16, *Cor.* 1), $FK \times F'R = \overline{OE}^2$, and, (P. 20, *Cor.*, B. 3), $KT \times RT = \overline{QT}^2$; hence,

$$\overline{HP}^2 : \overline{OE}^2 :: \overline{HT}^2 : \overline{QT}^2. \quad . \quad . \quad . \quad . \quad . \quad (4)$$

From the similar triangles QHT and OHQ we have

$$HT : QT :: HQ : OQ, \text{ or } OA,$$

or,

$$\overline{HT}^2 : \overline{QT}^2 :: \overline{HQ}^2 : \overline{OA}^2; \quad . \quad . \quad . \quad . \quad . \quad (5)$$

combining (4) and (5), and replacing $\overline{HQ}^2$ by its value $AH \times HC$, we have

$$\overline{HP}^2 : \overline{OE}^2 :: AH \times HC : \overline{OA}^2,$$

or,

$$\overline{HP}^2 = \frac{\overline{OE}^2}{\overline{OA}^2}(AH \times HC), \quad . \quad . \quad . \quad . \quad (6)$$

which was to be proved.

Cor. 1. If H is taken at either focus, HP is one-half of the parameter of the curve, (*Art.* 13). In this case, the rectangle of the corresponding segments of the transverse axis is equal to $\overline{OE}^2$, (P. 10, *Cor.* 3); denoting the *semi-parameter* by p, we have, from (6),

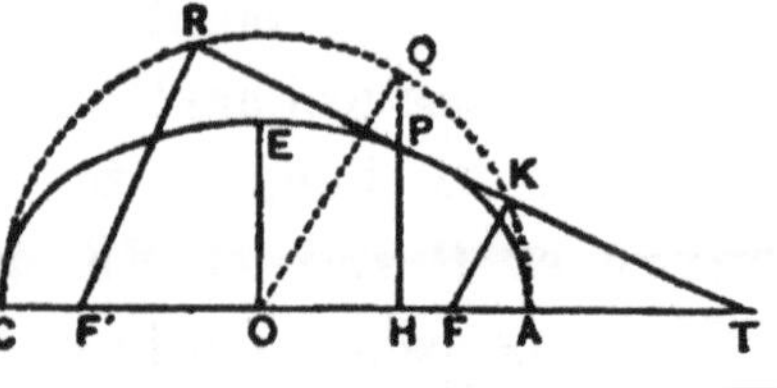

$$p^2 = \frac{\overline{OE}^2}{\overline{OA}^2} \times \overline{OE}^2, \text{ or } p^2 = \frac{\overline{OE}^4}{\overline{OA}^2};$$

extracting the square root of both members, we have,

$$p = \frac{\overline{OE}^2}{OA}, \text{ or } 2p = \frac{4\overline{OE}^2}{2OA};$$

hence, the proportion,

$$2OA : 2OE :: 2OE : 2p,$$

that is, *the parameter is a third proportional to the transverse and conjugate axes.*

Cor. 2. In passing from point to point of the curve the square of the ratio of the semi-axes does not change; hence, *the squares of the ordinates of any two points of the ellipse are to each other as the rectangles of the segments into which they divide the transverse axis.*

Scho. The ultimate basis of the preceding corollary is the definition of the ellipse; hence, the property there enunciated is a characteristic property of the ellipse. It is to be observed that the segments alluded to are both *internal*, that is, their common extremity lies between the extremities of the transverse axis.

PROPOSITION XX. THEOREM.

The ordinate of any point of an ellipse, is to the ordinate of the corresponding point of the circumscribing circle, as the semi-conjugate axis, is to the semi-transverse axis.

Let HP be the ordinate of any point, P, of an ellipse, and let HQ be the corresponding ordinate of the circumscribing circle. Let PT be tangent to the ellipse, at P, and QT tangent to the circle OA, at Q.

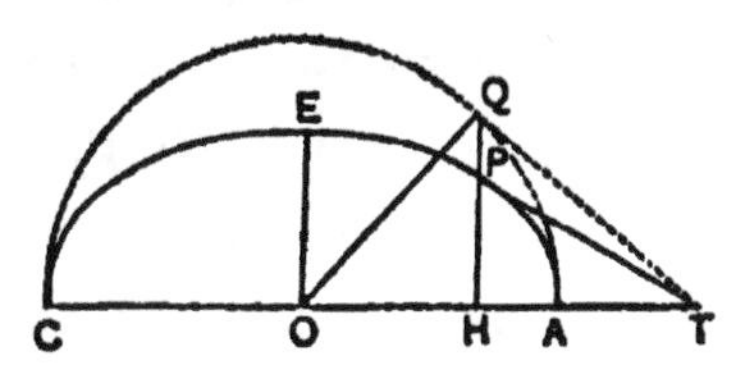

From P. 19, we have,

$$\overline{HP}^2 = \frac{\overline{OE}^2}{\overline{OA}^2}(AH \times HC), \quad . \quad . \quad . \quad . \quad (1)$$

and from P. 18, *Cor.*, B. 3, we have,

$$\overline{HQ}^2 = AH \times HC. \quad . \quad . \quad . \quad . \quad . \quad . \quad . \quad (2)$$

Dividing (1) by (2), member by member, we have,

$$\frac{\overline{HP}^2}{\overline{HQ}^2} = \frac{\overline{OE}^2}{\overline{OA}^2}, \text{ or } \frac{HP}{HQ} = \frac{OE}{OA},$$

whence, $\quad HP : HQ :: OE : OA,$

which was to be proved.

Scho. If OA remains unchanged, whilst OE increases, the ellipse approaches its circumscribing circle, and when OE becomes equal to OA, the ellipse coincides with the circle. Hence, the circle is a particular case of the ellipse; in this case, the foci of the ellipse coincide with the centre of the circle.

PROPOSITION XXI. THEOREM.

If a conic surface is cut by a plane, whose inclination to the plane of the base is less than that of an element of the surface, the section is an ellipse.

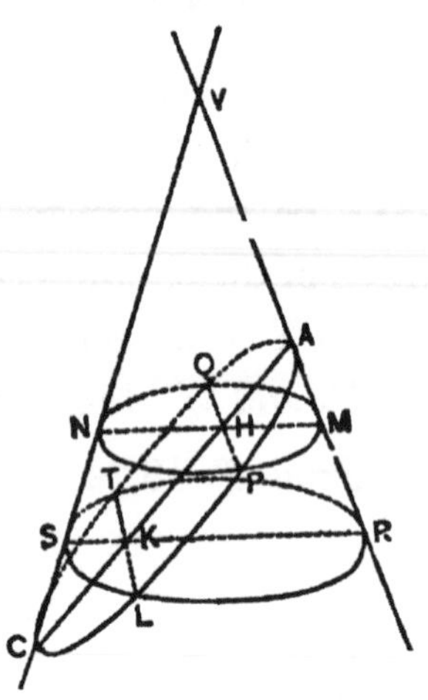

Let SVR be any meridian plane of the conic surface whose vertex is V, and whose base is RLT, and let MPQ be a section parallel to the base. Let APCT be a section of the surface by a plane perpendicular to SVR, and whose inclination, AKR, to the plane of the base, is less than VSR; and let the plane of this section intersect the circle MPQ in QP, and the circle RST in TL.

The plane SVR divides the surface into two symmetrical parts, and because it is perpendicular to the planes APCT, MNQ, and RST, it divides the corresponding curves symmetrically; hence, AC is the axis of the curve APCT, MN is a diameter of the circle MPQ, and RS is a diameter of the circle RLT. Furthermore, HP is perpendicular to AC, and MN, and TL is perpendicular to AC, and RS.

From P. 18, *Cor.*, B. 3, we have,

$$\overline{HP}^2 = NH \times HM, \quad . \quad . \quad . \quad . \quad . \quad . \quad . \quad . \quad (1)$$

and,
$$\overline{KL}^2 = SK \times KR; \quad . \quad . \quad . \quad . \quad . \quad . \quad . \quad . \quad (2)$$

whence,
$$\overline{HP}^2 : \overline{KL}^2 :: NH \times HM : SK \times KR. \quad . \quad . \quad (3)$$

From the similar triangles CNH and CSK, we have,

$$NH : SK :: HC : KC, \quad . \quad . \quad . \quad . \quad . \quad . \quad (4)$$

and from the similar triangles AHM and AKR, we have,

$$HM : KR :: AH : AK. \quad . \quad . \quad . \quad . \quad . \quad (5)$$

Multiplying (4) and (5), term by term, we have,

$$NH \times HM : SK \times KR :: AH \times HC : AK \times KC. \, . \, . \quad (6)$$

From (3) and (6), we have,

$$\overline{HP}^2 : \overline{KL}^2 :: AH \times HC : AK \times KC,$$

which is a characteristic property of the ellipse, (P. 19, *Scho.*); hence, the section APCT is an ellipse, *which was to be proved.*

Cor. If the cutting plane is parallel to the base, the section is a circle, which is a particular case of the ellipse.

III. THE HYPERBOLA.

Definitions.

14. An **hyperbola** is a plane curve that may be generated by a point, moving so that the difference of its distances from two fixed points is equal to a given line.

The moving point is called the **generatrix**; the fixed points are **foci**; the straight line through the foci and limited by the curve is the **transverse axis**; the point midway between the foci is the **centre**; the perpendicular to the transverse axis at the centre is the **conjugate axis**; and any line through the centre, limited by the curve, is a **diameter**.

15. The **ordinate** of a point is its distance from the transverse axis, indefinitely prolonged.

16. **Focal lines** of a point are lines from the foci to that point.

17. The **excentricity** is the distance from the centre to either focus, divided by half the transverse axis.

18. The **parameter of the curve** is the double ordinate through the focus.

PROPOSITION XXII. PROBLEM.

To construct an hyperbola when the foci and the difference of the focal lines of any point of the curve are given.

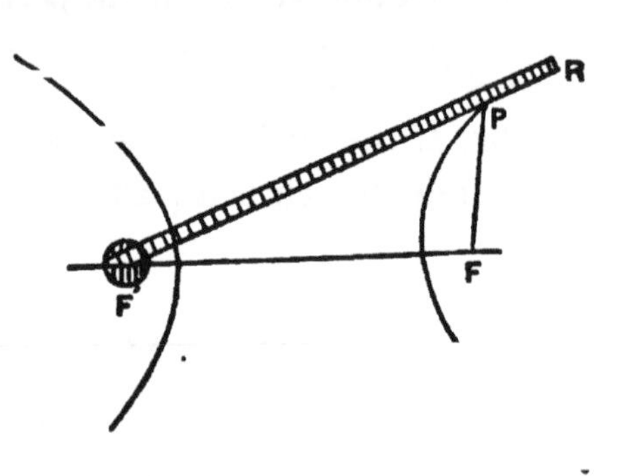

Let F and F′ be the foci. Fasten a ruler, FR, so that it can revolve about F′ as a centre; then take a string, whose length differs from that of the ruler by the given difference of the focal lines, and fasten one extremity at F and the other at R; press the string against the ruler by a pencil, P, and revolve the ruler around F′; the pencil will trace out a portion of the curve, for in every position we shall have F′P—FP equal to the given difference between the focal lines.

If the string is shorter than the ruler the pencil will describe a part of the right-hand branch; if longer, the pencil will describe a part of the left-hand branch.

Cor. It may be shown, as in P. 10, *Cor.* 1, that the transverse axis is bisected at the centre, and that the difference between the focal lines of any point of the curve is equal to the transverse axis.

Scho. Although the conjugate axis, DE, does not cut the curve, it is nevertheless a definite line. It is bisected at the centre, and each half is a mean proportional between the focal lines to either vertex of the tranverse axis.

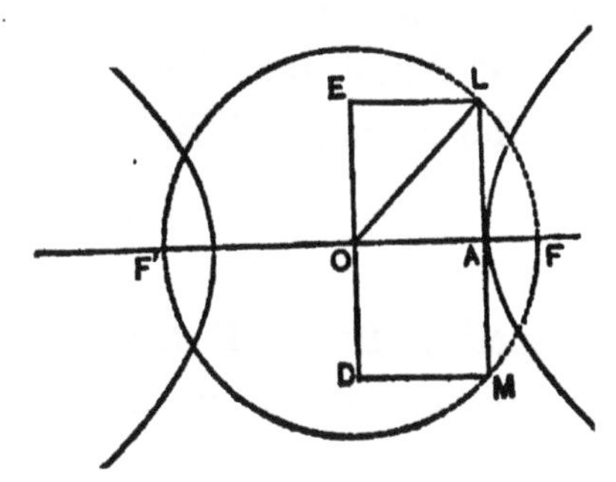

Let a circle be described on FF' as a diameter; then draw ML perpendicular to FF', through A, and it will be equal to the conjugate axis. For,

$$\overline{AL}^2 = \overline{AF} \times \overline{AF'}.$$

PROPOSITION XXIII. THEOREM.

If a point lies without the hyperbola, the difference of its focal distances is less than the transverse axis; if it lies within the curve, the difference of its focal distances is greater than the transverse axis.

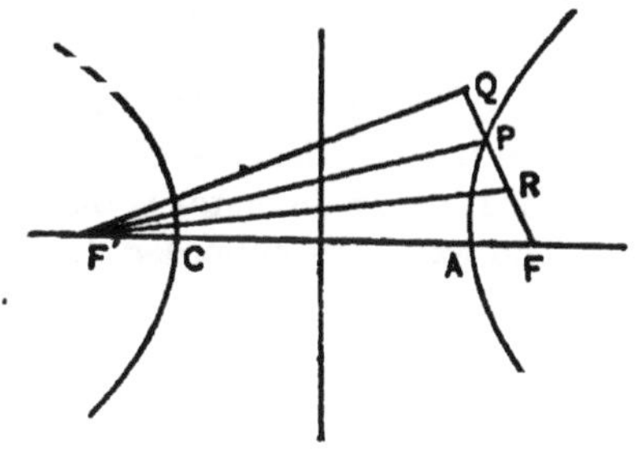

1°. Let Q lie without the hyperbola whose foci are F and F', and whose transverse axis is AC.

Draw F'Q and FQ; from P, the point in which FQ cuts the curve, draw PF'.

In the triangle PF'Q, we have,

$$QF' < QP + PF';$$

subtracting QF from the first member, and its equal QP + PF from the second member, we have,

$$QF' - QF < PF' - PF, \text{ or } AC,$$

which was to be proved.

2°. Let R lie within the curve.

Draw F′R and FR; prolong the latter till it meets the curve at P and draw PF′.

In the triangle PF′R, we have,

$$F'R + PR > PF',$$

subtracting PR+FR from the first member, and its equal PF from the second member, we have,

$$F'R - FR > PF' - PF, \text{ or, } AC,$$

which was to be proved.

Cor. If the difference of the focal lines of a point is less than the transverse axis, the point is without the curve; if this difference is greater than the transverse axis, the point is within the curve; and if this difference is equal to the transverse axis, the point is on the curve.

Scho. It may be shown by a course of reasoning similar to that employed in P. 12 that the curve is symmetrical with respect to both axes, and that every diameter is bisected at the centre.

The axes are supposed to be indefinitely prolonged in both directions.

PROPOSITION XXIV. THEOREM.

The bisectrix of the angle between the focal lines of a point of the hyperbola, is tangent to the curve at that point.

Let P be a point of the curve, F′P and FP, the corresponding focal lines, and PT, the bisectrix of the angle FPF′.

Lay off PQ equal to PF, and draw FQ cutting PT at K: then is F′Q equal to AC, by definition. Take any point, R, of PT, and draw RF, RQ, and RF′.

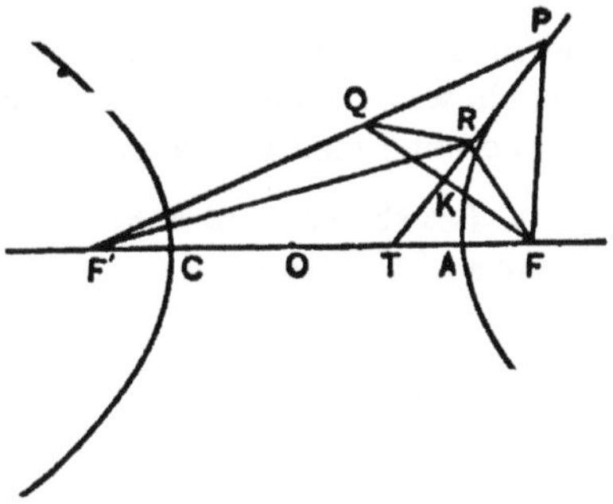

Because FP is equal to QP, and the angle FPK to QPK, PT is perpendicular to FQ, at its middle point K; consequently, RQ is equal to RF.

Now, in the triangle F′RQ, we have,

$$F'R - RQ < F'Q.$$

But, F′R−RQ is equal to F′R−FR, and F′Q is equal to AC; hence,

$$F'R - FR < AC,$$

that is, every point of PT, except P, lies without the curve, (P. 23, *Cor.*); hence, PT is tangent to the curve at P, *which was to be proved.*

Cor. 1. To draw a tangent to the curve at any point: draw the corresponding focal lines and bisect the angle between them; the bisectrix is the required tangent.

Cor. 2. Because PT bisects the angle FPF′, we have,

$$F'T : FT :: F'P : FP;$$

whence, by composition and division,

$$F'T + FT : F'T - FT :: F'P + FP : F'P - FP,$$

or,

$$2OF : 2OT :: F'P + FP : 2OA. \quad . \quad . \quad (1)$$

Scho. 1. It may be shown by a course of reasoning similar to that in P. 15, that the tangents at the extremities of any diameter are parallel.

Scho. 2. It may be proved as in P. 16, that a perpendicular from either focus to any tangent intersects the tangent on the circumference of a circle whose diameter is the transverse axis, and that the product of the perpendiculars is equal to the square of the semi-conjugate axis.

Note. Let the student demonstrate the principles expressed in the scholiums of this and the preceding theorem.

PROPOSITION XXV. THEOREM.

The semi-transverse axis of an hyperbola is a mean proportional between the distance from the centre to the foot of the tangent, and the distance from the centre to the foot of the ordinate of the point of contact.

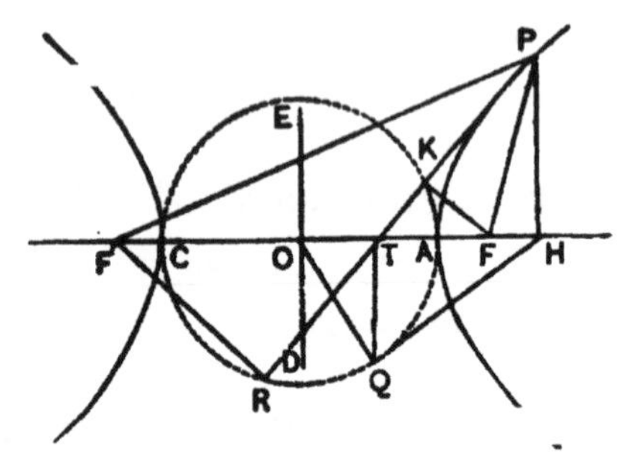

Let PT be tangent to the hyperbola at P; let O, be the centre, HP, the ordinate of P, and F′P and FP, its focal lines.

From the right-angled triangles F′HP and FHP, we have,

$$\overline{F'P}^2 = \overline{F'H}^2 + \overline{HP}^2, \quad . \quad . \quad . \quad . \quad . \quad (1)$$

and,

$$\overline{FP}^2 = \overline{FH}^2 + \overline{HP}^2; \quad . \quad . \quad . \quad . \quad . \quad (2)$$

subtracting (2) from (1), member from member, we have,

$$\overline{F'P}^2 - \overline{FP}^2 = \overline{F'H}^2 - \overline{FH}^2,$$

whence, by factoring, we have,

$$(F'P + FP)\,(F'P - FP) = (F'H + FH)\,(F'H - FH);$$

but, F′P — FP is equal to AC, or to 2OA, F′H + FH is equal to 2OH, and F′H — FH is equal to FF′, or to 2OF; hence, the preceding equation becomes,

$$(F'P + FP) \times 2OA = 2OH \times 2OF,$$

whence,

$$F'P + FP : 2OF :: 2OH : 2OA. \quad . \quad . \quad (3)$$

From P. 24, *Cor.* 2, we have, by alternation and inversion,

$$F'P + FP : 2OF :: 2OA : 2OT, \quad . \quad . \quad . \quad (4)$$

combining (3) and (4), and dividing each term by 2, we have,

$$OH : OA :: OA : OT, \quad . \quad . \quad . \quad . \quad . \quad (5)$$

which was to be proved.

Cor. Through T, draw TQ, perpendicular to AC, meeting the circumference OA, at Q; also draw QO and QH; then, because OQ is equal to OA, we have from (5),

$$OH : OQ :: OQ : OT;$$

but, the triangles OQH and OTQ have a common angle TOQ, and since the sides about that angle are proportional, they are similar, and consequently, QH is perpendicular to OQ, that is, it is tangent to the circumference OA, at Q. Also (P. 18, *Cor.*, and P. 19, B. 3),

$$\overline{TQ}^2 = AT \times TC = TK \times TR.$$

Proposition XXVI. Theorem.

The square of the ordinate of any point of an hyperbola, is equal to the square of the ratio of the semi-axes, multiplied by the rectangle of the distances from the foot of the ordinate to the extremities of the transverse axis.

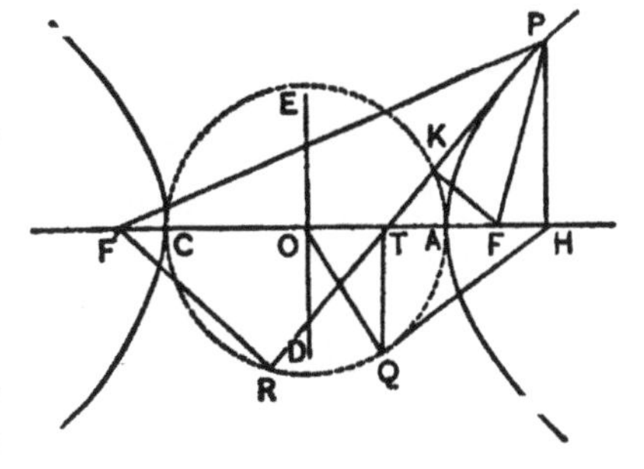

Let P be any point on the hyperbola whose foci are F and F'. Let PT be tangent to the curve at P, and let FK and F'R be perpendicular to PT.

The right-angled triangles THP, TKF, and TRF', have an acute angle in each equal; they are therefore similar; hence,

HP : KF :: TH : TK, . (1)

and,

HP : RF′ :: TH : TR, . (2)

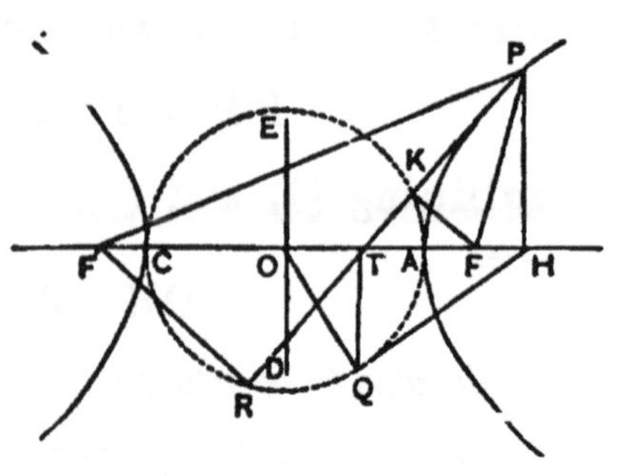

multiplying (1) by (2), term by term, we have,

$$\overline{HP}^2 : KF \times RF' :: \overline{TH}^2 : TK \times TR.$$

But, from P. 24, *Scho.* 2, KF × RF′ is equal to $\overline{OE}^2$; and from P. 25, *Cor.*, TK × TR is equal to $\overline{TQ}^2$; hence,

$$\overline{HP}^2 : \overline{OE}^2 :: \overline{TH}^2 : \overline{TQ}^2. \quad . \quad . \quad . \quad . \quad (3)$$

From the similar triangles HTQ and HQO, we have,

$$TH : TQ :: HQ : OQ, \text{ or } OA;$$

hence,

$$\overline{TH}^2 : \overline{TQ}^2 :: \overline{HQ}^2 : \overline{OA}^2. \quad . \quad . \quad . \quad . \quad (4)$$

But, $\overline{HQ}^2 = HA \times HC$, (P. 20, *Cor.*, B. 3); hence,

$$\overline{TH}^2 : \overline{TQ}^2 :: HA \times HC : \overline{OA}^2; \quad . \quad . \quad . \quad (5)$$

combining (3) and (5), we have,

$$\overline{HP}^2 : \overline{OE}^2 :: HA \times HC : \overline{OA}^2,$$

or,

$$\overline{HP}^2 = \frac{\overline{OE}^2}{\overline{OA}^2}(HA \times HC), \quad . \quad . \quad . \quad . \quad (6)$$

which was to be proved.

Cor. 1. If H coincides with either focus, HP is one-half of the parameter; in this case, the rectangle of the distances from H to the extremities of the transverse axis is equal to $\overline{OE}^2$, (P. 22, *Scho.*); denoting the semi-parameter by p, we have, from (6),

$$p^2 = \frac{\overline{OE}^2}{\overline{OA}^2} \times \overline{OE}^2, \text{ or } 2p = \frac{4\overline{OE}^2}{2OA};$$

hence,

$$2OA : 2OE :: 2OE : 2p,$$

that is, *the parameter is a third proportional to the transverse and the conjugate axes.*

Cor. 2. In passing from point to point of the curve the square of the ratio of the semi-axes does not change; hence, *the squares of the ordinates of any two points of the curve are to each other as the rectangles of the distances from the foot of each, to the extremities of the transverse axis.*

The distances HA and HC may be called *external segments* of CA, the point H being always on the prolongation of CA.

Scho. The preceding principle depends on the definition of the hyperbola; hence, the property in question is characteristic of the hyperbola.

Proposition XXVII. Theorem.

If a conic surface is cut by a plane whose inclination to the plane of the base is greater than that of an element of the surface, the section is an hyperbola.

Let SVR be a meridian plane of the conic surface whose vertex is V, and whose base is RLT, and let MPQ be a section parallel to the base. Let TQAPL be a portion of the section made by a plane perpendicular to SVR and whose inclination AKR, to the plane of the base, is greater than VSR; and let this plane intersect the circle MPQ in QP, and the circle RLT in TL. The plane will intersect the second nappe in the branch XCY, whose vertex is C.

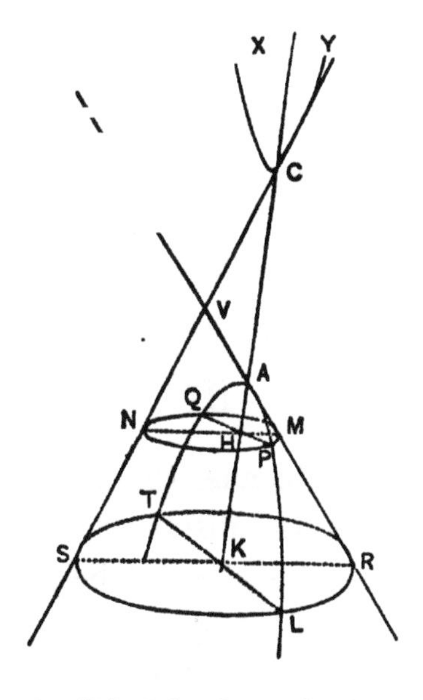

For the reason before given, (P. 21), CAK is the principal axis of the curve, MN is a diameter of the circle MPQ, RS is a diameter of the circle RLT, HP is perpendicular to AK and NM, and TL is perpendicular to AK and RS.

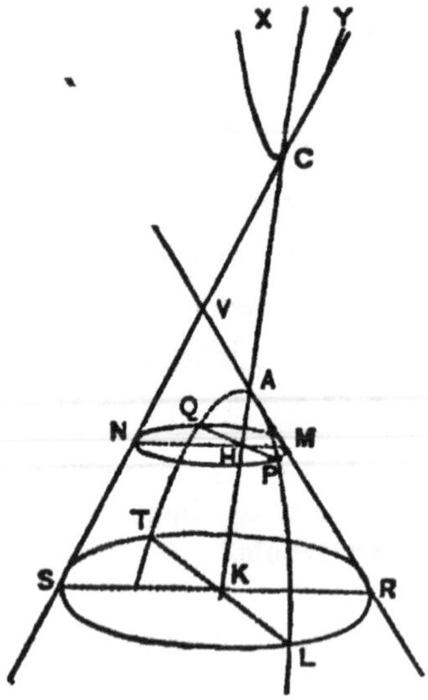

From the circles MPQ and RLT, we have, as in P. 21,

$$\overline{HP}^2 = NH \times HM,$$

and

$$\overline{KL}^2 = SK \times KR,$$

or,

$$\overline{HP}^2 : \overline{KL}^2 :: NH \times HM : SK \times KR. \quad . \quad . \quad (1)$$

From the similar triangles CNH and CSK, we have,

$$NH : SK :: HC : KC; \quad . \quad . \quad . \quad . \quad (2)$$

and from the similar triangles AHM and AKR, we have,

$$HM : KR :: HA : KA. \quad . \quad . \quad . \quad . \quad (3)$$

Multiplying (2) and (3), term by term, we have,

$$NH \times HM : SK \times KR :: HA \times HC : KA \times KC. \quad (4)$$

From (1) and (4) we have,

$$\overline{HP}^2 : \overline{KL}^2 :: HA \times HC : KA \times KC, \quad . \quad . \quad (5)$$

and because the points H and K are on the prolongation of CA, the distances HA, HC, KA and KC are *external segments* of CA. The property expressed by (5) is characteristic of the hyperbola, (P. 26, *Scho.*); hence, the section made by the plane TAL is an hyperbola, ***which was to be proved.***

General Scholium. The parabola, the ellipse, and the hyperbola are conic sections.

TRIGONOMETRY.

I. INTRODUCTION—USE OF LOGARITHMS.

DEFINITIONS AND EXPLANATIONS.

1. The common logarithm of a number is the exponent of the power to which it is necessary to raise 10 to produce the given number; thus, 2 is the logarithm of 100, because $10^2 = 100$.

NOTE.—The principles used in this introduction are fully explained in the Manual of Algebra, chapter XI.

2. A logarithm consists of two parts; an entire part, called the **characteristic**, and a decimal part, called the **mantissa**. Either of these parts may be 0; thus, the logarithm of 6 is 0.778151 and the logarithm of 1000 is 3.000000; in the former the characteristic is 0, and in the latter the mantissa is 0.

The characteristic can always be found by the following

RULE.

If the number has an entire part, the characteristic of its logarithm is 1 less than the number of places of figures in that part; if the number is purely decimal, the characteristic of its logarithm is negative and numerically greater by 1 than the number of ciphers that immediately follow the decimal point.

The characteristic of the logarithm of 425 is 2; that of .0425 is -2; if the characteristic is negative, the minus sign is written above the number; thus, -2 is written $\bar{2}$.

The mantissa, which is always positive, is entirely independent of the place of the decimal point; its value, in any given case, can be found from a table of logarithms, and in finding it, the decimal point may be moved to the right, or to the left, at pleasure.

Table of Logarithms.

3. The table appended to this volume gives the complete logarithms of all numbers from 1 to 100; it also gives the mantissas of the logarithms of all numbers from 100 to 10,000; by means of it, we can find approximate values for the mantissas of the logarithms of all numbers.

Use of the Table.

1°. *To find the logarithm of any number from 1 to 100.* Look for the number on the first page of the table in the column headed N; the number opposite in the column headed log., is the required logarithm; thus, log 16 = 1.204120.

2°. *To find the logarithm of any number.* Find the characteristic by the rule already given. Then, drop the decimal point and find the mantissa as follows:

1st. If the number, after dropping the decimal point, contains four places of figures, let it stand; if it contains less than four places, annex ciphers enough to make four places; then look for the first three figures of the result in the column headed N, and follow the corresponding line across the page to the column headed by the fourth figure; if the number there found contains *six* places of figures it is the required mantissa; if not, the *four* there found are the last four of the mantissa: to find the other two, follow the line back to the column headed 0, dropping to the line below if any *dots* are passed; then, if the number in that column has six places, prefix the first two to the four already found; if not, follow up the column headed 0 till a number is found which does contain six places, and prefix the first two of these. The result, preceded by a decimal point, is the required mantissa. It is to be observed that the *dots* stand for and are to be replaced by, ciphers.

Let the student verify the following equations:

1. $\log\ 2405 = 3.381115$;
2. $\log\ 380.6 = 2.580469$;
3. $\log\ 4.288 = 0.632255$;
4. $\log\ .0375 = \bar{2}.574031$;
5. $\log\ .005762 = \bar{3}.760573$;
6. $\log\ 37.11 = 1.569491$.

2d. If the number, after dropping the decimal point, contains more than four places of figures, place a new decimal point after the fourth figure, and find the mantissa of the entire part of the result, as just explained: then multiply the decimal part of the result by the number standing in the same line in the column D, and add the product to the mantissa already found. Thus, to find the logarithm of 478.632, we first find the characteristic, which is 2; then placing the decimal point after 6, we find the mantissa corresponding to 4786, which is .679973; we then multiply .32 by the tabular difference 91; the product is 29, and this we add to .679973, which gives .680002; hence, log 478.632 = 2.680002.

Let the student verify the following equations:

1. $\log 3723.41 = 3.570941$;
$\log .034621 = \bar{2}.539340$;
$\log 99435 = 4.997539$;
$\log 9.08085 = 0.958127$.

3°. *To find the number corresponding to a given logarithm.* This is done by reversing the preceding rules. The method of proceeding will be illustrated by examples.

Ex. 1. Find the number corresponding to the logarithm 2.314762.

OPERATION.

Given mantissa .314762
Next less in table .314710 . . Cor. number 2064.
D = 210) 52 (.25

∴ Hence, $\log^{-1} 2.314762 = 206.425$, *Ans.*

Explanation.—We find the next less mantissa in the table and also the corresponding number. We then divide the difference of the mantissas by the corresponding number in the table under D; we annex the quotient to the number taken out, dropping the decimal point: this gives 206425. We then point off three places from the left, because the characteristic is 2.

Note.—The symbol $\log^{-1}$ is read, *the number whose logarithm is.*

Ex. 2. Find the number whose logarithm is $\bar{2}.960168$.

OPERATION.

Given mantissa .960168
Next less mantissa .960138 . . number 9123
D . . . 48) 30 (.6

∴ $\log^{-1} \bar{2}.960168 = 0.091236$, *Ans.*

Explanation.—The operation is the same as before, except in finding the mantissa; on account of passing dots, the mantissa is found on two lines, the figures 96 being on one line, and the figures 0138 on the line above.

Let the student verify the following equations:

3. $\log^{-1} \bar{2}.573484 = 0.0374528$;
4. $\log^{-1} 0.342147 = 2.1986$;
5. $\log^{-1} 2.797948 = 627.984$;
6. $\log^{-1} \bar{3}.654210 = 0.00451034$;
7. $\log^{-1} \bar{2}.606032 = 0.040368$.

Multiplication by Logarithms.

4. To multiply two or more numbers together, we have the following

RULE.

Find the algebraic sum of the logarithms of the factors; this will be the logarithm of the product, and the corresponding number will be the product.

Ex. 1. Find the product of 3176 and 24.75.

OPERATION.

$$\begin{aligned} \log 3176 &= 3.501880 \\ \log 24.75 &= \underline{1.393575} \\ \therefore \log^{-1} 4.895455 &= 78605.82, \textit{ Ans.} \end{aligned}$$

Ex. 2. Find the product of 3.586, 2.0146, 0.8372, and 0.0294.

OPERATION.

$$\begin{aligned} \log 3.586 &= 0.554610 \\ \log 2.0146 &= 0.304189 \\ \log 0.8372 &= \bar{1}.922829 \\ \log 0.0294 &= \underline{\bar{2}.468347} \\ \therefore \log^{-1} \bar{1}.249975 &= 0.17782, \textit{ Ans.} \end{aligned}$$

Note. The sum of the mantissas gives 2 to carry, and this added to the sum of the characteristics, which is -3, gives -1, or $\bar{1}$.

Division by Logarithms.

5. To divide one number by another, we have the following

RULE.

Subtract (algebraically) the logarithm of the divisor from that of the dividend; the difference is the logarithm of the quotient, and the corresponding number is the quotient.

Ex. 1. Divide 4.317 by 516.3.

OPERATION.

$$\log 4.317 = 0.635182$$
$$\log 516.3 = 2.712902$$
$$\therefore \log^{-1} \bar{3}.922280 = 0.0083614, \textit{Ans.}$$

Ex. 2. Divide 37.149 by 523.76. *Ans.* 0.0709274.

Use of the Arithmetical Complement.

6. The **arithmetical complement** of a logarithm, denoted by the symbol $(a \cdot c)$, is the result obtained by subtracting that logarithm from 10; thus $(a \cdot c)$ 3.712745 is 6.287255. It may be found by beginning at the left and subtracting each figure from 9, to the last *significant* figure, which must be subtracted from 10.

If we denote the minuend by a, the subtrahend by b, and the remainder by r, we have,

$$r = a - b = a + (10 - b) - 10 = a + (a \cdot c)\, b - 10;$$

hence, we may divide one number by another by the following

RULE.

To the logarithm of the dividend add the arithmetical complement of the logarithm of the divisor and deduct 10; the number corresponding to the resulting logarithm is the required quotient.

Ex. 1. Divide 37.149 by 523.76.

OPERATION.

$$\begin{aligned} \log 37.149 &= 1.569947 \\ (a\cdot c)\log 523.76 &= \underline{7.280867} \\ \therefore\ \log^{-1} \bar{2}.850814 &= 0.0709274,\ \textit{Ans.} \end{aligned}$$

Note. The 10 is subtracted mentally, in connection with the addition.

The preceding rules may be extended to finding the quotient of one continued product, by another. It is to be observed that 10 is to be subtracted every time that the arithmetical complement is used.

Ex. 2. Divide 23.7×34.56, by 415.5×20.26.

OPERATION.

$$\begin{aligned} \log 23.7 &= 1.374748 \\ \log 34.56 &= 1.538574 \\ (a\cdot c)\log 415.5 &= 7.381429 \\ (a\cdot c)\log 20.26 &= \underline{8.693361} \\ \therefore\ \log^{-1} \bar{2}.988112 &= 0.0973,\ \textit{Ans.} \end{aligned}$$

Solution of proportions by Logarithms.

7. A proportion, in which the fourth term is to be found, may be solved by the following

RULE.

To the sum of the logarithms of the second and third terms, add the arithmetical complement of the logarithm of the first term, dropping 10 from the sum; the number corresponding to the resulting logarithm is the required term.

Ex. Solve the proportion $2.714 : 0.1096 :: 43.7 : x$.

OPERATION.

$$\begin{aligned} \log 0.1096 &= \bar{1}.039811 \\ \log 43.7 &= 1.640481 \\ (a\cdot c) \log 2.714 &= \underline{9.566390} \end{aligned}$$

$$\therefore \qquad \log^{-1} 0.246682 = 1.7647, \textit{ Ans.}$$

Other applications.

8. We may raise a number to any power by multiplying the logarithm of the number, by the exponent of the power, and then finding the number that corresponds to the resulting logarithm. We may extract any root of a number by dividing the logarithm of the number, by the index of the root, and then finding the number that corresponds to the resulting logarithm.

Ex. 1. What is the cube of 5.152? *Ans.* 136.75.

Ex. 2. What is the cube root of 5.152? *Ans.* 1.72713.

Note. All the results obtained by means of logarithms are *approximate.*

II. CIRCULAR FUNCTIONS.

General definitions and explanations.

9. If the vertex of an angle is at the centre of a circle, the angle is measured by the arc intercepted by its sides (see *measure of angles*, Art. 58, Geom.). Both the angle and the arc are expressed in *degrees*, *minutes*, and *seconds*, denoted by the symbols °, ′, ″.

10. The **complement of an arc, or angle**, is equal to 90° *minus* the arc, or angle; thus, 43° 15′ is the complement of 46° 45′.

11. The **supplement of an arc, or angle**, is equal to 180° *minus* the arc, or angle; thus, 133° 15′ is the supplement of 46° 45′.

12. Arcs are compared by means of certain lines which depend on the arcs, and vary with them; these lines are called **circular functions.**

13. For the purpose of defining the circular functions, we draw a circle whose radius CA is equal to 1; we then draw its horizontal and vertical diameters, AB and DE. These divide the circumference into four quadrants of which AE is the *first*, EB the *second*, BD the *third*, and DA the *fourth*.

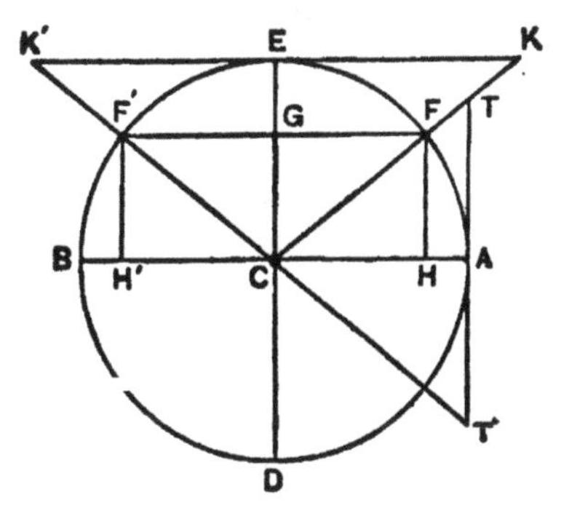

This circle, so divided, is called the **trigonometrical circle**; AB is called the **initial diameter**, and DE the **secondary diameter.**

The point A is called the **origin of arcs**, and all positive arcs are estimated from it around in the direction towards F. The point at which an arc terminates, is called its **extremity**. An arc is said to lie in that quadrant which contains its extremity; thus, the arc AF lies in the *first* quadrant, and AF′ in the *second* quadrant.

In applying the trigonometrical circle to the measure of angles, we suppose the vertex of the angle to be at the centre of the circle, and so placed that its initial side shall coincide with CA.

Definitions of the circular functions.

14. The **sine** of an arc is the distance from the initial diameter to the extremity of the arc; thus, HF is the sine of the arc AF.

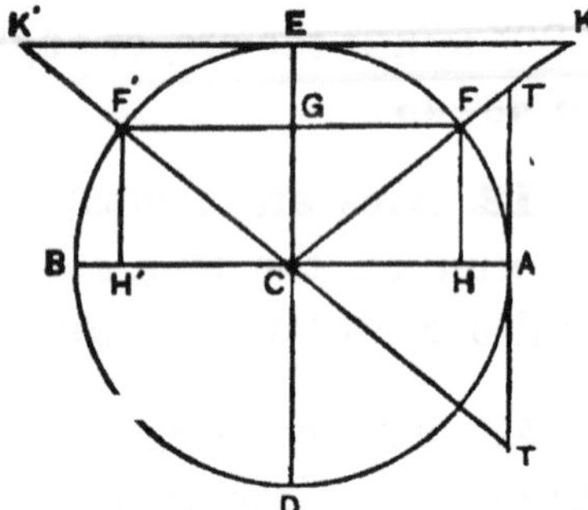

15. The **cosine** of an arc is the distance from the secondary diameter to the extremity of the arc; thus, GF is the cosine of AF.

The cosine is equal to the distance from the centre to the foot of the sine; thus, CH is equal to cos AF.

16. The **tangent** of an arc is a line that touches the arc at its origin, and is limited by the prolongation of the diameter through its extremity; thus, AT is the tangent of the arc AF.

17. The **cotangent** of an arc is a line that touches the circumference at the extremity of the secondary diameter, and is limited by the prolongation of the diameter through the extremity of the arc; thus, EK is the cotangent of the arc AF.

18. The **secant** is the distance from the centre of the arc to the extremity of the tangent; thus, CT is the secant of AF.

19. The **cosecant** is the distance from the centre to the extremity of the cotangent; thus, CK is the cosecant of AF.

20. The **versed sine** is the distance from the foot of the sine to the origin of arcs; thus, HA is the versed sine of AF.

21. The **coversed sine** is the distance from the foot of the cosine to the extremity of the secondary diameter; thus, GE is the coversed sine of AF.

It is obvious from the figure that the cosine, the cotangent, the cosecant, and the coversed sine of an arc, are respectively equal to the sine, the tangent, the secant, and the versed sine of the complement of the arc.

Because the supplement of an arc is as much greater than a quadrant as the arc itself is less, it follows that the sine of the supplement of an arc is equal to the sine of the arc, also that the tangent of the supplement of an arc is *numerically* equal to the tangent of the arc.

The circular functions of an arc, are the corresponding functions of the angle measured by that arc.

Tables of Circular Functions.

22. **A table of natural sines and tangents** is a table containing the values of those functions for all arcs, differing by 1 minute, from 0° to 90°, the radius being equal to 1.

23. **A table of logarithmic sines and tangents** is a table containing the logarithms of the natural sines and tangents, each logarithm being increased by 10.

The object of adding 10 to each logarithm is to avoid negative characteristics: thus, the natural sine of 37° 28′ is 0.6083 and its logarithm is $\bar{1}.784118$; by adding 10, we have the *tabular* logarithm 9.784118. The tens that are added are taken into account in the final result, as will be shown hereafter.

It is to be observed that the tabular logarithms, are the logarithms of the corresponding functions, in a circle whose radius is 10,000,000,000.

Explanation of the Table.

24. In the table appended to this work the logarithmic sines and tangents are given for every minute of the quadrant, and in the column D are the changes which these undergo for 1 second.

Because the cosine and the cotangent of an arc are respectively equal to the sine and the tangent of its complement, a table of sines and tangents may be used for finding the cosines and cotangents. To this end, the table is arranged in double columns, one corresponding to sines or tangents, and the other to cosines or cotangents.

For sines and cosines, the column of differences is placed to the right of each; for tangents and cotangents, it is placed between the two columns, and is common to both.

1°. *To find the logarithm of any function of an arc or angle.* If the arc, or angle, is less than 45°, look for the degrees at the top of the page and for the minutes in the left-hand column; then follow the corresponding line across the page to the column designated at *top* by the name of the function; or, if the arc or angle is greater than 45°, look for the degrees at the bottom of the page, and for the minutes in the right-hand column; then follow the corresponding line across the page to the column designated at *bottom* by the name of the function: take out the logarithm in this place, and set it aside. Also take out the corresponding number in the column D and multiply it by the number of seconds; then *add* the result to the logarithm set aside, for *sines* and *tangents*, and *subtract* it from that logarithm, for *cosines* and *cotangents*.

EXAMPLES.

1. Find the logarithmic sine of 18° 23′ 20″.

OPERATION.

$$\begin{array}{lr} & \log \sin\ 18^\circ\ 23' = 9.498825 \\ D = 6.34 & \\ \underline{\quad 20} & \\ 126.80\ \ .\ .\ .\ .\ .\ \ \textit{add} & \underline{127} \\ & \log \sin\ 18^\circ\ 23'\ 20'' = 9.498952,\ \textit{Ans.} \end{array}$$

2. Find the logarithmic cotangent of 62° 19′ 30″.

OPERATION.

$$\begin{array}{lr} & \log \cot\ 62^\circ\ 19' = 9.719862 \\ D = 5.12 & \\ \underline{\quad 30} & \\ 153.60\ \ .\ .\ .\ \ \textit{subtract} & \underline{154} \\ & \log \cot\ 62^\circ\ 19'\ 30'' = 9.719708,\ \textit{Ans.} \end{array}$$

Let the student verify the following equations:

3. log sin 41° 20′ 15″ = 9.819868;
4. log cos 40° 30′ 10″ = 9.881028;
5. log sin 66° 10′ 50″ = 9.961337;
6. log cos 54° 40′ 30″ = 9.762088;
7. log tan 54° 40′ 30″ = 10.149541;
8. log cot 5° 5′ 5″ = 11.050713;

2°. *To find the arc, or angle, corresponding to a given logarithmic function.* Look in the table for the next less logarithmic function of the same name: take out the corresponding degrees and minutes and set the result aside: also take out the corresponding value of D. Subtract the logarithm found in the table, from the given logarithm, and divide the remainder by the corresponding value of D; the quotient will be *seconds*, which must be *added* to the degrees and minutes set aside, in case of a *sine*, or *tangent*, and *subtracted*, in case of a *cosine*, or *cotangent*.

EXAMPLES.

1. Find the arc whose logarithmic sine is 9.516681.

OPERATION.

Given log sin . . . 9.516681
Next less in table. . 9.516657 . . arc 19° 11′
D . . . 6.05) 24 (4″
$\therefore$ log $\sin^{-1}$ 9.516681 = 19° 11′ 4″, *Ans.*

Note.—The symbols log $\sin^{-1}$, log $\tan^{-1}$, &c., are to be read *the arc whose logarithmic sine is, the arc whose logarithmic tangent is,* &c.

2. Find the arc whose logarithmic tangent is 10.321287.

OPERATION.

Given log tan . . 10.321287
Next less in table . 10.321179 . . arc 64° 29′.
D . . 5.41) 108 (20″
$\therefore$ log $\tan^{-1}$ 10.321287 = 64° 29′ 20″, *Ans.*

Let the student verify the following equations:

3. log $\sin^{-1}$ 9.634333 = 25° 31′ 19″;
4. log $\cos^{-1}$ 9.668135 = 62° 14′ 33″;
5. log $\tan^{-1}$ 10.364873 = 66° 39′ 10″;
6. log $\cot^{-1}$ 10.244220 = 29° 40′ 40″.

III. PLANE TRIGONOMETRY.

Definitions and explanations.

25. In every plane triangle there are six parts, three sides, and three angles. If three of these parts, one of which is a side, are given, or known, the other three may be found by computation.

26. The solution of a triangle is the operation of finding the remaining parts, when a sufficient number of parts are given.

27. Plane Trigonometry is that branch of Trigonometry which treats of the solution of plane triangles.

Method of proceeding.

28. The general method of proceeding is as follows.

We first deduce general equations expressing the relations between the parts of a triangle; we next transform these equations so as to adapt them to logarithmic computation; and then, we apply the resulting formulas to particular cases.

Principles used in solving right-angled triangles.

29. Let the triangle CAD be right-angled at A. Denote its hypothenuse, CD, by h; its base, CA, by b; and its altitude, or perpendicular, AD, by p.

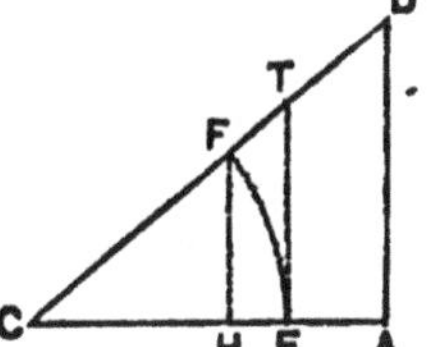

From C as a centre, with a radius CE, equal to 1, draw the arc EF; also, draw ET and FH perpendicular to CA; then is HF the sine of the angle C, CH is its cosine, and ET its tangent.

Because the triangles CHF, CET, and CAD, are similar, we have,

$$\text{CF} : \text{FH} :: \text{CD} : \text{DA}; \quad \text{or,} \quad 1 : \sin \text{C} :: h : p; \quad . \quad (1)$$

$$\text{CF} : \text{CH} :: \text{CD} : \text{CA}; \quad \text{or,} \quad 1 : \cos \text{C} :: h : b; \quad . \quad (2)$$

$$\text{CE} : \text{ET} :: \text{CA} : \text{AD}; \quad \text{or,} \quad 1 : \tan \text{C} :: b : p; \quad . \quad (3)$$

whence,

$$p = h \sin \text{C}; \quad \text{or,} \quad \sin \text{C} = \frac{p}{h}; \quad . \quad . \quad . \quad (4)$$

$$b = h \cos C; \quad \text{or,} \quad \cos C = \frac{b}{h}; \quad . \quad . \quad . \quad (5)$$

$$p = b \tan C; \quad \text{or,} \quad \tan C = \frac{p}{b}; \quad . \quad . \quad . \quad (6)$$

From equations (4), (5) and (6), we have the following general principles:

1°. *The perpendicular, is equal to the hypothenuse, into the sine of the angle at the base; and the sine of the angle at the base, is equal to the perpendicular, divided by the hypothenuse.*

2°. *The base, is equal to the hypothenuse, into the cosine of the angle at the base; and the cosine of the angle at the base, is equal to the base, divided by the hypothenuse.*

3°. *The perpendicular, is equal to the base, into the tangent of the angle at the base; and the tangent of the angle at the base, is equal to the perpendicular, divided by the base.*

In what precedes the *sine, cosine,* and *tangent* of C are *natural* functions. It is to be observed that either one of the acute angles may be taken as the angle at the base, in which case, the side opposite is to be regarded as the perpendicular, and the side adjacent as the base.

Logarithmic Formulas.

30. By the rule for multiplication by logarithms, (*Art.* 4), we have from the first of equations (4),

$$\log p = \log h + \log \text{nat} \sin C;$$

adding 10 to, and subtracting 10 from, the second member, we have,

$$\log p = \log h + (\log \text{nat} \sin C + 10) - 10;$$

but, log nat sin C + 10 is the tabular logarithm of sin C, (*Art.* 23), and is denoted by the symbol log sin; hence,

$$\log p = \log h + \log \sin C - 10. \quad . \quad . \quad . \quad \text{(A)}$$

By the rule for division by logarithms, (*Arts.* 5 and 6), we have from the second of equations (4),

$$\log \text{nat} \sin C = \log p + (a \cdot c) \log h - 10;$$

adding 10 to both members, remembering that the first member increased by 10 is the tabular logarithm of sin C, we have,

$$\log \sin C = \log p + (a \cdot c) \log h; \quad . \quad . \quad . \quad \text{(B)}$$

in like manner, we have, from equations (5) and (6), the following formulas:

$$\log b = \log h + \log \cos C - 10; \quad . \quad . \quad . \quad \text{(C)}$$
$$\log \cos C = \log b + (a \cdot c) \log h; \quad . \quad . \quad . \quad \text{(D)}$$
$$\log p = \log b + \log \tan C - 10; \quad . \quad . \quad \text{(E)}$$
$$\log \tan C = \log p + (a \cdot c) \log b. \quad . \quad . \quad \text{(F)}$$

From (C) we have, by transposition,

$$\log h = \log b + (10 - \log \cos C)$$

or,

$$\log h = \log b + (a \cdot c) \log \cos C. \quad . \quad . \quad . \quad \text{(G)}$$

In applying the preceding formulas, only approximate results are to be expected. In what follows we shall compute angles to within 5″ and sides to within .01, in accordance with the principle laid down in Art. 90, COMPLETE ARITHMETIC. In accordance with this principle the seconds in any result will be some multiple of 10″.

Applications.

31. In the solution of right-angled triangles there are two cases: in the *first*, we have given one acute angle, and one side; in the *second*, we have given any two sides.

CASE 1°. *Given, an acute angle, and any side, to find the remaining parts.*

The other acute angle is found by subtracting the given one from 90°. The remaining sides may be found from formulas (A), (C), (E), and (G).

EXAMPLES.

1. Given, the angle at the base 40° 20′, and the hypothenuse 450 *yds.*, to find the other parts.

OPERATION.

1°. Angle at the vertex $= 90° - 40°\ 20' = 49°\ 40'$.

2°. From formula (A),

$\log h = \log 450$ 2.653213

$\log \sin C = \log \sin 40°\ 20'$. 9.811061

$\therefore$ *perpendicular* $= \log^{-1} 2.464274 = 291.26$ *yds.*

3°. From formula (C),

$\log h = \log 450$ 2.653213

$\log \cos C = \log \cos 40°\ 20'$. 9.882121

$\therefore$ *base* $= \log^{-1} 2.535334 = 343.03$ *yds.*

2. Given, the angle at the vertex 39° 50′, and the hypothenuse 497 *ft.*, to find the other parts.

OPERATION.

1°. Angle at base $= 90° - 39°\ 50' = 50°\ 10'$.

2°. From formula (A),

$\log h = \log 497$ 2.696356

$\log \sin C = \log \sin 50°\ 10'$. 9.885311

$\therefore$ *perpendicular* $= \log^{-1} 2.581667 = 381.65$ *ft.*

4°. From formula (C),

$\log h = \log 497$ 2.696356

$\log \cos C = \log \cos 50°\ 10'$. 9.806557

$\therefore$ *base* $= \log^{-1} 2.502913 = 318.36$ *ft.*

3. Given, the angle at the base 50° 23′, and the base 290 *yds.*, to find the other parts.

OPERATION.

1°. Angle at the vertex $= 90° - 50°\ 23' = 39°\ 37'$.

2°. From formula (E),

$\log b = \log 290$	. . . .	2.462398
$\log \tan C = \log \tan 50°\ 23'$	.	10.082095

$\therefore$ *perpendicular* $= \log^{-1} 2.544493 = 350.34$ *yds.*

3°. From formula (G),

log 290	. . .	2.462398
$(a \cdot c) \log 50°\ 23'$	. .	0.195419

$\therefore$ *hypothenuse* $= \log^{-1} 2.657817 = 369.58$ *yds.*

4. Given, the angle at the vertex, 44° 2′, and the perpendicular, 84 *yds.*, to find the other parts.

Ans. $C = 45°\ 58'$, $b = 81.21$ *yds.*, $h = 116.84$ *yds.*

5. Given, the angle at the vertex, 36° 15′ 30″, and the hypothenuse, 112 *ft.*, to find the other parts.

Ans. $C = 53°\ 44'\ 30''$, $b = 66.24$ *ft.*, $p = 90.31$.

6. Given, the angle at the base, 41° 24′ 30″, and the hypothenuse, 397 *ft.*, to find the other parts.

Ans. $D = 48°\ 35'\ 30''$ $b = 297.76$ *ft.*, $p = 262.58$ *ft.*

7. Given, $C = 14°\ 37'\ 40''$, and $b = 111$ *yds.*, to find the other parts.

Ans. $D = 75°\ 22'\ 20''$, $p = 28.97$ *yds.*, $h = 114.72$ *yds.*

8. Given, $D = 27°\ 0'\ 25''$, and $p = 114$ *in.*, to find the other parts.

Ans. $C = 62°\ 59'\ 35''$, $b = 58.1$ *in.*, $h = 127.95$ *in.*

CASE 2°. *Given any two sides, to find the remaining parts.*

The angle at the base may be found by one of the formulas (B), (D), or (F); the solution is then completed as in Case 1°.

EXAMPLES.

1. Given, the hypothenuse, 640 *yds.*, and the perpendicular, 300 *yds.*, to find the other parts.

OPERATION.

1°. From formula (B),

$$\log p = \log 300 \quad . \quad . \quad . \quad 2.477121$$
$$(a \cdot c) \log h = (a \cdot c) \log 640 \quad . \quad \underline{7.193820}$$
$$\therefore \quad C = \log \sin^{-1} 9.670941 = 27° \ 57' \ 10''.$$

2°. $D = 90° - 27° \ 57' \ 10'' = 62° \ 2' \ 50''$.

3°. From formula (C), we find $b = 565.33$ *yds.*

2. Given, the hypothenuse, 392 *ft.*, and the base, 256 *ft.*, to find C, D, and p.

OPERATION.

1°. From formula (D),

$$\log 256 \quad . \quad . \quad . \quad . \quad . \quad 2.408240$$
$$(a \cdot c) \log 392 \quad . \quad . \quad . \quad . \quad . \quad \underline{7.406714}$$
$$\therefore \quad C = \log \cos^{-1} \ 9.814954 = 49° \ 13' \ 40''$$

2°. $D = 90° - 49° \ 13' \ 40'' = 40° \ 46' \ 20''$.

3°. From formula (B), we find $p = 296.87$.

3. Given, the perpendicular, 360 *yds.*, and the base, 270 *yds.*, to find C, D, and h.

OPERATION.

1°. From formula (F),

$$\log 360 \quad . \quad . \quad . \quad . \quad . \quad 2.556303$$
$$(a \cdot c) \log 270 \quad . \quad . \quad . \quad . \quad . \quad \underline{7.568636}$$
$$\therefore \quad C = \log \tan^{-1} 10.124939 = 53° \ 7' \ 50''.$$

2°. $D = 90° - 53° \ 7' \ 50'' = 36° \ 52' \ 10''$.

3°. From formula (G), we find $h = 450$ *yds.*

Note.—The hypothenuse may be found without the use of logarithms from the formula, $h^2 = p^2 + b^2$.

4. Given, $h = 718$ *yds.*, and $b = 425$ *yds.*, to find the other parts.

Ans. $C = 53° 42' 20''$, $D = 36° 17' 40''$, $p = 578.7$ *yds.*

5. Given, $h = 420$ *ft.*, and $p = 320$ *ft.*, to find the other parts. *Ans.* $C = 49° 38'$, $D = 40° 22'$, $b = 272.02$ *ft.*

6. Given, $p = 99$ *yds.*, and $b = 132$ *yds.*, to find the other parts.

Ans. $C = 36° 52' 10''$, $D = 53° 7' 50''$, $h = 164.99$ *yds.*

7. Given, $p = 446$ *ft.*, and $b = 810$ *ft.*, to find the other parts.

Ans. $C = 28° 50' 20''$, $D = 61° 9' 40''$, $h = 924.68$ *ft.*

Principles used in solving oblique triangles.

32. In what follows the angles of a triangle will be denoted by A, C, and D; and their opposite sides by a, c, and d.

First. From the vertex D, of the triangle ACD, draw DH perpendicular to the base, AC; H may lie on CA, or it may lie on the prolongation of CA; in the latter case, the angle DAH is the supplement of A, and consequently, sin DAH is equal to sin A; in both cases, we have, (Pr. 1°, *Art.* 29),

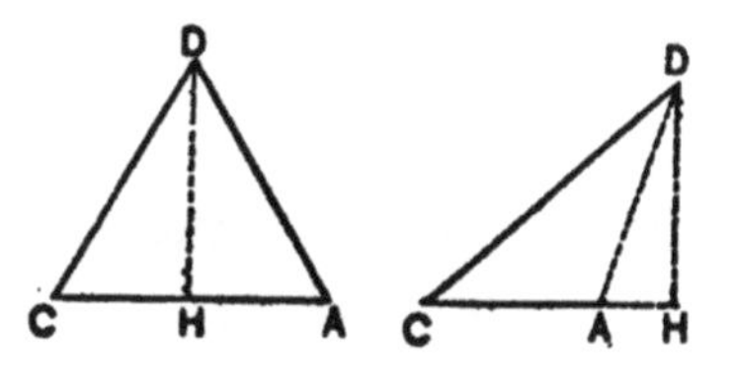

$$DH = a \sin C, \quad \text{and} \quad DH = c \sin A,$$

whence,

$$a \sin C = c \sin A,$$

or,

$$a : c :: \sin A : \sin C; \quad . \quad . \quad . \quad . \quad . \quad (1)$$

hence, the following principle:

1°. *Any two sides of a plane triangle are to each other as the sines of their opposite angles.*

Secondly. Let ACD be a triangle, in which CD is less than DA.

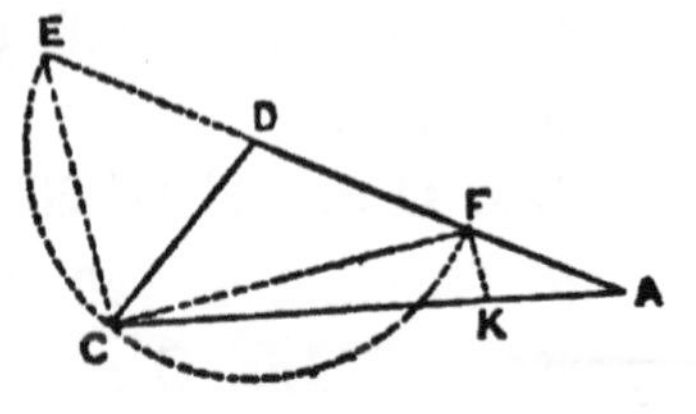

From D as a centre, with the radius DC, describe the arc FCE, cutting AD at F, and the prolongation of AD at E; draw EC and CF; also, through F, draw FK parallel to EC; then will ECF and CFK be right angles.

Because CDE is an exterior angle of the triangle, it is equal to C + A; but, CDE is an angle at the centre, and CFE is an inscribed angle, subtended by the same arc; hence,

$$\text{CFE} = \tfrac{1}{2}(\text{C} + \text{A}). \quad . \quad . \quad . \quad . \quad . \quad (2)$$

Because CFE is an exterior angle of the triangle ACF, we have,

$$\tfrac{1}{2}(\text{C} + \text{A}) = \text{ACF} + \text{A},$$

or,

$$\text{ACF} = \tfrac{1}{2}(\text{C} - \text{A}). \quad . \quad . \quad . \quad . \quad . \quad (3)$$

From the right-angled triangles FCE and CFK, we have, (*Prin.* 3°, *Art.* 29),

$$\text{EC} = \text{CF} \tan \tfrac{1}{2}(\text{C} + \text{A}),$$

and,

$$\text{FK} = \text{CF} \tan \tfrac{1}{2}(\text{C} - \text{A});$$

whence,

$$\text{EC} : \text{FK} :: \tan \tfrac{1}{2}(\text{C} + \text{A}) : \tan \tfrac{1}{2}(\text{C} - \text{A}). \quad . \quad . \quad (4)$$

Because DF and DE are each equal to DC, AE is equal to $c + a$, and AF is equal to $c - a$. From the similar triangles AEC and AFK, we have,

$$\text{AE} : \text{AF} :: \text{EC} : \text{FK},$$

or,

$$c + a : c - a :: \text{EC} : \text{FK}; \quad . \quad . \quad . \quad . \quad . \quad (5)$$

from (4) and (5), we have,

$$c + a : c - a :: \tan \tfrac{1}{2}(C + A) : \tan \tfrac{1}{2}(C - A); \quad (6)$$

hence, the following principle:

2°. *The sum of two sides of a plane triangle is to the difference of those sides, as the tangent of half the sum of their opposite angles, is to the tangent of half the difference of those angles.*

Thirdly. From the vertex D, of the triangle ACD, draw DH perpendicular to the base AC, dividing it into the segments CH and HA; the vertex D can always be chosen so that H shall fall between A and C. Denote CH by S, and HA by S′.

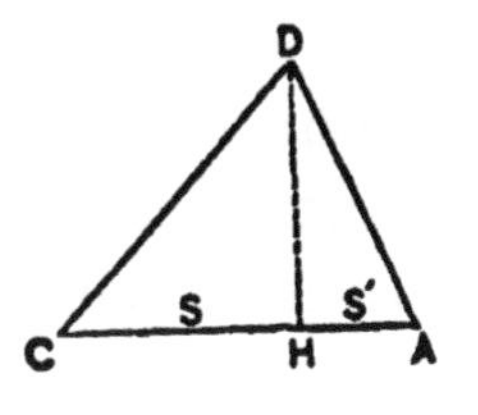

From the right-angled triangles CHD and AHD, we have,

$$a^2 = S^2 + \overline{DH}^2, \quad \text{and} \quad c^2 = S'^2 + \overline{DH}^2;$$

whence, by subtraction,

$$a^2 - c^2 = S^2 - S'^2, \text{ or, } (a+c)(a-c) = (S+S')(S-S'),$$

or,

$$a + c : S + S' :: S - S' : a - c; \quad . \quad . \quad (7)$$

hence, we have the following principle:

3°. *The sum of two sides, is to the sum of the corresponding segments of the third side, as the difference of these segments, is to the difference of the two sides.*

Logarithmic Formulas.

33. From proportion (1), *Art.* 32, we have,

$$a = \frac{c \sin A}{\sin C}, \quad \text{and} \quad \sin A = \frac{a \sin C}{c};$$

whence, from the rules for multiplication and division by logarithms, we have,

$$\log a = \log c + \log \text{ nat } \sin A + (a \cdot c) \log \text{ nat } \sin C - 10, . \quad (1)$$

and,

$$\log \text{ nat } \sin A = \log a + \log \text{ nat } \sin C + (a \cdot c) \log c - 10. \quad (2)$$

Because the logarithm of a natural function is less by 10 than the corresponding tabular logarithm, the arithmetical complement of that logarithm is greater by 10 than the arithmetical complement of the corresponding tabular logarithm; hence, the form of equation (1) will not change in passing to tabular logarithms, and we shall have,

$$\log a = \log c + \log \sin A + (a \cdot c) \log \sin C - 10. \quad (H)$$

If we pass to tabular logarithms, by adding 10 to both members of (2), we have,

$$\log \sin A = \log a + \log \sin C + (a \cdot c) \log c - 10. . \quad (I)$$

From proportion (6), *Art.* 32, and by the rule for solving proportions by logarithms, we have,

$$\log \text{ nat } \tan \tfrac{1}{2}(C - A) = \log (c - a) + \log \text{ nat } \tan \tfrac{1}{2}(C + A) + (a \cdot c) \log (c + a) - 10;$$

and passing to tabular logarithms, we have,

$$\log \tan \tfrac{1}{2}(C - A) = \log (c - a) + \log \tan \tfrac{1}{2}(C + A) + (a \cdot c) \log (c + a) - 10. . . . . . \quad (K)$$

From proportion (7), *Art.* 32, we have,

$$\log (S - S') = \log (a + c) + \log (a - c) + (a \cdot c) \log (S + S') - 10. . . . . \quad (L)$$

34. In the solution of oblique triangles there are *four* cases.

CASE 1°. *Given, one side and two angles, to find the other parts.*

The third angle may be found by subtracting the sum of the given angles from 180°; after which, the required sides may be found by formula (H).

EXAMPLES.

1. Given, $d = 350$ *yds.*, $A = 39° 40'$, and $C = 47° 20'$, to find the other parts.

OPERATION.

1°. $D = 180° - (39° 40' + 47° 20') = 93°$.

2°. From formula (H), by changing c and C, into d and D, we have,

$$\log a = \log d + \log \sin A + (a \cdot c) \log \sin D - 10.$$

$\log d = \log 350$	2.544068
$\log \sin A = \log \sin 39° 40'$ . . .	9.805039
$(a \cdot c) \log \sin D = (a \cdot c) \log \sin 93°$.	0.000596

$$\therefore\ a = \log^{-1} 2.349703 = 223.72 \textit{ yds.}$$

Note—Because the sine of an angle is equal to the sine of its supplement, log sin 93° is equal to log sin 87°.

3°. From formula (H), by changing a and A, into c and C, also c and C, into d and D, we have,

$$\log c = \log d + \log \sin C + (a \cdot c) \log \sin D - 10.$$

$\log d = \log 350$ 2.544068
$\log \sin C = \log \sin 47° 20'$ 9.866470
$(a \cdot c) \log \sin D = (a \cdot c) \log \sin 93°$ . . 0.000596

$$\therefore c = \log^{-1} 2.411134 = 257.71.$$

2. Given, $d = 40.8$ *yds.*, $A = 58° 07'$, and $C = 22° 37'$, to find the other parts.

Ans. $D = 99° 16'$, $a = 35.1$ *yds.*, $c = 15.9$ *yds.*

3. Given, $c = 160$, $C = 83° 53'$, and $A = 38° 25'$, to find the other parts.

Ans. $D = 57° 42'$, $a = 99.99$, $d = 136$.

CASE 2°. *Given, two sides, and the angle opposite one of them, to find the other parts.*

The angle opposite the second given side can be found by formula (I), after which, the solution is completed as in Case 1°.

EXAMPLES.

1. Given, $a = 137$, $c = 84$, and $C = 30° 40$, to find the other parts.

OPERATION.

1°. From formula (I), we have,

$\log a = \log 137$ 2.136721
$\log \sin C = \log \sin 30° 40'$ 9.707606
$(a \cdot c) \log c = (a \cdot c) \log 84$ 8.075721

$$\therefore A = \log \sin^{-1} 9.920048 = 56° 17' 20'',$$

or $123° 42' 40''$.

Note.—This case corresponds to Prob. 11, B. 2. There may be *two* solutions, *one* solution, or *no* solution. The perpendicular from the vertex D, to the opposite side CA, (see figure Art. 32), is equal to $a \sin C$, and consequently, its value may be computed; let it be denoted by p. We shall then have the following results:

First. If C is acute, and if c is less than a, but greater than p, there will be *two* solutions.

Secondly. If C is acute, and if c is equal to p; or, if C is either acute or obtuse, and if c is greater than a, there will be but *one* solution.

Thirdly. If c is less than p, there is *no* solution. In the present example p is equal to 69.88, and consequently, there are two solutions.

2°. $D = 180° - (30° 40' + 56° 17' 20'') = 93° 2' 40''$,
and, $D' = 180° - (30° 40' + 123° 42' 40'') = 25° 37' 20''$.

3°. From formula (H), we find,

$$d = 164.46, \text{ and } d' = 71.22.$$

2. Given, $A = 32°$, $a = 200$, and $c = 250$, to find the other parts.

Ans. $\begin{cases} C = 41° 29', & D = 106° 31', d = 361.84 \\ C' = 138° 31', & D' = 9° 29', \quad d' = 62.18. \end{cases}$

3. Given, $C = 36° 30'$, $a = 64$, and $c = 96$, to find the remaining parts.

Ans. $A = 23° 21' 50''$, $D = 120° 8' 10''$, $d = 139.58$.

CASE 3°. *Given two sides and their included angle, to find the other parts.*

The half sum of the required angles is found by subtracting the given angle from 180°, and dividing the result by 2; the half difference of these angles is found by means of formula (K); and the remaining side is then found by formula (H).

EXAMPLES.

1. Given, $a = 90$, $c = 108$, and $D = 80°$, to find the other parts.

1°. $\frac{1}{2}(C + A) = \frac{1}{2}(180° - 80°) = 50°$;

$(c + a) = 108 + 90 = 198$; and $(c - a) = 108 - 90 = 18$.

From formula (K),

log 18	1.255273
log tan 50°	10.076187
$(a \cdot c)$ log 198	7.703335

$\frac{1}{2}(C - A) = \tan^{-1}\ 9.034795 = 6°\ 11'$.

Adding the half difference to the half sum, we find C, and subtracting the half difference from the half sum, we have A; hence,

$C = 50° + 6°\ 11' = 56°\ 11'$, and $A = 50° - 6°\ 11' = 43°\ 49'$;

2°. From formula (H), we find $d = 128$.

2. Given, $c = 843$, $a = 480$, and $D = 128°\ 4'$, to find the other parts.

Ans. $A = 18°\ 21'\ 20''$, $C = 33°\ 34'\ 40''$, $d = 1200$.

3. Given, $c = 16.96$, $a = 11.96$, and $D = 60°\ 43'\ 40''$, to find the other parts.

Ans. $C = 76°\ 4'\ 40''$, $A = 43°\ 11'\ 40''$, $d = 15.24$.

CASE 4°. *Given the three sides, to find the other parts.*

In this case we take the longest side as a base, and from the opposite vertex, we draw a perpendicular to this base, dividing it into two segments; the sum of these segments is equal to the base, and their difference is found by formula L; then, the greater segment is equal to half their sum, *plus* half their difference, and the less segment is equal to half their sum, *minus* half their difference. It is to be remembered that the greater segment lies next the greater side, and the less segment next the less side. After the segments are found, the angles at the base may be determined by formula (D), and the angle at the vertex may be found as before.

EXAMPLES.

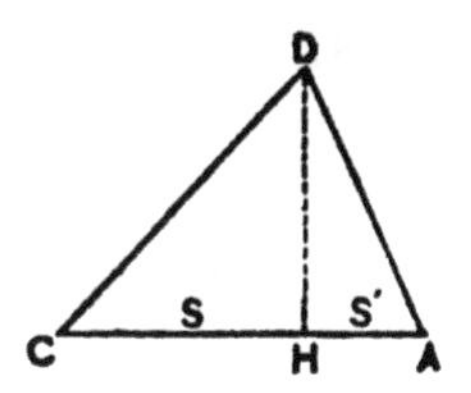

1. Given $a = 144$, $c = 108$, and $d = 200$, to find the other parts.

1°. $a + c = 252$, $a - c = 36$, $S + S' = d = 200$.

$\therefore$ from formula (L), we have,

$$\begin{array}{ll} \log 252 \ldots\ldots & 2.401401 \\ \log 36 \ldots\ldots & 1.556303 \\ (a \cdot c) \log 200 \ldots\ldots & \underline{7.698970} \end{array}$$

$$\therefore \quad S - S' = \log^{-1} 1.656674 = 45.36.$$

$$\therefore \quad S = \tfrac{1}{2}(200 + 45.36) = 122.68,$$

and,

$$S' = \tfrac{1}{2}(200 - 45.36) = 77.32.$$

2°. From formula (D), we have,

$$\begin{array}{ll} \log S = \log 122.68 \;\;.\;. & 2.088774 \\ (a \cdot c) \log a = (a \cdot c) \log 144 \;\;.\;. & \underline{7.841638} \end{array}$$

$$C = \log \cos^{-1} 9.930412 = 31^\circ\, 34'\, 30''.$$

and,

$$\begin{array}{ll} \log S' = \log 77.32 \;\;.\;. & 1.888292 \\ (a \cdot c) \log c = (a \cdot c) \log 108 \;\;.\;. & \underline{7.966576} \end{array}$$

$$\therefore \quad A = \log \cos^{-1} 9.854868 = 44^\circ\, 16'\, 50''.$$

3°. $D = 180^\circ - (31^\circ\, 34'\, 30'' + 44^\circ\, 16'\, 50'') = 104^\circ\, 8'\, 40''$.

2. Given, $a = 55$, $c = 44$, and $d = 66$, to find the other parts.

Ans. $A = 55^\circ\, 46'\, 20''$, $C = 41^\circ\, 24'\, 30''$, $D = 82^\circ\, 49'\, 10''$.

3. Given, $a = 136$, $c = 100$, and $d = 160$, to find the other parts.

Ans. A = 57° 41′ 20″, C = 38° 25′ 20″, D = 83° 53′, 20″.

4. Given, $a = 648$, $c = 374$, and $d = 712$, to find the other parts.

Ans. A = 64° 46′ 50″, C = 31° 28′ 30″, D = 83° 44′ 40″.

IV. ANALYTICAL TRIGONOMETRY.

Definitions and explanations.

35. Analytical trigonometry treats of the general relations of *circular functions.*

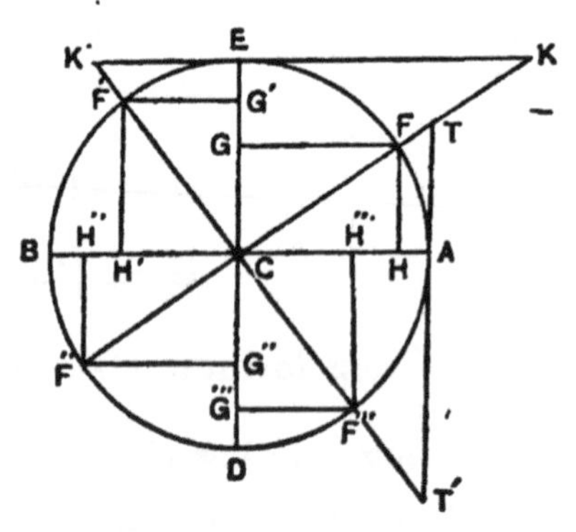

In considering the general relations of circular functions, the arc is supposed to have any value from 0° to 360°; the figure shows these functions for arcs, terminating in each of the four quadrants. Sines and tangents are estimated from the initial diameter AB, either *upward,* or *downward;* cosines and cotangents are estimated from the secondary diameter ED, either *to the right,* or *to the left.* The algebraic signs of these functions are determined in accordance with the following

RULE.

Distances estimated upward are plus, and those estimated downward are minus; distances estimated toward the right are plus, and those estimated toward the left are minus.

General relations.

36. Assume the figure and notation of *Art.* 13, and denote the arc AF by x.

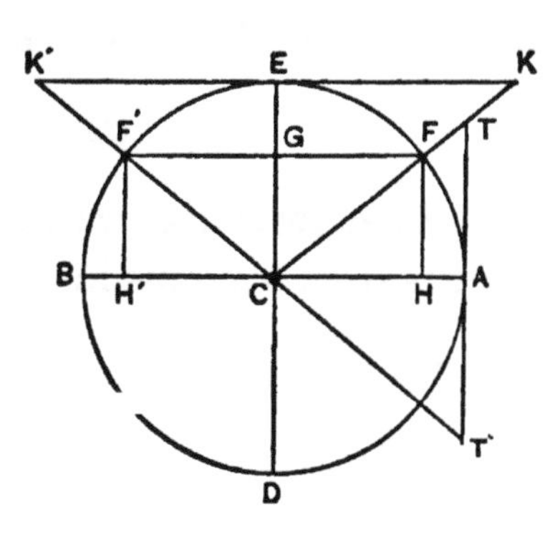

From the right-angled triangles CHF and CAT, we have the following relations:

$$\overline{CH}^2 + \overline{HF}^2 = \overline{CF}^2,$$

or,

$$\sin^2 x + \cos^2 x = 1;$$

whence,

$$\sin^2 x = 1 - \cos^2 x, \quad \text{and} \quad \cos^2 x = 1 - \sin^2 x; \quad . \ (1)$$

Note.—The symbols $\sin^2 x$, $\cos^2 x$, &c., denote the square of the sine of x, the square of the cosine of x, &c.; they are read, *sine square of* x, *cosine square of* x, &c.

$$\text{CH} : \text{HF} :: \text{CA} : \text{AT}, \quad \text{or,} \quad \cos x : \sin x :: 1 : \tan x;$$

whence,

$$\tan x = \frac{\sin x}{\cos x}; \quad . \ . \ . \ . \ . \ . \ . \ (2)$$

$$\text{CH} : \text{CA} :: \text{CF} : \text{CT}, \quad \text{or,} \quad \cos x : 1 :: 1 : \sec x;$$

whence,

$$\sec x = \frac{1}{\cos x}; \quad . \ . \ . \ . \ . \ . \ . \ (3)$$

From the triangles CGF and CEK, we have the following relations:

$$\text{CG} : \text{GF} :: \text{CE} : \text{EK}, \quad \text{or,} \quad \sin x : \cos x :: 1 : \cot x;$$

whence,

$$\cot x = \frac{\cos x}{\sin x}; \quad . \ . \ . \ . \ . \ . \ . \ (4)$$

$$\text{CG} : \text{CF} :: \text{CE} : \text{CK}, \quad \text{or,} \quad \sin x : 1 :: 1 : \text{cosec}\, x;$$

whence,
$$\operatorname{cosec} x = \frac{1}{\sin x}. \quad . \quad . \quad . \quad . \quad . \quad . \quad (5)$$

If we multiply (2) by (4), member by member, we have,

$$\tan x \cot x = 1. \quad . \quad . \quad . \quad . \quad . \quad . \quad . \quad (6)$$

To these formulas, we may add the following, which are obvious from a simple inspection of the figure:

$$\text{ver sin } x = 1 - \cos x, \text{ and coversin } x = 1 - \sin x; \quad . \quad (7)$$

$$\sec^2 x = 1 + \tan^2 x, \text{ and } \operatorname{cosec}^2 x = 1 + \cot^2 x. \quad . \quad . \quad (8)$$

The following method of expressing the relations between the functions of any angle, and also the relations between the parts of a right-angled triangle, is due to Prof. J. W. Nicholson, of Louisiana.

Let ACD be a triangle, right-angled at A. Denote the angle at the base by C, the hypothenuse by h, the perpendicular by p, and the base by b. Then, let the sides of the triangle, and the several functions of C, be arranged as shown in the diagram. The nine elements, thus expressed, are called *parts*; the two parts that lie next to a given part, whether counted along a chord, or around the circumference, are said to be adjacent to it; and any two parts that are separated by two intervening parts are said to be opposite to each other. In all cases we have the following relations:

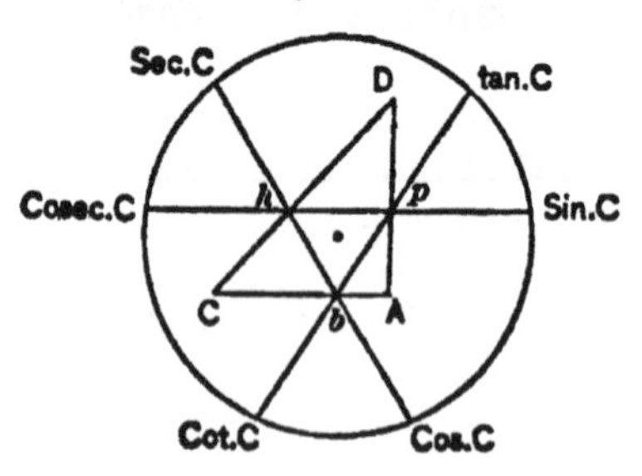

1°. *Any part is equal to the product of its adjacent parts;*

2°. *Any part is equal to the reciprocal of its opposite part.*

Signs of circular functions in different quadrants.

37. In accordance with the rule for signs, in *Art.* 35, it is plain that the sine of an arc in the first, or in the second quadrant, is *plus;* and that the sine of an arc in the third, or in the fourth quadrant, is *minus;* also, the cosine of an arc in the first, or in the fourth quadrant, is *plus;* and the cosine of an arc in the second, and in the third quadrant, is *minus*. If due regard is paid to the rule for signs, the formulas of *Art.* 36 will hold true for all values of the arc from O to 360°. Applying what precedes to the formulas of *Art.* 36, we may form the following

TABLE.

ARC.	SIN.	COS.	TAN.	COT.	SEC.	COSEC.
1st quad.	+	+	+	+	+	+
2d quad.	+	−	−	−	−	+
3d quad.	−	−	+	+	−	−
4th quad.	−	+	−	−	+	−

Functions of negative arcs.

38. If an arc is estimated from the origin A, around toward F‴, it is said to be *negative*. It is obvious that the sine of a negative arc is equal to that of an equal positive arc, with its algebraic sign changed, and that the cosine of a negative arc is equal to that of an equal positive arc; hence, from the formulas of *Art.* 36, we have,

$$\sin(-x) = -\sin x,\ \cos(-x) = \cos x,\ \tan(-x) = -\tan x,$$
$$\text{and } \cot(-x) = -\cot x.$$

Limiting Values.

39. The **limiting values** of a circular function are its values when the arc is some multiple of a quadrant. In determining the sign of a limiting value, it is to be remembered that the sign of a varying magnitude up to the limit, is its sign at the limit. If we now suppose the arc to commence at 0°, and to increase by insensible increments up to 360°, we shall have the following results:

Sin $0° = +0$, sin $90° = +1$, sin $180° = +0$, sin $270° = -1$, sin $360° = -0$; cos $0° = +1$, cos $90° = +0$; cos $180° = -1$, cos $270° = -0$, cos $360° = +1$. Substituting these results in the formulas of *Art.* 36, we have the following

TABLE.

ARC.	SIN.	COS.	TAN.	COT.	SEC.	COSEC.
0°	+0	+1	+0	+∞	+1	+∞
90°	+1	+0	+∞	+0	+∞	+1
180°	+0	−1	−0	−∞	−1	+∞
270°	−1	−0	+∞	+0	−∞	−1
360°	−0	+1	−0	−∞	+1	−∞

Functions of Certain Angles.

40. 1°. Let F′AF be an arc of 60°, and let CA be drawn perpendicular to the chord F′F; then AF will be an arc of 30°, HF will be the sine of 30°, and CH will be the cosine of 30°.

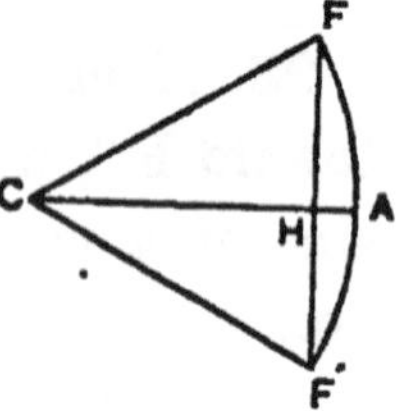

The chord FF′ is equal to 1; consequently, HF is equal to $\frac{1}{2}$; that is, sin $30° = \frac{1}{2}$. From formula (1), *Art.* 36, we have,

$$\cos 30° = \sqrt{1 - \tfrac{1}{4}} = \tfrac{1}{2}\sqrt{3}.$$

2°. Let AF be an arc of 45°; HF is its sine, and CH its cosine.

Then, because ACF is equal to 45°, HFC is also equal to 45°, and consequently, CH is equal to HF. But, $\overline{CH}^2 + \overline{HF}^2 = \overline{CF}^2$, or $2\overline{CH}^2 = \overline{CF}^2$, or, $2 \sin^2 45° = 1$; hence,

$$\sin 45° = \sqrt{\tfrac{1}{2}} = \tfrac{1}{2}\sqrt{2}, \text{ and } \cos 45° = \sqrt{\tfrac{1}{2}} = \tfrac{1}{2}\sqrt{2}.$$

From formulas (2) and (4), *Art.* 36, we have,

$$\tan 45° = 1, \text{ and } \cot 45° = 1.$$

3°. Because the sine of an arc is equal to the cosine of its complement, and the reverse, we have, from principle 1°,

$$\sin 60° = \tfrac{1}{2}\sqrt{3} \text{ and } \cos 60° = \tfrac{1}{2},$$

Functions of the Sum and Difference of Two Arcs.

41. Let AF and FG be two arcs of the trigonometric circle; denote the former by x, and the latter by y.

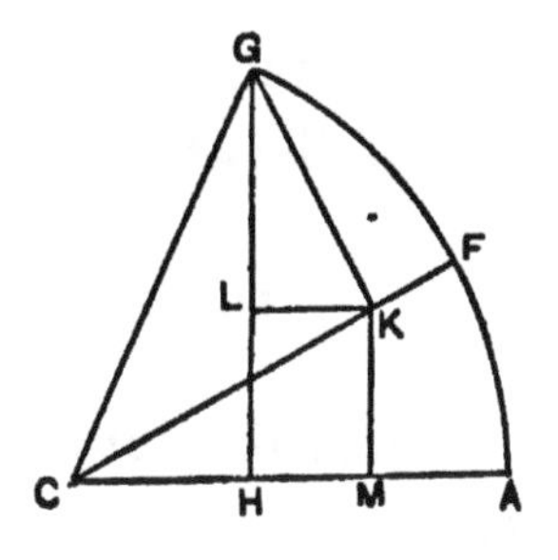

Draw the radii CA, CF, and CG; from G draw GK perpendicular to CF, and GH perpendicular to CA; from K, draw KM perpendicular to CA, and KL parallel to CA; then, HG will be the sine of $x + y$, CH the cosine of $x + y$, KG the sine of y, and CK the cosine of y.

From the figure, we have,

$$\text{HG} = \text{HL} + \text{LG}, \text{ or, } \sin(x + y) = \text{LG} + \text{HL}. \quad . \quad . \quad (1)$$

Now, the triangle KLG is similar to the triangle KMC, (P. 15, *Cor.* 2, B. 3); hence, the angle KGL is equal to x. From the right-angled triangle KLG, we have, (*Art.* 29),

$$\text{LG} = \text{KG} \cos x = \cos x \sin y,$$

and from the right-angled triangle CMK, we have,

$$\text{HL} = \text{MK} = \text{CK} \sin x = \sin x \cos y.$$

Substituting these values in (1), we have,

$$\sin(x + y) = \sin x \cos y + \cos x \sin y. \quad . \quad . \quad (a)$$

Formula (a) is true for all values of x and y; if we therefore substitute $-y$ for y and reduce, (*Art.* 38), we have,

$$\sin(x - y) = \sin x \cos y - \cos x \sin y. \quad . \quad . \quad (b)$$

If we substitute $90° - x$, for x, in (b), we have,

$$\sin[90° - (x + y)] = \sin(90° - x)\cos y - \cos(90° - x)\sin y; \quad (2)$$

but, the sine of 90° *minus* an arc, is equal to the cosine of the arc, and the cosine of 90° *minus* an arc, is equal to the sine of the arc; hence, equation (2) becomes

$$\cos(x + y) = \cos x \cos y - \sin x \sin y. \quad . \quad . \quad (c)$$

If we substitute $-y$, for y, in (c), and reduce (*Art.* 38), we have,

$$\cos(x - y) = \cos x \cos y + \sin x \sin y. \quad . \quad . \quad (d)$$

If we divide (a) by (c), member by member, we have,

$$\frac{\sin(x + y)}{\cos(x + y)} = \frac{\sin x \cos y + \cos x \sin y}{\cos x \cos y - \sin x \sin y};$$

dividing both terms of the second member by $\cos x \cos y$ and reducing, by the principle that the sine, divided by the cosine, is equal to the tangent, we have,

$$\tan(x + y) = \frac{\tan x + \tan y}{1 - \tan x \tan y}. \quad . \quad . \quad . \quad . \quad (e)$$

If we substitute $-y$, for y, in (e), and reduce, (*Art.* 38), we have,

$$\tan(x - y) = \frac{\tan x - \tan y}{1 + \tan x \tan y}. \quad . \quad . \quad . \quad . \quad (f)$$

Functions of Double Arcs.

42. If we make $y = x$ in the formulas (a), (c), and (e) of *Art.* 41, and reduce, we have,

$$\sin 2x = 2 \sin x \cos x \quad . \quad . \quad . \quad . \quad . \quad . \quad (a')$$

$$\cos 2x = \cos^2 x - \sin^2 x \quad . \quad . \quad . \quad . \quad . \quad . \quad (c')$$

$$\tan 2x = \frac{2 \tan x}{1 - \tan^2 x}. \quad . \quad . \quad . \quad . \quad . \quad . \quad . \quad . \quad (e')$$

If we substitute $1 - \sin^2 x$, for $\cos^2 x$, in (c'), and then substitute $1 - \cos^2 x$, for $\sin^2 x$, and reduce, we have,

$$\cos 2x = 1 - 2 \sin^2 x, \text{ or, } \sin x = \sqrt{\tfrac{1}{2}(1 - \cos 2x)}, . \; . \; (k)$$

$$\cos 2x = 2 \cos^2 x - 1, \text{ or, } \cos x = \sqrt{\tfrac{1}{2}(1 + \cos 2x)}. \; . \; . \; (l)$$

Additional Formulas.

43. If we add (b) to (a), member to member, and then subtract (b) from (a), member from member, we have,

$$\sin (x + y) + \sin (x - y) = 2 \sin x \cos y; \quad . \quad . \quad (1)$$

$$\sin (x + y) - \sin (x - y) = 2 \cos x \sin y. \quad . \quad . \quad (2)$$

In like manner, we deduce from formulas (d) and (c),

$$\cos (x + y) + \cos (x - y) = 2 \cos x \cos y; \quad . \quad . \quad (3)$$

$$\cos (x - y) - \cos (x + y) = 2 \sin x \sin y. \quad . \quad . \quad (4)$$

If we now make

$$x = \tfrac{1}{2}(p + q), \text{ and, } y = \tfrac{1}{2}(p - q),$$

or, $\quad x + y = p$ and $x - y = q$,

we have from formulas (1), (2), (3), and (4),

$$\sin p + \sin q = 2 \sin \tfrac{1}{2}(p + q) \cos \tfrac{1}{2}(p - q); \quad . . \quad (m)$$

$$\sin p - \sin q = 2 \cos \tfrac{1}{2}(p + q) \sin \tfrac{1}{2}(p - q); \quad . . \quad (n)$$

$$\cos p + \cos q = 2 \cos \tfrac{1}{2}(p + q) \cos \tfrac{1}{2}(p - q); \quad . . \quad (o)$$

$$\cos q - \cos p = 2 \sin \tfrac{1}{2}(p + q) \sin \tfrac{1}{2}(p - q); \quad . . \quad (p)$$

Dividing (p) by (o), member by member, remembering that the *sine*, divided by the *cosine*, is equal to the *tangent*, we have,

$$\frac{\cos q - \cos p}{\cos q + \cos p} = \tan \tfrac{1}{2}(p + q) \tan \tfrac{1}{2}(p - q). \quad . . . \quad (q)$$

If we substitute $\tfrac{1}{2}x$, for x, in (k) and (l), and reduce, we have,

$$\sin \tfrac{1}{2}x = \sqrt{\tfrac{1}{2}(1 - \cos x)}, \text{ or, } 1 - \cos x = 2 \sin^2 \tfrac{1}{2}x, \quad . \quad (r)$$

$$\cos \tfrac{1}{2}x = \sqrt{\tfrac{1}{2}(1 + \cos x)}, \text{ or, } 1 + \cos x = 2 \cos^2 \tfrac{1}{2}x. \quad . \quad (s)$$

Functions of Arcs Greater than 90°.

44. Every arc greater than 90° is made up of one or more quadrants, plus or minus an arc less than 90°. Denote this auxiliary arc by z.

If we take an arc in the second quadrant it may be denoted by $90° + z$, or by $180° - z$, an arc in the third quadrant may be denoted by $180° + z$ or by $270° - z$, and an arc in the fourth quadrant may be denoted by $270° + z$, or by $360° - z$.

The functions of these arcs can be found in terms of z by formulas (a), (b), (c), &c., as follows:

$$\sin (90° + z) = \sin 90° \cos z + \sin z \cos 90° = \cos z,$$

$$\sin (180° - z) = \sin 180° \cos z - \sin z \cos 180° = + \sin z, \text{ \&c.}$$

By proceeding in this manner we may form the following

TABLE.

Arc = 90° + z.	*Arc* = 270° − z.
sin = cos z, cos = − sin z.	sin = − cos z, cos = − sin z,
tan = − cot z, cot = − tan z.	tan = cot z, cot = tan z.
Arc = 180° − z.	*Arc* = 270° + z.
sin = sin z, cos = − cos z.	sin = − cos z, cos = sin z.
tan = − tan z, cot = − cot z.	tan = − cot z, cot = − tan z.
Arc = 180° + z.	*Arc* = 360° − z.
sin = − sin z, cos = − cos z,	sin = − sin z, cos = cos z.
tan = tan z, cot = cot z.	tan = − tan z, cot = − cot z.

V. SPHERICAL TRIGONOMETRY.

Definition and Explanations.

45. Spherical Trigonometry is that branch of Trigonometry which treats of the solution of Spherical Triangles.

In every spherical triangle there are six *parts*, three sides and three angles. If any three of these parts are given, the others can be found by computation.

The method of solving spherical triangles is entirely similar to that explained in *Art.* 28. In applying this method, we shall suppose that all the triangles considered lie on the surface of a sphere whose radius is equal to 1; we shall also suppose that each part of every triangle is less than 180°.

Formulas, used in Solving Right-angled Spherical Triangles.

46. Let ACD be a spherical triangle, right-angled at A, and let O be the centre of the sphere on which it is situated.

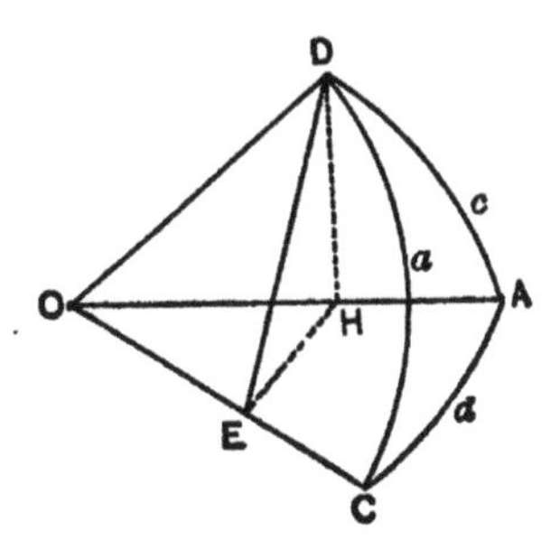

Draw the radii OA, OC, and OD; from D, draw DH perpendicular to OA and from H, draw HE perpendicular to OC; then draw ED. The line ED is perpendicular to OC, (P. 5, B. 6), and consequently, the angle HED is equal to the diedral angle whose edge is OC, (*Art.* 92, *Geom.*), that is, it is equal to the spherical angle ACD. Because A is a right angle, the planes OAC and OAD are perpendicular to each other; hence, DH is perpendicular to the plane AOC, (P. 14, *Cor.* 1, B. 6), and consequently, to the line HE; that is, the triangle EHD is right-angled at H.

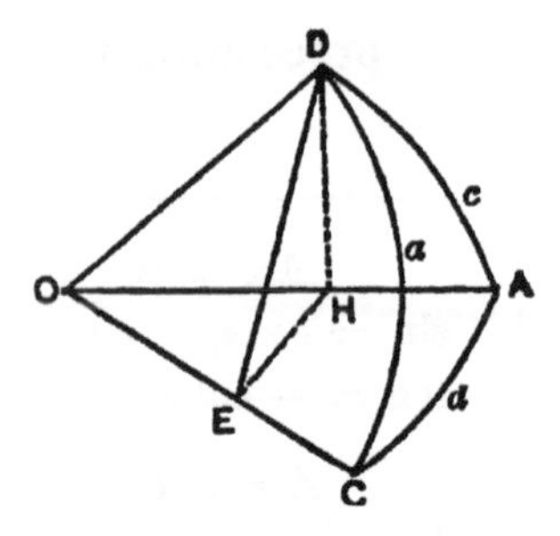

Denote the angles of the triangle by A, D, and C, and the sides opposite these angles by a, d, and c. Then, is HD equal to $\sin c$, OH, to $\cos c$, ED, to $\sin a$, and OE, to $\cos a$.

It is to be observed, that we may interchange C and D, provided we also interchange c and d.

From the right-angled triangle EHD, we have,

$$\text{DH} = \text{ED} \sin \text{HED},$$

or,
$$\sin c = \sin a \sin \text{C}; \quad \ldots \ldots \quad (1)$$

changing c into d, and C into D, we have,

$$\sin d = \sin a \sin \text{D}. \quad \ldots \ldots \quad (2)$$

From the right-angled triangle OEH, we have,

$$\text{OE} = \text{OH} \cos \text{EOH},$$

or,
$$\cos a = \cos c \cos d. \quad \ldots \ldots \quad (3)$$

From the triangles EHD and OEH, we have,

$$\cos \text{HED} = \frac{\text{EH}}{\text{ED}} = \frac{\text{OH} \sin \text{EOH}}{\text{ED}},$$

or, $$\cos C = \frac{\cos c \sin d}{\sin a};$$

substituting for sin d, its value taken (2) and reducing, we have,

$$\cos C = \cos c \sin D; \quad . \quad . \quad . \quad . \quad . \quad . \quad . \quad (4)$$

changing C into D, c into d, and D into C, we have,

$$\cos D = \cos d \sin C. \quad . \quad . \quad . \quad . \quad . \quad . \quad . \quad (5)$$

From the same triangles as before, we have,

$$\tan HED = \frac{HD}{EH} = \frac{HD}{OH \sin EOH},$$

or, $$\tan C = \frac{\sin c}{\cos c \sin d} = \frac{\tan c}{\sin d},$$

whence, by reduction, remembering that the cotangent is equal to the reciprocal of the tangent, we have,

$$\sin d = \tan c \cot C; \quad . \quad . \quad . \quad . \quad . \quad . \quad . \quad (6)$$

changing d into c, c into d, and C into D, we have,

$$\sin c = \tan d \cot D. \quad . \quad . \quad . \quad . \quad . \quad . \quad . \quad (7)$$

From the same triangles as before, we have,

$$\cos HED = \frac{EH}{ED} = \frac{OE \tan EOH}{ED},$$

or, $$\cos C = \frac{\cos a \tan d}{\sin a} = \cot a \tan d, \quad . \quad . \quad . \quad (8)$$

changing C into D, and d into c, we have,

$$\cos D = \cot a \tan c. \quad . \quad . \quad . \quad . \quad . \quad . \quad . \quad (9)$$

Multiplying (7) by (6), member by member, we have,

$$\sin c \sin d = \tan c \tan d \cot C \cot D;$$

dividing both members by $\tan c$ $\tan d$, remembering that the sine divided by the tangent is equal to the cosine, we have,

$$\cos c \cos d = \cot C \cot D;$$

substituting for $\cos c$ $\cos d$, its value taken from (3), we have,

$$\cos a = \cot C \cot D. \quad . \quad . \quad . \quad . \quad . \quad (10)$$

Formulas (1) to (10) enable us to solve a right-angled spherical triangle, when any two parts besides the right angle are given.

Napier's Rules for circular parts.

47. In any right-angled triangle, the two sides about the right angle, the complements of the opposite angles, and the complement of the hypothenuse, are called **Napier's circular parts.**

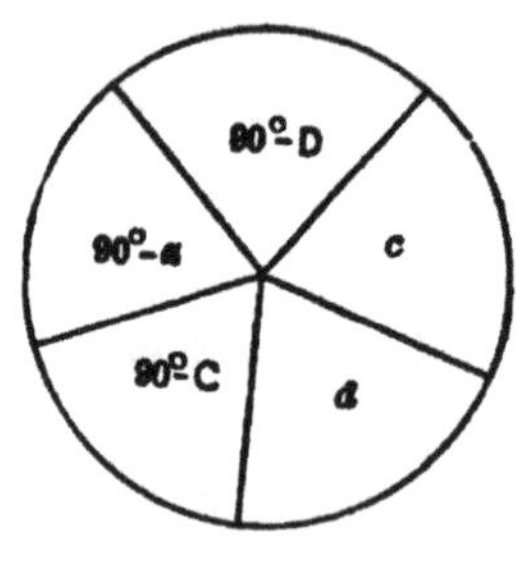

Let these parts, taken in order, be arranged in a circle, as shown in the diagram. If we take any three of these parts, they will either be adjacent to each other, or one of them will be separated from each of the other two by an intervening part. In the former case, the one lying between the other two, is called the **middle part**, and the other two, are called **adjacent parts**; in the latter case the one that is separated from the other two is called the **middle part**, and the other two are called **opposite parts.** Now let us take each part in succession as a middle part, first regarding only the *opposite parts*, and then regarding only the *adjacent* parts.

1°. From formulas (1), (2), (4), (3), and (5), of *Art.* 46, we have, after a slight change of form,

$$\sin c = \cos (90° - a) \cos (90° - C); \quad (1)$$
$$\sin d = \cos (90° - a) \cos (90° - D); \quad (2)$$
$$\sin (90° - C) = \cos c \cos (90° - D); \quad . \quad . \quad . \quad (3)$$
$$\sin (90° - a) = \cos c \cos d; \quad . \quad . \quad . \quad . \quad . \quad . \quad . \quad . \quad (4)$$
$$\sin (90° - D) = \cos d \cos (90° - C). \quad . \quad . \quad . \quad (5)$$

Comparing these formulas with the diagram, we see that the following rule is always true:

1st. *The sine of the middle part is equal to the rectangle of the cosine of the opposite parts.*

2°. From formulas (7), (6), (8), (10), and (9), of *Art.* 46, we have,

$$\sin c = \tan d \tan (90° - D); \quad . \quad . \quad . \quad (6)$$
$$\sin d = \tan c \tan (90° - C); \quad . \quad . \quad . \quad (7)$$
$$\sin (90° - C) = \tan (90° - a) \tan d; \quad . \quad . \quad . \quad (8)$$
$$\sin (90° - a) = \tan (90° - C) \tan (90° - D); \quad (9)$$
$$\sin (90° - D) = \tan (90° - a) \tan c. \quad . \quad . \quad . \quad (10)$$

Comparing these formulas with the diagram, we see that the following rule is always true:

2d. *The sine of the middle part is equal to the rectangle of the tangents of the adjacent parts.*

Discussion.

48. In applying Napier's rules, the required part is always determined by means of its sine. Now, the same sine corresponds to two different arcs, which are supplements of each other; it is, therefore, important to ascertain which of these is to be taken.

Two arcs, or angles, are said to be of the **same species**, when both are less than 90°, or both greater than 90°;

they are of different species, when one is greater, and the other less than 90°.

From formula (4), *Art.* 46, we have,

$$\sin D = \frac{\cos C}{\cos c}. \quad . \quad . \quad . \quad . \quad . \quad . \quad (1)$$

Now, the first member of (1) is always positive, because D is always less than 180°; consequently, cos C must have the same sign as cos c; this can only be the case when C and c are of the *same species;* hence, we have the following principle:

1°. *Each side about the right angle must be of the same species as its opposite angle.*

Formula (3), *Art.* 46, is

$$\cos a = \cos d \cos c. \quad . \quad . \quad . \quad . \quad . \quad (2)$$

If a is less than 90°, cos a is *positive;* consequently cos d and cos c must have the same sign, that is, d and c must be of the same species; if a is greater than 90°, cos a is *negative;* consequently, cos d and cos c must have contrary signs, that is, d and c must be of different species; hence, the following principle:

2°. *If the hypothenuse is less than 90°, the sides about the right angle must be of the same species; if the hypothenuse is greater than 90°, the sides about the right angle must be of different species.*

These principles enable us to decide as to the value of the arc, or angle, in every instance, except when we have given an oblique angle and its opposite side. Then, there may be several cases.

Let ACD be right-angled at A, and let the angle C, and the side DA be given. Prolong CD and CA till they intersect at C′. Lay off C′D′ equal to CD, C′A′ equal to CA, and draw the arc A′D′. The triangles ACD and A′C′D′ have two sides and their included angle equal, each to each; hence, the remaining parts are equal, each to each, that is, the angle A′ is equal to A, the angle D′ to D, and the side D′A′ to DA. The triangles ACD and A′CD′ have, therefore, the angle C common, and the opposite sides AD and A′D′ equal; consequently, whenever the triangle A′CD′ can be formed, there are two solutions. But in order that this triangle may be formed, the given parts must satisfy equation (4), *Art.* 46, which gives,

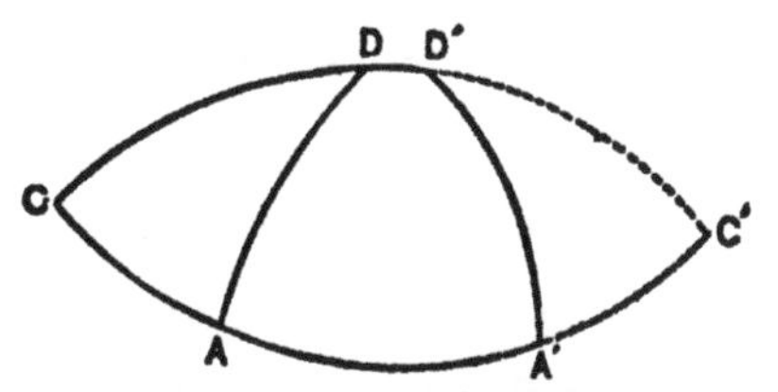

$$\sin D = \frac{\cos C}{\cos c}, \quad . \quad . \quad . \quad . \quad . \quad . \quad . \qquad (3)$$

in which sin D cannot be greater than 1, nor less than 0.

First. Let C and c be of the same species, and let the difference between C and 90° be *less* than the difference between c and 90°. In this case, cos C is *numerically less* than cos c; condition (3) is satisfied; and there are *two* solutions.

Secondly. Let C be equal to c. In this case, sin D is equal to 1, that is, D is equal to 90°; the two triangles ACD and A′CD′, are bi-rectangular and equal; and there is but *one* solution.

Thirdly. Let C and c be of the same species, and let the difference between C and 90° be greater than the difference between c and 90°. In this case, cos C is *numerically* greater than $\cos c$; condition (3) is not satisfied; and there is no *solution.*

Fourthly. Let C and c be of different species. In this case $\cos C$ and $\cos c$ have different signs; condition (3) is not satisfied; and there is no solution.

Applications of Napier's Rules.

49. In applying Napier's rules, two of the circular parts must be given. Then, whatever these may be, we select that formula of *Art.* 47, whose second member contains the given parts, and by means of this formula we find a third part. In like manner, we find each of the other parts in succession.

It is to be observed that we can find any one of the unknown parts, without finding the others. To this end, we select the formula which contains the required and the given parts, and, if necessary, solve it with respect to the required part. From this, we find the numerical value of the required part by means of logarithms.

Note.—In solving the following examples, let seconds of arcs be neglected at every step of the operation. In taking out arcs corresponding to logarithmic functions, stop at the one next lower than the given logarithm.

EXAMPLES.

1. Given, $a = 86° \, 51'$ and $d = 18° \, 2'$, to find the remaining parts.

OPERATION.

Applying logarithms to formula (8), *Art.* 47, we have,

$$\log \sin (90° - C) = \log \tan (90° - a) + \log \tan d - 10;$$

substituting the proper values of a and d, we have,

$$\begin{array}{ll} \log \tan 3^\circ\ 9' \ . \ . & 8.740626 \\ \log \tan 18^\circ\ 2' \ . & \underline{9.512635} \end{array}$$

$$\therefore\ C = 90^\circ - \log \sin^{-1} 8.253261 = 90^\circ - 1^\circ\ 1' = 88^\circ\ 59'.$$

Applying logarithms to formula (5), *Art.* 47, we have,

$$\log \sin (90^\circ - D) = \log \cos d + \log \cos (90^\circ - C) - 10;$$

$$\begin{array}{ll} \log \cos 18^\circ\ 2' \ . & 9.978124 \\ \log \cos 1^\circ\ 1' \ . \ . & \underline{9.999932} \end{array}$$

$$\therefore\ D = 90^\circ - \log \sin^{-1} 9.978056 = 90^\circ - 71^\circ\ 56' = 18^\circ\ 4'.$$

Applying logarithms to formula (1), *Art.* 47, we have,

$$\log \sin c = \log \cos (90^\circ - a) + \log \cos (90^\circ - C) - 10,$$

or,
$$\log \sin c = \log \sin a + \log \sin C - 10;$$

$$\begin{array}{ll} \log \sin 86^\circ\ 51' \ . \ . & 9.999343 \\ \log \sin 88^\circ\ 59' \ . \ . & \underline{9.999932} \end{array}$$

$$\therefore\ c = \log \sin^{-1} 9.999275 = 86^\circ\ 41'.$$

2. Given, $a = 142^\circ\ 9'$, and $C = 54^\circ\ 1'$, to find the remaining parts.

In this example, we use formulas (1), (7), and (9), which give the logarithmic formulas

$$\log \sin c = \log \sin a + \log \sin C - 10, \quad . \ . \quad (a)$$

$$\log \sin d = \log \tan c + \log \cot C - 10, \quad . \ . \quad (b)$$

$$\log \cos D = \log \cot a + \log \tan c - 10. \quad . \ . \quad (c)$$

From (a), (b), and (c), we find,

$$c = 29^\circ\ 46', \quad d = 155^\circ\ 28', \quad \text{and } D = 137^\circ\ 23'. \ \textit{Ans.}$$

NOTE.—The side c must be less than 90°, (*Prin.* 1°, *Art.* 48); and because a is greater than 90°, c and d must be of different species, (*Prin.* 2°, *Art.* 48); from *Prin.* 1°, *Art.* 48, D and d must be of the same species.

3. Given, $c = 115^\circ\ 20'$, and $C = 91^\circ\ 2'$, to find the remaining parts.

$$d = \begin{cases} 2^\circ\ 11', \\ 177^\circ\ 49', \end{cases} \quad D = \begin{cases} 2^\circ\ 24', \\ 177^\circ\ 36', \end{cases} \quad a = \begin{cases} 115^\circ\ 19'. \\ 64^\circ\ 41'. \end{cases} \quad \textit{Ans.}$$

NOTE.—Because the difference between C and 90°, is less than the difference between c and 90°, there are two solutions, (*Art.* 48).

4. Given, $a = 78^\circ\ 20'$, and $C = 37^\circ\ 25'$, to find d.

Applying logarithms to formula (8), *Art.* 47, we have,

$$\log \cos C = \log \cot a + \log \tan d - 10;$$

whence, $\log \tan d = \log \cos C + (10 - \log \cot a)$,

or, $$\log \tan d = \log \cos C + (a{\cdot}c) \log \cot a. \quad . \quad . \quad (1)$$

From (1) we find, $d = 75^\circ\ 25'$.

Solution of Quadrantal Triangles.

50. We may solve a quadrantal triangle, that is, a triangle in which one side is a quadrant, by finding its polar triangle, which will be a right-angled spherical triangle; this triangle may be solved by Napier's rules, *Art.* 47; then, by passing back to *its* polar triangle, we shall have the parts of the given quadrantal triangle.

Formulas used in solving oblique spherical triangles.

51. In what follows, the angles of a spherical triangle will be denoted by A, C, and D; and their opposite sides by a, c, and d.

First. From the vertex D, of the spherical triangle ACD, draw the arc DE, perpendicular to the side AC, and denote it by p.

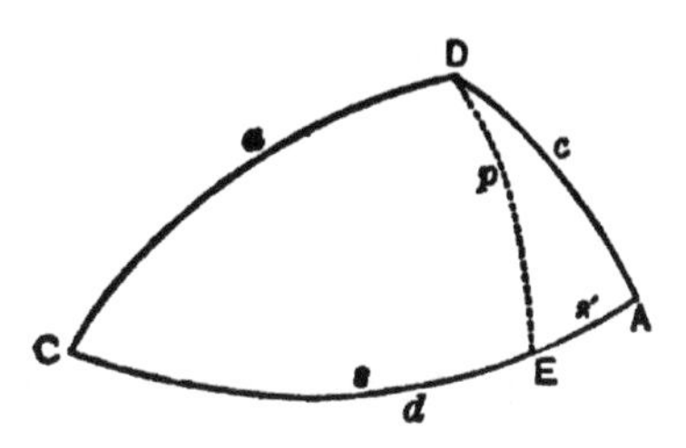

From formula (1), *Art.* 46, we have,

$$\sin p = \sin c \sin \text{A}, \quad . \quad . \quad . \quad . \quad . \quad . \quad (1)$$

and,

$$\sin p = \sin a \sin \text{C}. \quad . \quad . \quad . \quad . \quad . \quad . \quad (2)$$

Equating the second members of (1) and (2), we have,

$$\sin c \sin \text{A} = \sin a \sin \text{C},$$

or,

$$\sin a : \sin c :: \sin \text{A} : \sin \text{C}. \quad . \quad . \quad . \quad . \quad . \quad (3)$$

Hence, we have the following principle:

1°. *In any spherical triangle, the sines of any two sides, are to each other, as the sines of their opposite angles.*

NOTE.—If the point E falls on the prolongation of CA, the preceding formulas, and their consequent principle, will hold true.

Secondly. If we denote the distances from E to A, and

from E to C respectively, by s' and s, we have from formula (3), *Art.* 46,

$$\cos c = \cos p \cos s', \text{ and } \cos a = \cos p \cos s,$$

whence, the proportion,

$$\cos c : \cos a :: \cos s' : \cos s; \quad . \quad . \quad . \quad . \quad (4)$$

from which, by composition and division, we have,

$$\cos c + \cos a : \cos c - \cos a :: \cos s' + \cos s : \cos s' - \cos s,$$

or,

$$\frac{\cos c - \cos a}{\cos c + \cos a} = \frac{\cos s' - \cos s}{\cos s' + \cos s}. \quad . \quad . \quad . \quad (5)$$

Substituting for each member of (5), its value taken from formula (q), *Art.* 43, we have,

$$\tan \tfrac{1}{2}(a + c) \tan \tfrac{1}{2}(a - c) = \tan \tfrac{1}{2}(s + s') \tan \tfrac{1}{2}(s - s');$$

or,

$$\tan \tfrac{1}{2}(a+c) : \tan \tfrac{1}{2}(s+s') :: \tan \tfrac{1}{2}(s-s') : \tan \tfrac{1}{2}(a-c). \quad . \quad (6)$$

Hence, the following principle:

2°. *If a perpendicular is drawn from the vertex of any angle of a spherical triangle, the tangent of half the sum of the including sides, is to the tangent of half the sum of the segments of the other side, as the tangent of half the difference of the segments is to the tangent of half the difference of the including sides.*

NOTE.—If the point E falls on the prolongation of CA, the segments EA and EC are called *external* segments; in this case, the segment EA is to be regarded as *negative*. With this understanding, the preceding formulas, and their consequent principle, will hold true.

Solution of oblique spherical triangles.

52. Every spherical triangle contains *six* parts, *three* sides and *three* angles; if any three of these parts are given, the other three may be found by computation. There may be three cases.

FIRST CASE. *Given, two sides and an angle opposite one of them, to find the remaining parts. Or, given, two angles and a side opposite one, to find the other parts.*

1°. Let the sides a, c, and the angle C, be given. The remaining parts can then be found by means of the formulas of *Arts.* 47 and 51.

The angle A is determined by means of its sine; and because the sine of an angle is equal to the sine of its supplement, it may happen that there will be two solutions. To ascertain when there will be two solutions, let the arcs CD and CA be prolonged till they meet at C′. Lay off EA′ equal to EA, and draw the arc DA′. The right-angled triangles AED and A′ED, have two sides and the included angle of the one, equal to two sides and the included angle of the other, each to each; hence, they are equal in all their parts; therefore, the angle EA′D is equal to EAD, and the side DA′ to DA. But, the sine of CA′D is equal to the sine of its supplement EA′D; hence, the sines of a, c, A, and C belong equally to the two triangles ACD and A′CD. There will, therefore, be two solutions whenever these two triangles can be formed; that is, when both A and A′ fall between C and C′. Hence, if $s - s' > 0°$, that is, if $s > s'$, and if $s + s' < 180°$, there will be *two* solutions.

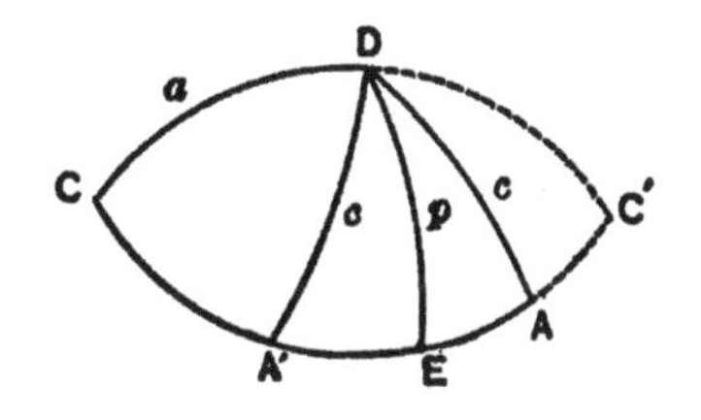

If $s - s' < 0°$, that is, if $s < s'$, and if $s + s' < 180°$, it is obvious that there will be but *one* solution. If $s' = 0$, ACD and A'CD will coincide and there will be but *one* solution.

From formula (1), *Art.* 46, we have,

$$\sin p = \sin c \sin A, \text{ or } \sin A = \frac{\sin p}{\sin c}. \quad . \quad . \quad (1)$$

If $\sin p > \sin c$, equation (1) will not be satisfied, and there will be *no* solution.

EXAMPLES.

1. Given, $a = 72°\ 10'$, $c = 79°$, and $C = 32°\ 20'$; to find the remaining parts.

We first find s and s'.

From the right-angled triangle CED, we have,

$$\sin(90° - C) = \tan(90° - a) \tan s, \quad . \quad (2)$$

from which we deduce the logarithmic formula,

$$\log \tan s = \log \cos C + \log \tan a - 10.$$

log cos 32° 20′	. .	9.926831
log tan 72° 10′	. .	10.492540

$$s = \log \tan^{-1} 10.419371 = 69°\ 9'.$$

Because $\sin(90° - C)$, and $\tan(90° - a)$, are positive in (2), $\tan s$ must be positive; hence, s is less than 90°.

Applying logarithms to proportion (4), *Art.* 51, we have,

$$\log \cos s' = \log \cos c + \log \cos s + (a \cdot c) \log \cos a - 10.$$

	log cos 79°	9.280599
	log cos 69° 9′ . . .	9.551356
$(a \cdot c)$	log cos 72° 10′ . .	0.513925

$$s' = \log \cos^{-1} 9.345880 = 77° \, 13'.$$

Because the first, second, and fourth terms of proportion (4), *Art.* 51, are positive, the third term is positive; hence, s' is less than 90°.

We see that $s < s'$, and that $s + s' < 180°$; hence, there is but one solution.

Applying logarithms to proportion (3), *Art.* 51, we have,

$$\log \sin \mathrm{A} = \log \sin a + \log \sin \mathrm{C} + (a \cdot c) \log \sin c - 10.$$

From this formula, we have,

	log sin 72° 10′ . .	9.978615
	log sin 32° 20′ . .	9.728227
$(a \cdot c)$	log sin 79°	0.008053

$$\therefore \mathrm{A} = \log \sin^{-1} 9.714895 = 31° \, 14'.$$

Because C < 90°, it is obvious that p is less than 90°; but p and A are of the same species; hence, we take the acute value of A.

The side AC is equal to $s + s'$; hence, $d = 146° \, 22'$.

From Napier's rules, we have,

$$\log \cos \mathrm{EDC} = \log \sin \mathrm{C} + \log \cos s - 10,$$

and,

$$\log \cos \mathrm{EDA} = \log \sin \mathrm{A} + \log \cos s' - 10,$$

From the first of these, we find $\text{EDC} = 79° \ 2'$, and from the second, we find $\text{EDA} = 83° \ 24'$. Hence,

$$\text{CDA} = \text{EDC} + \text{EDA} = 162° \ 26'.$$

2°. Let A, C, and c, be given; to find the remaining parts.

If we subtract each of the given parts from 180°, we shall have two sides and an angle opposite one of them, in the polar triangle; we can solve the polar triangle in the manner just explained, and then by subtracting each of the parts found, from 180°, we have the required parts of the given triangle.

EXAMPLE.

1. Given, $A = 58° \ 8'$, $C = 50° \ 12'$, and $c = 62° \ 42'$, to find the remaining parts.

Subtracting A, C, and c, from 180° and denoting the results by a', c', and C', we have in the polar triangle $a' = 121° \ 52'$, $c' = 129° \ 48'$, and $C' = 117° \ 18'$, to find the remaining parts.

The polar triangle has two solutions giving

$$A' = \begin{cases} 100° \ 48', \\ 79° \ 12'; \end{cases} \quad D' = \begin{cases} 60° \ 56', \\ 27° \ 46'; \end{cases} \quad d' = \begin{cases} 49° \ \ 5', \\ 23° \ 45'. \end{cases}$$

Passing to the given triangle by subtracting each of these parts from 180°, we have,

$$a = \begin{cases} 79° \ 12', \\ 100° \ 48'; \end{cases} \quad d = \begin{cases} 119° \ \ 4', \\ 152° \ 14'; \end{cases} \quad D = \begin{cases} 130° \ 55', \\ 156° \ 15'. \end{cases} \ \textit{Ans.}$$

SECOND CASE. *Given, two sides and their included angle, to find the remaining parts. Or, given, two angles and their included side, to find the remaining parts.*

1°. Let the sides a, d, and the included angle C, be given, to find the other parts.

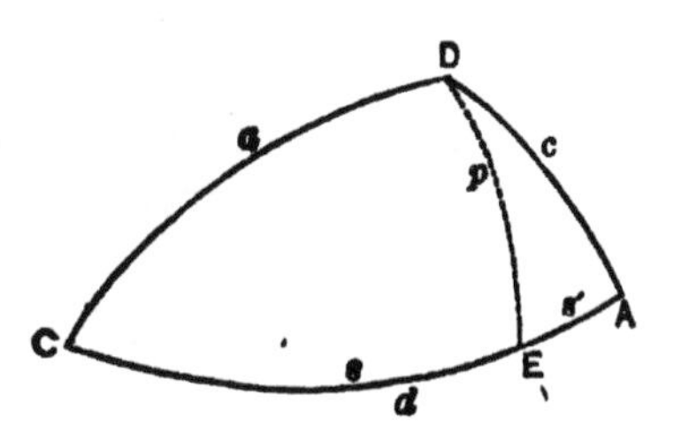

We first find the perpendicular p, and the segments s and s'; we then complete the solution by Napier's rules.

EXAMPLES.

1. Given, $a = 68° 46'$, $d = 37° 10'$, and $C = 39° 23'$, to find the remaining parts.

From the right-angled triangle CED, we have,

$$\log \sin p = \log \sin C + \log \sin a - 10,$$

and, $$\log \sin s = \log \tan p + \log \cot C - 10.$$

From these formulas we find,

$$p = 36° 15', \text{ and } s = 63° 16'.$$

Because s is greater than d, the segments are external, and s' is therefore equal to $d - s$, or to $-26° 6'$.

From the right-angled triangles CED and AED, we have,

$$\cos c = \cos p \cos s'; \quad . \quad . \quad . \quad . \quad . \quad . \quad (1)$$
$$\cos A = \tan s' \cot c; \quad . \quad . \quad . \quad . \quad . \quad (2)$$
$$\cos EDC = \tan p \cot a; \quad . \quad . \quad . \quad . \quad (3)$$
$$\cos EDA = \tan p \cot c. \quad . \quad . \quad . \quad . \quad (4)$$

From these formulas, by means of logarithms, we find,

$$c = 43° 36', \; A = 120° 57', \; EDC = 73° 27', \text{ and}$$
$$EDA = 39° 39'.$$

Because s' is negative, tan s' is negative in (2), and consequently cos A is negative; hence, A is greater than 90°.

Because s and s' are external segments, the angle D is equal to EDC — EDA; hence,

$$D = 73° 27' - 39° 39' = 33° 48'.$$

2°. Let the angles A, D, and the included side c, be given, to find the other parts.

We pass to the polar triangle as before; we then solve the polar triangle, as just explained; we next pass back to the given triangle, which will then be completely known.

EXAMPLES.

1. Given, $A = 51° 30'$, $D = 131° 30'$, and $c = 80° 19'$; to find the remaining parts.

Ans. $C = 59° 16'$, $a = 63° 50'$, and $d = 120° 46'$.

2. Given, $A = 34° 15'$, $D = 42° 15'$, and $c = 76° 37'$; to find the remaining parts.

Ans. $a = 40° 3'$, $d = 50° 6'$, and $C = 121° 37'$.

THIRD CASE. *Given, the three sides, to find the three angles.* Or, *given, the three angles, to find the three sides.*

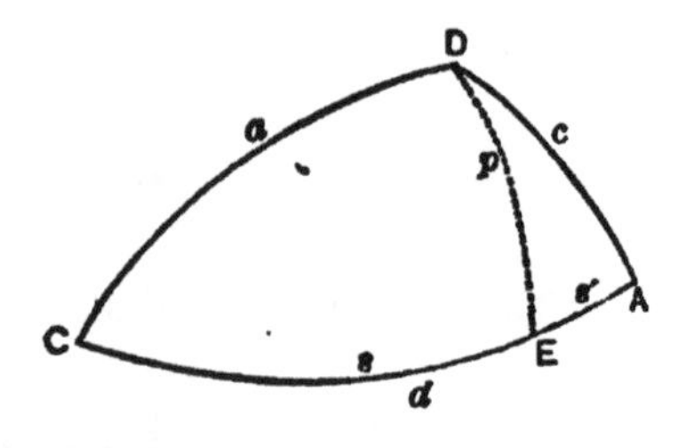

We find the half difference of s and s' by formula (6), *Art.* 51; we then add the half difference to $\frac{1}{2}(s+s')$, or $\frac{1}{2}d$, to find the greater segment, and subtract the half difference from $\frac{1}{2}d$ to find the less segment; the solution is then completed as in previous cases.

EXAMPLE.

1. Given, $a = 83° 14'$, $c = 56° 40'$, and $d = 114° 30'$; to find A, C, and D.

Applying logarithms to formula (6), *Art.* 51, remembering that $s + s'$ is equal to d, we have,

$$\log \tan \tfrac{1}{2}(s - s') = \log \tan \tfrac{1}{2}(a + c) + \log \tan \tfrac{1}{2}(a - c) - \log \tan \tfrac{1}{2}d.$$

log tan 69° 57′	10.437756
log tan 13° 17′	9.373064
	19.810820
log tan 57° 15	10.191639

$$\tfrac{1}{2}(s - s') = \log \tan^{-1} 9.619181 = 22° 35'.$$

$$\therefore s = \tfrac{1}{2}(s + s') + \tfrac{1}{2}(s - s') = 57° 15' + 22° 35' = 79° 50'.$$
$$s' = \tfrac{1}{2}(s + s') - \tfrac{1}{2}(s - s') = 57° 15' - 22° 35' = 34° 40'.$$

From the right-angled triangles CED and AED, we have,

$$\cos C = \tan s \cot a; \quad \ldots \quad (1)$$
$$\cos A = \tan s' \cot c; \quad \ldots \quad (2)$$
$$\cos EDA = \sin A \cos s'; \quad \ldots \quad (3)$$
$$\cos EDC = \sin C \cos s. \quad \ldots \quad (4)$$

From these formulas, by the aid of logarithms, we find,

C=48° 35′, A=62° 57′, EDA=42° 55′, and EDC=82° 24′;

hence, D=EDA+EDC=125° 19′.

2°. Let the angles A, C, and D, be given, to find the sides a, c, and d.

If we subtract each of the given angles from 180°, we have the corresponding sides of a polar triangle, which can be solved by the method just explained. If we then subtract each angle of this triangle from 180°, we have the sides of the primitive triangle.

EXAMPLE.

1. Given, $A = 48° 30'$, $C = 62° 54'$, and $D = 125° 20'$, to find the remaining parts.

Ans. $a = 56° 41'$, $c = 83° 12'$, and $d = 114° 30'$.

MENSURATION.

INTRODUCTION.

Definitions.

1. The **measure** of a quantity is an expression for the quantity in terms of a quantity of the same kind taken as a *unit*.

2. Mensuration is that branch of mathematics which treats of the measurement of Geometrical magnitudes.

The rules of Mensuration are simple deductions from the principles and formulas of Geometry.

Explanatory remarks.

3. The **unit of measure of any line** is a straight line; as 1 foot, 1 yard, &c. The measure of the line is a denominate number, whose unit is the assumed *linear unit.*

4. The **unit of measure of a surface** is a square, one of whose sides is the linear unit; as 1 square foot, 1 square yard, &c. The measure of a surface is a denominate number, whose unit is the assumed *superficial unit.*

By the **product of two lines**, we mean the product obtained by multiplying the number of linear units in one, by the number of linear units in the other. The unit of the product is a *superficial unit.*

5. The **unit of measure of a volume** is a cube, one of whose edges is the linear unit; as, 1 cubic foot, 1 cubic yard, &c. The measure or **content** of a volume is a denominate number, whose unit is the assumed *cubic unit.*

By the **product of three lines**, we mean the continued product of the number of linear units in each line. The unit of the product is a *cubic unit,* or a *unit of volume.*

By the **product of a line and a surface**, we mean the product obtained by multiplying the number of linear units in the line, by the number of superficial units in the surface. The unit of the product is a *cubic unit,* or a *unit of volume.*

Note.—In what follows, the results are only approximate; as a general rule, numbers will be carried out to two places of decimals and angles to minutes of arc; the last figure in either case will generally be of doubtful value.

I. MENSURATION OF LINES.

Lines considered.

6. The only lines treated of in Elementary Geometry are straight lines, and arcs of circles.

1°. *To find the hypothenuse of a right-angled triangle.*

RULE.

Find the sum of the squares of the sides about the right angle and extract the square root of the result, (P. 8, B. 4.)

EXAMPLES.

1. The sides about a right angle are 400 *yds.,* and 600 *yds.*; what is the length of the corresponding hypothenuse?

$$h = \sqrt{(400)^2 + (600)^2} = 721.11 \text{ } yds., \text{ } Ans.$$

2. The sides about a right angle are 13 *ft.*, and 25 *ft.*; what is the corresponding hypothenuse? *Ans.* 28.18 *ft.*

3. What is the diagonal of a rectangle whose sides are 14 *rds.*, and 19 *rds.*? *Ans.* 23.6 *rds.*

4. The radius of the base of a cone is $4\frac{1}{2}$ *ft.*, and its altitude is 6 *ft.*; what is its slant height? *Ans.* 7.5 *ft.*

5. The hypothenuse of a right-angled triangle is 45 *yds.*, and one side about the right angle is 37 *yds.*; what is the other side?

$$side = \sqrt{(45)^2 - (37)^2} = 25.61 \text{ } yds., \text{ } Ans.$$

6. The radius of the upper base of a conic frustum is 8 *ft.*, the radius of the lower base is 14 *ft.*, and the altitude is 6 *ft.*; what is the slant height?

$$s = \sqrt{(14 - 8)^2 + (6)^2} = 8.49 \text{ } ft., \text{ } Ans.$$

2°. *To find the circumference of a circle whose radius is given.*

RULE.

Multiply twice the radius by 3.1416, (P. 11, B. 5, *Cor.* 2).

EXAMPLES.

1. The radius of a circle is 3.75 *ft.*, what is its circumference?

$$c = 7.5 \times 3.1416 = 23.56 \text{ } ft., \text{ } Ans.$$

2. What is the circumference of a circle whose radius is 15.375 *rds.*? *Ans.* 96.60 *rds.*

3. Find the circumference whose radius is 30.25.

Ans. 190.07.

3°. *To find the length of an arc of a circle whose radius is given.*

RULE.

Multiply the length of the circumference, by the number of degrees in the arc, and divide the product by 360, (P. 11, B. 5, *Cor.* 4).

EXAMPLES.

1. What is the length of an arc of 30°, in a circle whose radius is 20.5 *yds.*?

$$a = 41 \times 3.1416 \times \frac{30}{360} = \frac{128.81}{12} = 10.73 \textit{ yds., Ans.}$$

2. Find the length of an arc of 17°, the radius being equal to 25.5. *Ans.* 7.57.

3. The radius of a circle is 25.75; what is the length of an arc of 66° 30′? *Ans.* 29.89.

4. The length of a circumference is 179.86 *ft.*; what is the length of its radius?

$$r = \frac{179.86}{2\pi} = 28.63 \textit{ ft., Ans.}$$

5. The radius of a circle is 31¼ *ft.*, and the length of an arc of its circumference, is 16.36 feet; what is the corresponding angle at the centre?

$$n = \frac{16.36}{2\pi r} \times 360° = 30°, \textit{ nearly, Ans.}$$

II. MENSURATION OF SURFACES.

Surfaces considered.

7. The surfaces treated of in Elementary Geometry are plane, spherical, conical, and cylindrical.

4°. *To find the area of a plane triangle when the base and altitude are given.*

RULE.

Multiply the base by the altitude, and divide the result by 2, (P. 4, B. 4).

EXAMPLES.

1. What is the area of a triangle whose base is 1300 *ft.*, and whose altitude is 520 *ft.*?

$$A = \frac{1300 \times 520}{2} = 338{,}000 \ sq. ft., \ Ans.$$

2. Find the area of a triangle whose base is $13\frac{1}{3}$ *yds.*, and whose altitude is 10 *yds.*? *Ans.* $66\frac{2}{3}$ *sq. yds.*

3. The base of a triangle is 49 *yds.*, and its altitude is $21\frac{1}{7}$ *yds.*; what is its area? *Ans.* 518 *sq. yds.*

5°. *To find the area of a plane triangle, when two sides and their included angle are given.*

8. Let the sides AC, AD, and the angle A, of the triangle ACD, be given. Call AC the base.

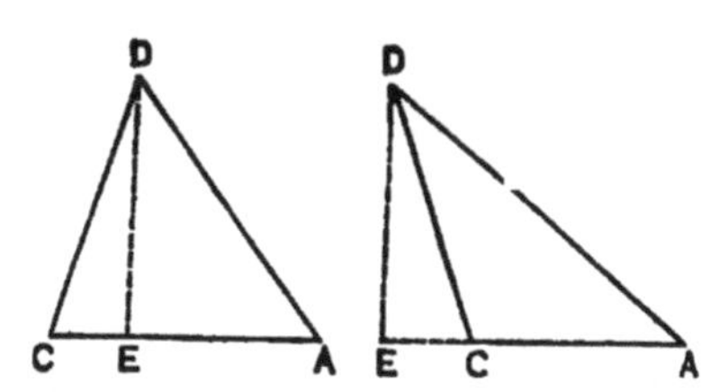

From D, draw DE perpendicular to the base; DE will be the altitude. Let the angles of the triangle be denoted by A, C, and D; and let their opposite sides be denoted by a, c, and d.

Then, whether E falls on AC, or on its prolongation, we shall have, (*Trig.*, *Art.* 29),

$$DE = c \sin A.$$

If we denote the area of the triangle by T, we shall have, from the preceding article,

$$2T = cd \sin A; \quad . \quad . \quad . \quad . \quad . \quad . \quad . \quad (1)$$

applying logarithms to formula (1), we have,

$$\log (2T) = \log c + \log d + \log \sin A - 10; \quad . \quad (2)$$

hence, we have the following

RULE.

Find the logarithms of the given sides and the logarithmic sine of their included angle, and take their sum; from this sum subtract 10; the result will be the logarithm of twice the area of the triangle. Find the number corresponding to this logarithm and divide it by 2; the quotient will be the required area.

EXAMPLES.

1. Given, $c = 153$ *ft.*, $d = 211$ *ft.*, and $A = 62° 20'$, to find the area of the triangle.

OPERATION.

log 153	2.184691
log 211	2.324282
log sin 62° 20′	9.947269

$$\therefore 2T = \log^{-1} 4.456242 = 28{,}591.84;$$

hence, $\quad T = 14{,}295.92$ *sq. ft.*, *Ans.*

2. What is the area of a triangle, two of whose sides are 60 *yds.* and 80 *yds.*, their included angle being 28° 57′?

Ans. 1161.71 *sq. yds.*

3. Two sides of a triangular field are respectively 25 *rds.*, and 21¼ *rds.* in length, and their included angle is 45°; what is the area of the field?

Ans. 187.82 *sq. rds.* = 1 *A.* 27.82 *sq. rds.*

6°. *To find the area of a plane triangle when its three sides are given.*

LEMMA.

To find expressions for the sine and for the cosine of half an angle of a plane triangle.

9. Assume the figure and the notation of the last article; and let it be required to find the sine and the cosine of $\frac{1}{2}$C.

From P. 9, and P. 10, B. 4, we have,

$\overline{DA}^2 = \overline{AC}^2 + \overline{CD}^2 \mp 2\overline{AC} \times \overline{CE}$

or, $c^2 = d^2 + a^2 \mp 2d \times \overline{CE}$. (1)

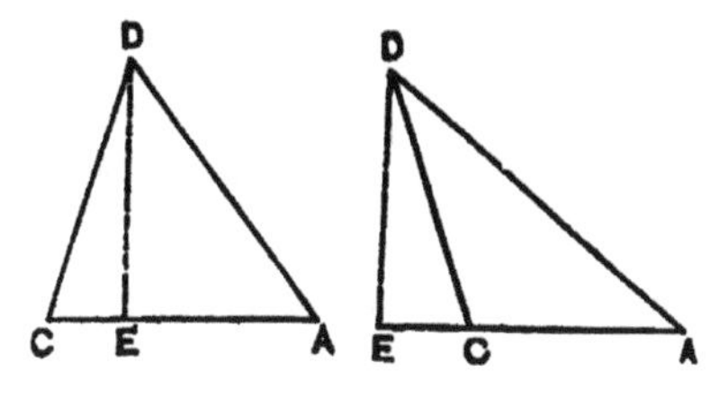

The upper sign corresponds to the case in which C is *acute*, and the lower sign to the case in which C is obtuse.

Now, when C is acute, we have,

$$CE = CD \cos C, \text{ or, } CE = a \cos C;$$

when C is obtuse, we have,

$$CE = CD \cos DCE = -CD \cos C, \text{ or, } CE = -a \cos C.$$

If we substitute the first value of CE in (1), taking the upper sign of the last term; or if we substitute the second value taking the lower sign, we shall have, in both cases,

$$c^2 = d^2 + a^2 - 2ad \cos C;$$

whence,

$$\cos C = \frac{d^2 + a^2 - c^2}{2ad} \quad . \quad . \quad . \quad . \quad . \quad . \quad . \quad . \quad (2)$$

If we add 1 to both members of equation (2), remembering that $1 + \cos C = 2 \cos^2 \frac{1}{2}C$, (*Trig.*, *Art.* 43), we shall have, after reduction,

$$\cos^2 \tfrac{1}{2}C = \frac{d^2 + a^2 + 2ad - c^2}{4ad} = \frac{(a + d)^2 - c^2}{4ad} \quad . \quad (3)$$

The numerator of the final member of (3) can be factored, giving,

$$(a + d)^2 - c^2 = (a + d + c)\,(a + d - c);$$

hence, equation (3) may be written,

$$\cos^2 \tfrac{1}{2}C = \frac{\tfrac{1}{2}(a + c + d)\,\tfrac{1}{2}(a + d - c)}{ad}. \quad . \quad (4)$$

If we denote the sum of the sides by s, we shall have,

$$\tfrac{1}{2}(a + c + d) = \tfrac{1}{2}s, \text{ and } \tfrac{1}{2}(a + d - c) = \tfrac{1}{2}s - c.$$

which substituted in (4) give,

$$\cos^2 \tfrac{1}{2}C = \frac{\tfrac{1}{2}s\,(\tfrac{1}{2}s - c)}{ad}, \text{ or, } \cos \tfrac{1}{2}C = \sqrt{\frac{\tfrac{1}{2}s\,(\tfrac{1}{2}s - c)}{ad}}. \quad (5)$$

If we now subtract both members of equation (2) from 1, remembering that $1 - \cos C = 2 \sin^2 \tfrac{1}{2}C$, (*Trig., Art.* 43), we shall have, after reduction,

$$\sin^2 \tfrac{1}{2}C = \frac{c^2 + 2ad - d^2 - a^2}{4ad} = \frac{c^2 - (a - d)^2}{4ad}. \quad . \quad (6)$$

Factoring the final member of (6), we have,

$$\sin^2 \tfrac{1}{2}C = \frac{\tfrac{1}{2}(c + a - d)\,\tfrac{1}{2}(c + d - a)}{ad}. \quad . \quad . \quad (7)$$

But we have,

$$\tfrac{1}{2}(c + a - d) = \tfrac{1}{2}s - d, \text{ and } \tfrac{1}{2}(c + d - a) = \tfrac{1}{2}s - a.$$

Substituting these in (7), we have,

$$\sin^2 \tfrac{1}{2}C = \frac{(\tfrac{1}{2}s-d)\,(\tfrac{1}{2}s-a)}{ad}, \text{ or } \sin \tfrac{1}{2}C = \sqrt{\frac{(\tfrac{1}{2}s-d)\,(\tfrac{1}{2}s-a)}{ad}}. \quad (8)$$

Equations (5) and (8) give the required functions of $\tfrac{1}{2}C$.

To find the area of ACD.

10. Let the area of ACD be denoted by T. We have from formula (1), *Art.* 8,

$$2\,T = ad \sin C. \quad . \quad . \quad . \quad . \quad . \quad . \quad . \quad (9)$$

But, $\sin C = 2 \sin \tfrac{1}{2}C \cos \tfrac{1}{2}C$, (*Trig., Art.* 42);

whence, by substitution and reduction,

$$T = ad \sin \tfrac{1}{2}C \cos \tfrac{1}{2}C. \quad . \quad . \quad . \quad . \quad . \quad (11)$$

Substituting the values of $\sin \tfrac{1}{2}C$, and $\cos \tfrac{1}{2}C$, taken from (8) and (5), we have

$$T = \sqrt{\tfrac{1}{2}s\,(\tfrac{1}{2}s-a)\,(\tfrac{1}{2}s-c)\,(\tfrac{1}{2}s-d)}. \quad . \quad . \quad (12)$$

Hence, we may find the area of a plane triangle by the following

RULE.

Find half the sum of the three sides, and from it subtract each side separately; then find the continued product of the half sum and the three remainders; extract the square root of the product.

If we apply logarithms to formula (12), we have,

$$\log T = \tfrac{1}{2}[\log \tfrac{1}{2}s + \log(\tfrac{1}{2}s-a) + \log(\tfrac{1}{2}s-b) + \log(\tfrac{1}{2}s-c)].$$

EXAMPLES.

1. Given, $a = 75$ *ft.*, $b = 115$ *ft.*, and $c = 140$ *ft.*, to find the area of the triangle.

OPERATION.

$\frac{1}{2}s = 165$, $\frac{1}{2}s - a = 90$, $\frac{1}{2}s - b = 50$, and $\frac{1}{2}s - c = 25$.

log 165	. . .	2.217484
log 90	. . .	1.954243
log 50	. . .	1.698970
log 25	. . .	1.397940
	2)	7.268637

$T = \log^{-1} 3.634318 = 4{,}308.42$ *sq. ft.*, *Ans.*

2. Given, $a = 90$ *ft.*, $b = 120$ *ft.*, and $c = 150$ *ft.*, to find the area of the triangle. *Ans.* 5400 *sq. ft.*

3. The sides of a triangular field are 196 *rds.*, 201 *rds.*, and 102.76 *rds.*; how may acres in the field?

Ans. 61 *A.* 79.68 *sq. rds.*

4. Find the area of an equilateral triangle, each of whose sides is equal to 75 *ft.* *Ans.* 2435.7 *sq. ft.*

7°. *To find the area of a parallelogram whose base and altitude are given.*

RULE.

Multiply the base by the altitude, (P. 2, B. 4).

EXAMPLES.

1. What is the area of a parallelogram whose base is 25.5 *ft.*, and whose altitude is 17 *ft.*? *Ans.* 433.5 *sq. ft.*

2. The base of a parallelogram is equal to 34.8 *rds.*, and its altitude is 24 *rds.*; how many acres in the field?

Ans. 5.22 *A.*

3. How many square yards of carpeting will be required to carpet a room 42 *ft.* long and 19 *ft.* wide?

Ans. $88\frac{2}{3}$ *yds.*

If two adjacent sides and their included angle are given, we have the following

RULE.

Find the continued product of the two sides and the sine of their included angle.

4. The adjacent sides of a parallelogram are 20 *yds.*, and 37 *yds.*, and their included angle is 43°; what is the area of the parallelogram?

$$\log(area) = \log 20 + \log 37 + \log \sin 43° - 10;$$

$$\therefore \text{ area} = 504.7 \textit{ sq. yds., Ans.}$$

5. The adjacent sides of a parallelogram are 125.81, and 57.65; and their included angle is 57° 25′; what is the area of the parallelogram? *Ans.* 6111.4.

8°. *To find the area of a trapezoid.*

RULE.

Multiply the half sum of its parallel bases by the altitude, (P. 5, B. 4).

EXAMPLES.

1. The parallel sides of a trapezoid are 20.25 *ft.*, and 12.75 *ft.*, and the altitude is 9.75 *ft.*; what is the area of the trapezoid?

$$\tfrac{1}{2}(20.25 + 12.75) \times 9.75 = 160.88 \textit{ sq. ft., Ans.}$$

2. The parallel sides of a trapezoidal field measure 30 *rds.*, and 49 *rds.*, and the perpendicular distance between them is 61.6 *rds.*; how many acres does the field contain?

Ans. 15¼ *A.*, *nearly.*

3. The length of a board is 14.5 *ft.*; its breadth at the

wider end is 16 *in.*, and at the narrower end 11.6 *in.*; what is its superficial content? *Ans.* 16⅔ *sq. ft.*

9°. *To find the area of a regular polygon.*

11. Let ACDEF be a regular polygon, whose centre is O; let OP be its apothem, OA, OC, OD, OE, and OF, radii, dividing it into equal isosceles triangles. Then, its area is equal to the length of one side multiplied by the number of sides, *into* one-half of the apothem, (P. 4, B. 5).

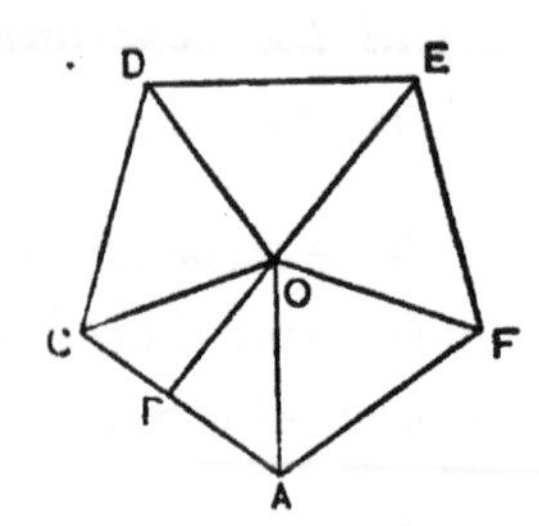

If we denote the area by A, the length of one side by s, the apothem by a, and the number of sides by n, we shall have

$$A = s \times n \times \tfrac{1}{2}a, \text{ or } A = \tfrac{1}{2}sna. \quad . \quad . \quad . \quad (1)$$

The angle AOP is equal to $\frac{360°}{2n}$, or to $\frac{180°}{n}$, (Art. 87, *Geom.*). From the right-angled triangle APO, we have,

$$PA = OP \tan AOP, \text{ or, } \tfrac{1}{2}s = a \tan \frac{180°}{n};$$

$$\therefore a = \tfrac{1}{2}s \cot \frac{180°}{n}. \quad . \quad . \quad . \quad . \quad . \quad . \quad (2)$$

Substituting in (1), and reducing, we have,

$$A = \tfrac{1}{4}ns^2 \cot \frac{180°}{n}. \quad . \quad . \quad . \quad . \quad . \quad . \quad (3)$$

Hence, the following

RULE.

Multiply one-fourth the square of one side by the cotangent of half the angle at the centre into the number of sides.

EXAMPLES.

1. Find the area of a regular hexagon, one of whose sides is 20 *yds.*

Applying logarithms to for nula (3), we have,

$$\log 4A = \log n + 2 \log s + \log \cot \frac{180^\circ}{n} - 10.$$

Substituting for n and s their values as given above, we have,

$$\log 4A = \log 6 + 2\log 20 + \log \cot 30^\circ - 10 = 3.618772.$$

$\therefore$ $4A = 4156.91$ *sq. yds.*, or, $A = 1039.23$ *sq. yds.*, *Ans.*

2. The side of a regular octagon is 25 *ft.*; what is its area? *Ans.* 3017.77 *sq. ft.*

3. The side of a regular pentagon is 22 *yds.*; what is its area? *Ans.* 832.71 *sq. yds.*

4. The side of a regular decagon is 15 *ft.*; what is its area? *Ans.* 1731.2 *sq. ft.*

Note.—To find the area of any polygon, we divide it into triangles, by diagonals drawn from any vertex; we then determine the length of each diagonal; we next compute the area of each triangle by the method explained in *Art.* 10; the sum of the areas of these triangles will be the required area.

10°. *To find the area of a circle whose radius is given.*

RULE.

Multiply the square of the radius by 3.1416, (P. 11, *Cor.* 3, B. 5).

EXAMPLES.

1. What is the area of a circle whose radius is 30 *in.*?

$$A = (30)^2 \times 3.1416 = 2827.44 \text{ sq. in., Ans.}$$

2. Find the area of a circle whose radius is 4.75 *rds.*

Ans. 70.88 *sq. rds.*

3. The radius of a circle is 8.25 *yds.*; what is its area?

Ans. 213.83 *sq. yds.*

4. What is the area of a circular pond, whose diameter is 12.75 *rds.*? *Ans.* 127.68 *sq. rds.*

5. The diameter of a circular park is 77.75 *rds.*; how many acres does it contain? *Ans.* 29.67 *A.*

To find the area of a circular sector, we first find the area of the circle to which it belongs; we then multiply this area by the quotient obtained, by dividing the angle of the sector by 360°.

6. What is the area of a sector of 60°, in a circle whose radius is 42 *yds.*?

Circle $= 5541.78$ *sq. yds.*, $\therefore$ sector $= \frac{1}{6} \times 5541.78$, or 923.63 *sq. yds.*, *Ans.*

7. The angle of a sector is 45°, and its radius 4.75 *ft.*; what is its area? *Ans.* 8.86 *sq. ft.*

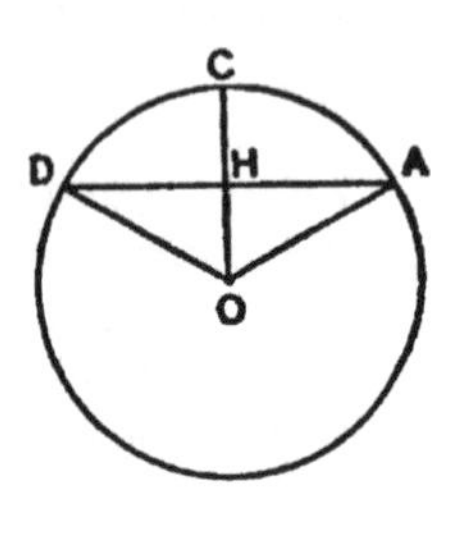

To find the area of a circular segment ACD, we first find the area of the sector AODC; we then subtract from it the area of the triangle AOD; the difference is the required area.

8. What is the area of a circular segment corresponding to an angle of 120° at the centre, the radius of the circle being 20 *ft.*?

Ans. 245.68 *sq. ft.*

11°. *To find the area of the lateral surface of a right prism, or of a cylinder.*

RULE.

Multiply the perimeter of either base, by the altitude, (P. 3, B. 7, and P 1, *Cor.* 2, B. 8).

EXAMPLES.

1. The perimeter of the base of a right prism is 20 *ft.* and its altitude is 10 *ft.*; what is the area of its lateral surface? *Ans.* 200 *sq. ft.*

2. The base of a right prism is a regular pentagon, one of whose sides is 25 *in.* and the altitude of the prism is 12 *ft.*; what is the entire surface of the prism, including the bases? *Ans.* 139.93 *sq. ft.*

3. The radius of the base of a right cylinder is 10 *in.*, and the altitude of the cylinder is 6 *in.*; what is its entire surface? *Ans.* 1005.31 *sq. in.*

12°. *To find the area of the lateral surface of a right pyramid, or of a cone.*

RULE.

Multiply the perimeter of the base by one-half the slant height, (P. 4, B. 7, and P. 2, *Cor.* 4, B. 8).

EXAMPLES.

1. The perimeter of the base of a right pyramid is 40 *ft.*, and the slant height is 4 *ft.*; what is the area of its lateral surface? *Ans.* 80 *sq. ft.*

2. The radius of the base of a right cone is 3 *ft.*, and the altitude of the cone is 4 *ft.*; what is its lateral surface? *Ans.* 47.12 *sq. ft.*

3. The radius of the base of a cone is 2 *ft.*, and the altitude of the cone is 6 *ft.*; what is the area of its entire surface, including the base? *Ans.* 52.31 *sq. ft.*

To find the lateral surface of a frustum of a right pyramid, or of a cone, we multiply the half sum of the perimeters of the upper and lower bases, by the slant height, (P. 4, *Cor.*, B. 7, and P. 3, *Cor.* 2, B. 8.)

4. The perimeter of the upper base of a frustum of a right

pyramid is 18 *ft.*, the perimeter of its lower base is 28 *ft.*, and its slant height is 6 *ft.*; what is the area of its lateral surface? *Ans.* 138 *sq. ft.*

5. The radius of the upper base of a conic frustum is 3 *ft.*, the radius of the lower base is 4 *ft.*, and the slant height is 8 *ft.*; what is its entire surface, including both bases?
Ans. 254.46 *sq. ft.*

13°. *To find the area of the surface of a sphere.*

RULE.

Multiply 4 times the square of the radius by 3.1416, (P. 7, *Cor.* 1, B. 8.)

EXAMPLES.

1. What is the area of the surface of a sphere whose radius is 20 *in.*?

$$4 \times 20^2 \times 3.1416 = 5026.56 \text{ sq. in., Ans.}$$

2. What is the area of the surface of a sphere whose diameter is 5 *ft.*? *Ans.* 78.54 *sq. ft.*

To find the area of a zone, we multiply twice the radius, by the altitude of the zone, and that result by π, (P. 7, *Cor.* 3, B. 8.)

3. Find the area of a zone, on a sphere, whose radius is 7 *ft.*, the altitude of the zone being 3 *ft.* *Ans.* 131.95 *sq. ft.*

4. The radius of the earth, considered as a sphere, is equal to 3,956 miles, the latitude of the tropic of cancer is 23° 30′, and that of the arctic circle is 66° 30′; what is the area of the northern temperate zone? *Ans.* 50,966,183.59 *sq. mi.*

The altitude is equal to 3956 (sin 66° 30′ − sin 23° 30′); the natural sines of 66° 30′ and 22° 30′ are .91706 and .39875, hence the altitude is 2050.43 miles.

14°. *To find the area of a spherical triangle.*

RULE.

Diminish the sum of its angles by 180°, and divide the result by 90°; this will give the spherical excess. Multiply one-half the square of the radius by 3.1416; this will give the area of the tri-rectangular rectangle. Then, multiply the area of the tri-rectangular triangle by the spherical excess, (P. 16, B. 9).

EXAMPLES.

1. The angles of a spherical triangle are 85°, 105°, and 50°; the radius of the sphere on which the triangle is situated is 22.25 *ft.*; what is the area of the triangle?

Ans. 518.43 *sq. ft.*

The spherical excess is $\frac{2}{3}$; the area of the tri-rectangular triangle is 777.64 *sq. ft.*

2. The angles of a spherical triangle are 140°, 92°, and 68°; the radius of the sphere is 30 *ft.*; what is the area of the triangle? *Ans.* 1884.96 *sq. ft.*

III. MENSURATION OF VOLUMES.

Volumes considered.

12. The volumes treated of in elementary Geometry are the prism, and the cylinder; the pyramid, the cone, and their frustums; and the sphere.

15°. *To find the volume of a prism, or of a cylinder.*

RULE.

Multiply the area of its base by its altitude, (P. 14, B. 7, and P. 1, *Cor.* 3, B. 8.)

EXAMPLES.

1. Find the volume of a block of stone whose length is $3\frac{1}{6}$ *ft.*, whose breadth is $2\frac{2}{3}$ *ft.*, and whose altitude is $2\frac{1}{2}$ *ft.*

Base $= 3\frac{1}{6} \times 2\frac{2}{3} = 8\frac{4}{9}$ *sq. ft.*; $\therefore 8\frac{4}{9} \times 2\frac{1}{2} = 21\frac{1}{9}$ *cu. ft.*, *Ans.*

2. The sides of the base of a triangular prism are 3 *yds.*, 4 *yds.*, and 5 *yds.*; and the altitude of the prism is 6 *yds.*; what is the volume of the prism? *Ans.* 36 *cu. yds.*

3. What is the volume of a cylinder, the radius of whose base is 4.5 *yds.*, and whose altitude is 11 *yds.*?

Ans. 699.79 *cu. yds.*

4. The radius of the base of cylindrical log is 9 *in.*, and its length is 21 *ft.*; what are its cubic contents?

Ans. 37.1 *cu. ft.*

5. What is the content of a cylindrical vat 15 *ft.* in diameter, and 12 *ft.* deep? *Ans.* 2120.58 *cu. ft.*

16°. *To find the volume of a right pyramid, or of a cone.*

RULE.

Multiply the area of its base by one-third of its altitude, (P. 17, B. 7, and P. 2, *Cor.* 5, B. 8).

EXAMPLES.

1. The area of the base of a pyramid is 74.3 *sq. yds.*, and its altitude is 21 *yds.*; what is its volume?

Ans. 520.1 *cu. yds.*

2. The base of a right pyramid is a regular pentagon, each side of which is equal to 10 *ft.*; and its altitude is 18 *ft.*; what is its volume? *Ans.* 1032.29 *cu. ft.*

3. The radius of the base of a right cone is 8.5 *yds.*, and its altitude is 15.75 *yds.*; what is its content?

Ans. 1191.65 *cu. yds.*

4. The radius of the base of a cone is 3 *ft.*, and its slant height is 5 *ft.*; what is its volume? *Ans.* 37.7 *cu. ft.*

17°. *To find the volume of a frustum of a right pyramid, or of a frustum of a cone.*

RULE.

Find the sum of the upper base, the lower base, and a mean proportional between the two bases; then multiply this sum by one-third of the altitude, (P. 18, *Cor.* 2, B. 7, and P. 3, *Cor.* 3, B. 8.)

EXAMPLES.

1. The upper base of a pyramidal frustum is 7.2 *sq. yds.*, the lower base is 12.5 *sq. yds.*, and the altitude of the frustum is 6 *yds.*; what is its cubic content?

Ans. $(7.2 + 12.5 + \sqrt{7.2 \times 12.5}) \times 2 = 58.37$ *cu. yds.*

2. The bases of a pyramidal frustum are regular hexagons; one side of the upper base is 2 *ft.*, one side of the lower base is 3 *ft.*, and the altitude of the frustum is 7 *ft.*; what is the volume of the frustum? *Ans.* 115.178 *cu. ft.*

3. The radius of the upper base of a conic frustum is 3.5 *ft.*, the radius of its lower base is 14 *ft.*, and the altitude of the frustum is 33 *ft.*; what is its volume?

Ans. 8,889.92 *cu. ft.*

4. The radius of the upper base of a conic frustum is 2.5 *ft.*, the radius of its lower base is 3.5 *ft.*, and its altitude is 6 *ft.*; what is the volume of the frustum? *Ans.* 171.22 *cu. ft.*

18°. *To find the volume of a sphere.*

RULE

Multiply four-thirds of the cube of the radius by 3.1416, (P. 8, *Cor.* 1, B. 8.)

The factor $\frac{4}{3}\pi$ is equal to 4.1888; $\therefore$ $V = 4.1888R^3$.

EXAMPLES.

1. Find the volume of a sphere whose radius is 3 *ft.*

$$V = 4.1888 \times 27 = 113.1 \textit{ cu. ft., nearly, Ans.}$$

2. What is the volume of a sphere whose radius is 2 *yds.?*

Ans. 33.51 *cu. yds.*

To find the volume of a spherical sector, we first find the area of the zone which forms its base; we then multiply this by one-third the radius.

3. The altitude of a zone, on a sphere whose radius is 10 *ft.*, is 3 *ft.*; what is the volume of the corresponding spherical sector?

Ans. 628.32 *cu. ft.*

The area of the zone is found by Rule 13°.

4. The area of a zone, on the surface of a sphere whose radius is 11 *yds.*, is equal to 276 *sq. yds.*; what is the volume of the corresponding spherical sector?

Ans. 1012 *cu. yds.*

5. What is the radius of a sphere whose volume is 417 *cu. yds.?*

Ans. 4.63 *yds.*

19°. *To find the volume of a wedge.*

Definitions.

3. A **wedge** is a polyedron, bounded by a parallelogram, two trapezoids, and two triangles; as, ACDE–PQ.

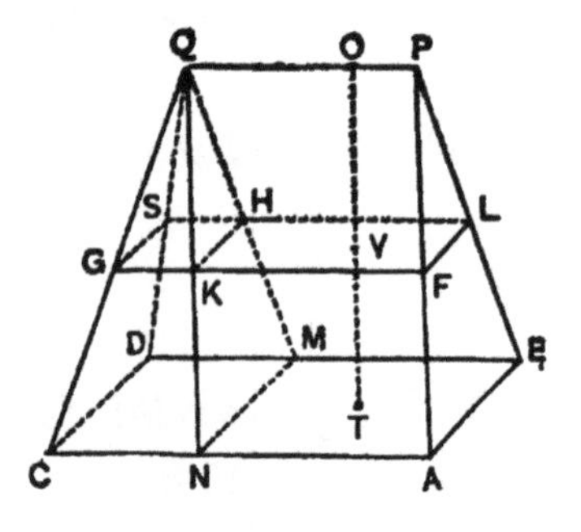

The parallelogram AD, is called the **back**; the trapezoids AQ and EQ, are called **faces**; the triangles APE and CQD, are called **ends**; the line PQ, in which the faces meet and which is parallel

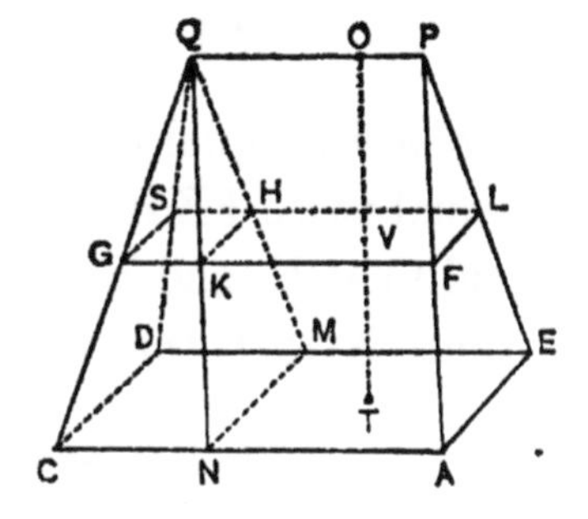

to the back, is called the **edge**; and a line OT, drawn from any point of the edge, perpendicular to the back, is called the **altitude of the wedge.**

14. If a plane is passed through V, the middle of the altitude, and parallel to the back of the wedge, its intersection with the wedge is called the **middle section** of the wedge: thus, FGSL is the middle section of the wedge ACDE–PQ.

15. A **frustum of a wedge** is the part included between the back and any secant plane parallel to the back; thus, ACDE–FGSL is a frustum of the wedge ACDE–PQ.

Expression for the volume.

16. Let us first suppose that the length of the edge is less than the length of the back. Through the extremity Q, of the edge PQ, pass a plane QNM parallel to PAE; this plane will divide the wedge into a triangular prism QNM–A, and a pyramid, NCDM–Q. Through the middle of the altitude, pass a plane parallel to ACDE; it will intersect the prism and the pyramid in the *middle sections* FH and KS.

Denote the area of ANME by b, of NCDM by b', of FKHL by m, and of KGSH by m'; also denote the altitude by $6h$.

The triangular prism ANME–P, is the half of a parallelopipedon whose edges are ME, MN, and MQ; hence, the volume of this prism is equal to ANME $\times$ $\frac{1}{2}$OT, that is, to $b \times 3h$. The altitude of the pyramid is equal to OT; hence, its volume is equal to NCDM $\times$ $\frac{1}{3}$OT, that is, to $b' \times 2h$. Denoting the volume of the wedge by V, and the area of its back by B, we have,

$$V = h\,[3b+2b'] = h\,[(b+b')+2b+b'] = h\,[B+2b+b']. \quad (1)$$

Now, KF is equal to AN, and KH to $\frac{1}{2}$NM; hence, ANME is equal twice FKHL, that is, b is equal to $2m$; also, GK is equal to $\frac{1}{2}$CN, and HK to $\frac{1}{2}$NM; hence, NCDM is equal to 4 times KGSH, that is, b' is equal to $4m'$; hence,

$$2b + b' = 4m + 4m' = 4(m + m');$$

denoting the sum of m and m', that is, the middle section of the wedge, by M, and substituting in (1), we have,

$$V = h\,[B + 4M]. \quad . \quad . \quad . \quad . \quad . \quad . \quad (2)$$

If the length of the edge is greater than the length of the back, the volume of the wedge is equal to the difference of the volumes of the prism and the pyramid, and the expression for the volume is of the same form as before.

We may therefore find the volume of a wedge by the following

RULE.

To the area of the back add four times the middle section, and multiply the sum by one-sixth of the altitude.

EXAMPLES.

1. The length of the back of a wedge is 30 *ft.*, the breadth of the back is 20 *ft.*, the length of the edge is 40 *ft.*, and the altitude is 18 *ft.*; what is its volume? *Ans.* 6000 *cu. ft.*

The length of the middle section is $\frac{1}{2}$(30 *ft* + 40 *ft.*), or to 35 *ft.*, and the breadth of that section $\frac{1}{2}$(20) *ft.*, or 10 *ft.*

2. The back of a wedge is 18 *ft.* long, and 9 *ft.* wide; the edge is 20 *ft.* long; and the altitude is 6 *ft.*: what is its volume? *Ans.* 504 *cu. ft.*

20°. *To find the volume of a prismoid.*

Definitions.

17. A **prismoid** is a frustum of a wedge, (*Art.* 15); thus, ACDE–PQRT is a prismoid.

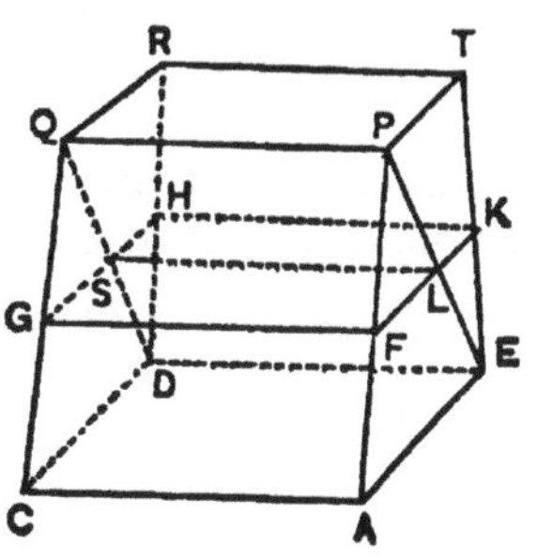

The parallel faces, ACDE and PQRT, are called **bases**; the perpendicular distance between the bases is called the **altitude** of the prismoid; and the section FGHK, through the middle of the altitude and parallel to the bases, is the **middle section** of the prismoid.

Expression for the Volume.

18. Let the volume of the prismoid be denoted by V, its lower base by B, its upper base by B′, its middle section by M, and its altitude by $6h$.

Pass a plane through PQ and DE; this will divide the prismoid into two wedges, having the same altitude as the prismoid. The back of the first wedge is equal to B, and its middle section is FGSL; the back of the second wedge is B′ and its middle section is LSHK.

Because the prismoid is equal to the sum of the two wedges, we have, (*Art.* 16),

$$V = h[B + 4FGSL] + h[B' + 4LSHK] = h[B + B' + 4M]. \quad (1)$$

We may therefore find the volume of a prismoid by the following

RULE.

Find the sum of the end sections together with four times the middle section, and multiply the result by one-sixth of the altitude.

EXAMPLES.

1. The length of the lower base of a prismoid is 24 *ft.*, its breadth 18 *ft.*; the length of the upper base is 16 *ft.*, and its breadth 14 *ft.*; and the altitude of the prismoid is 27 *ft.*: what is its volume? *Ans.* 8,712 *cu. ft.*

The length of the middle section is $\frac{1}{2}$(24 *ft.* + 16 *ft.*), or 20 *ft.*, and the breadth of that section is $\frac{1}{2}$(18 *ft.* + 14 *ft.*) = 16 *ft.*

2. The end sections of a volume of earth are respectively 340 *sq. ft.*, and 590 *sq. ft.*; the middle section is 475 *sq. ft.*; and the length of the section is 100 *ft.*: what is the volume of earth? *Ans.* 47,166$\frac{2}{3}$ *cu. ft.*

3. What is the volume of a stick of hewn timber, whose ends are 30 *in.* by 27 *in.*, and 24 *in.* by 18 *in.*, its length being 24 *ft.*? *Ans.* 102 *cu. ft.*

Formula (1), *Art.* 18, is called the *prismoidal formula*; it is used by engineers in computing the volumes of earth-work in railroad cuttings and embankments.

It may be shown that the prismoidal formula holds good for the cone, the cylinder, the pyramid, the frustum of a cone, the frustum of a pyramid, and for the sphere.

In the cone, one of the end sections is 0; in the sphere, both of the end sections are 0 and the middle section is a great circle.

Let the student show that the formula gives the correct volume in each of the cases just named.

A TABLE

OF

LOGARITHMS OF NUMBERS

From 1 to 10,000

N.	Log.	N.	Log.	N.	Log.	N.	Log.
1	0·000000	26	1·414973	51	1·707570	76	1·880814
2	0·301030	27	1·431364	52	1·716003	77	1·886491
3	0·477121	28	1·447158	53	1·724276	78	1·892085
4	0·602060	29	1·462398	54	1·732394	79	1·897627
5	0·698970	30	1·477121	55	1·740363	80	1·903090
6	0·778151	31	1·491362	56	1·748188	81	1·908485
7	0·845098	32	1·505150	57	1·755875	82	1·913814
8	0·903090	33	1·518514	58	1·763428	83	1·919078
9	0·954243	34	1·531479	59	1·770852	84	1·924279
10	1·000000	35	1·544068	60	1·778151	85	1·929419
11	1·041393	36	1·556303	61	1·785330	86	1·934498
12	1·079181	37	1·568202	62	1·792392	87	1·939519
13	1·113943	38	1·579784	63	1·799341	88	1·944483
14	1·146128	39	1·591065	64	1·806180	89	1·949390
15	1·176091	40	1·602060	65	1·812913	90	1·954243
16	1·204120	41	1·612784	66	1·819544	91	1·959041
17	1·230449	42	1·623249	67	1·826075	92	1·963788
18	1·255273	43	1·633468	68	1·832509	93	1·968483
19	1·278754	44	1·643453	69	1·838849	94	1·973128
20	1·301030	45	1·653213	70	1·845098	95	1·977724
21	1·322219	46	1·662758	71	1·851258	96	1·982271
22	1·342423	47	1·672098	72	1·857333	97	1·986772
23	1·361728	48	1·681241	73	1·863323	98	1·991226
24	1·380211	49	1·690196	74	1·869232	99	1·995635
25	1·397940	50	1·698970	75	1·875061	100	2·000000

Remark.—In the following table, in the nine right-hand columns of each page, where the first or leading figures change from 9's to 0's, points or dots are introduced instead of the 0's, to catch the eye, and to indicate that from thence the two figures of the Logarithm to be taken from the second column, stand in the next line below.

N.	0	1	2	3	4	5	6	7	8	9	D.
100	000000	0434	0868	1301	1734	2166	2598	3029	3461	3891	432
101	4321	4751	5181	5609	6038	6466	6894	7321	7748	8174	428
102	8600	9026	9451	9876	•300	•724	1147	1570	1993	2415	424
103	012837	3259	3680	4100	4521	4940	5360	5779	6197	6616	419
104	7033	7451	7868	8284	8700	9116	9532	9947	•361	•775	416
105	021189	1603	2016	2428	2841	3252	3664	4075	4486	4896	411
106	5306	5715	6125	6533	6942	7350	7757	8164	8571	8978	408
107	9384	9789	•195	•600	1004	1408	1812	2216	2619	3021	404
108	033424	3826	4227	4628	5029	5430	5830	6230	6629	7028	400
109	7426	7825	8223	8620	9017	9414	9811	•207	•602	•998	396
110	041393	1787	2182	2576	2969	3362	3755	4148	4540	4932	393
111	5323	5714	6105	6495	6885	7275	7664	8053	8442	8830	389
112	9218	9606	9993	•380	•766	1153	1538	1924	2309	2694	386
113	053078	3463	3846	4230	4613	4996	5378	5760	6142	6524	382
114	6905	7286	7666	8046	8426	8805	9185	9563	9942	•320	379
115	060698	1075	1452	1829	2206	2582	2958	3333	3709	4083	376
116	4458	4832	5206	5580	5953	6326	6699	7071	7443	7815	372
117	8186	8557	8928	9298	9668	••38	•407	•776	1145	1514	369
118	071882	2250	2617	2985	3352	3718	4085	4451	4816	5182	366
119	5547	5912	6276	6640	7004	7368	7731	8094	8457	8819	363
120	079181	9543	9904	•266	•626	•987	1347	1707	2067	2426	360
121	082785	3144	3503	3861	4219	4576	4934	5291	5647	6004	357
122	6360	6716	7071	7426	7781	8136	8490	8845	9198	9552	355
123	9905	•258	•611	•963	1315	1667	2018	2370	2721	3071	351
124	093422	3772	4122	4471	4820	5169	5518	5866	6215	6562	349
125	6910	7257	7604	7951	8298	8644	8990	9335	9681	••26	346
126	100371	0715	1059	1403	1747	2091	2434	2777	3119	3462	343
127	3804	4146	4487	4828	5169	5510	5851	6191	6531	6871	340
128	7210	7549	7888	8227	8565	8903	9241	9579	9916	•253	338
129	110590	0926	1263	1599	1934	2270	2605	2940	3275	3609	335
130	113943	4277	4611	4944	5278	5611	5943	6276	6608	6940	333
131	7271	7603	7934	8265	8595	8926	9256	9586	9915	•245	330
132	120574	0903	1231	1560	1888	2216	2544	2871	3198	3525	328
133	3852	4178	4504	4830	5156	5481	5806	6131	6456	6781	325
134	7105	7429	7753	8076	8399	8722	9045	9368	9690	••12	323
135	130334	0655	0977	1298	1619	1939	2260	2580	2900	3219	321
136	3539	3858	4177	4496	4814	5133	5451	5769	6086	6403	318
137	6721	7037	7354	7671	7987	8303	8618	8934	9249	9564	315
138	9879	•194	•508	•822	1136	1450	1763	2076	2389	2702	314
139	143015	3327	3639	3951	4263	4574	4885	5196	5507	5818	311
140	146128	6438	6748	7058	7367	7676	7985	8294	8603	8911	309
141	9219	9527	9835	•142	•449	•756	1063	1370	1676	1982	307
142	152288	2594	2900	3205	3510	3815	4120	4424	4728	5032	305
143	5336	5640	5943	6246	6549	6852	7154	7457	7759	8061	303
144	8362	8664	8965	9266	9567	9868	•168	•469	•769	1068	301
145	161368	1667	1967	2266	2564	2863	3161	3460	3758	4055	299
146	4353	4650	4947	5244	5541	5838	6134	6430	6726	7022	297
147	7317	7613	7908	8203	8497	8792	9086	9380	9674	9968	295
148	170262	0555	0848	1141	1434	1726	2019	2311	2603	2895	293
149	3186	3478	3769	4060	4351	4641	4932	5222	5512	5802	291
150	176091	6381	6670	6959	7248	7536	7825	8113	8401	8689	289
151	8977	9264	9552	9839	•126	•413	•699	•985	1272	1558	287
152	181844	2129	2415	2700	2985	3270	3555	3839	4123	4407	285
153	4691	4975	5259	5542	5825	6108	6391	6674	6956	7239	283
154	7521	7803	8084	8366	8647	8928	9209	9490	9771	••51	281
155	190332	0612	0892	1171	1451	1730	2010	2289	2567	2846	279
156	3125	3403	3681	3959	4237	4514	4792	5069	5346	5623	278
157	5899	6176	6453	6729	7005	7281	7556	7832	8107	8382	276
158	8657	8932	9206	9481	9755	••29	•303	•577	•850	1124	274
159	201397	1670	1943	2216	2488	2761	3033	3305	3577	3848	272
N.	0	1	2	3	4	5	6	7	8	9	D.

N.	0	1	2	3	4	5	6	7	8	9	D.
160	204120	4391	4663	4934	5204	5475	5746	6016	6286	6556	271
161	6826	7096	7365	7634	7904	8173	8441	8710	8979	9247	269
162	9515	9783	••51	•319	•586	•853	1121	1388	1654	1921	267
163	212188	2454	2720	2986	3252	3518	3783	4049	4314	4579	266
164	4844	5109	5373	5638	5902	6166	6430	6694	6957	7221	264
165	7484	7747	8010	8273	8536	8798	9060	9323	9585	9846	262
166	220108	0370	0631	0892	1153	1414	1675	1936	2196	2456	261
167	2716	2976	3236	3496	3755	4015	4274	4533	4792	5051	259
168	5309	5568	5826	6084	6342	6600	6858	7115	7372	7630	258
169	7887	8144	8400	8657	8913	9170	9426	9682	9938	•193	256
170	230449	0704	0960	1215	1470	1724	1979	2234	2488	2742	254
171	2996	3250	3504	3757	4011	4264	4517	4770	5023	5276	253
172	5528	5781	6033	6285	6537	6789	7041	7292	7544	7795	252
173	8046	8297	8548	8799	9049	9299	9550	9800	••50	•300	250
174	240549	0799	1048	1297	1546	1795	2044	2293	2541	2790	249
175	3038	3286	3534	3782	4030	4277	4525	4772	5019	5266	248
176	5513	5759	6006	6252	6499	6745	6991	7237	7482	7728	246
177	7973	8219	8464	8709	8954	9198	9443	9687	9932	•176	245
178	250420	0664	0908	1151	1395	1638	1881	2125	2368	2610	243
179	2853	3096	3338	3580	3822	4064	4306	4548	4790	5031	242
180	255273	5514	5755	5996	6237	6477	6718	6958	7198	7439	241
181	7679	7918	8158	8398	8637	8877	9116	9355	9594	9833	239
182	260071	0310	0548	0787	1025	1263	1501	1739	1976	2214	238
183	2451	2688	2925	3162	3399	3636	3873	4109	4346	4582	237
184	4818	5054	5290	5525	5761	5996	6232	6467	6702	6937	235
185	7172	7406	7641	7875	8110	8344	8578	8812	9046	9279	234
186	9513	9746	9980	•213	•446	•679	•912	1144	1377	1609	233
187	271842	2074	2306	2538	2770	3001	3233	3464	3696	3927	232
188	4158	4389	4620	4850	5081	5311	5542	5772	6002	6232	230
189	6462	6692	6921	7151	7380	7609	7838	8067	8296	8525	229
190	278754	8982	9211	9439	9667	9895	•123	•351	•578	•800	228
191	281033	1261	1488	1715	1942	2169	2396	2622	2849	3075	227
192	3301	3527	3753	3979	4205	4431	4656	4882	5107	5332	226
193	5557	5782	6007	6232	6456	6681	6905	7130	7354	7578	225
194	7802	8026	8249	8473	8696	8920	9143	9366	9589	9812	223
195	290035	0257	0480	0702	0925	1147	1369	1591	1813	2034	222
196	2256	2478	2699	2920	3141	3363	3584	3804	4025	4246	221
197	4466	4687	4907	5127	5347	5567	5787	6007	6226	6446	220
198	6665	6884	7104	7323	7542	7761	7979	8198	8416	8635	219
199	8853	9071	9289	9507	9725	9943	•161	•378	•595	•813	218
200	301030	1247	1464	1681	1898	2114	2331	2547	2764	2980	217
201	3196	3412	3628	3844	4059	4275	4491	4706	4921	5136	216
202	5351	5566	5781	5996	6211	6425	6639	6854	7068	7282	215
203	7496	7710	7924	8137	8351	8564	8778	8991	9204	9417	213
204	9630	9843	••56	•268	•481	•693	•906	1118	1330	1542	212
205	311754	1966	2177	2389	2600	2812	3023	3234	3445	3656	211
206	3867	4078	4289	4499	4710	4920	5130	5340	5551	5760	210
207	5970	6180	6390	6599	6809	7018	7227	7436	7646	7854	209
208	8063	8272	8481	8689	8898	9106	9314	9522	9730	9938	208
209	320146	0354	0562	0769	0977	1184	1391	1598	1805	2012	207
210	322219	2426	2633	2839	3046	3252	3458	3665	3871	4077	206
211	4282	4488	4694	4899	5105	5310	5516	5721	5926	6131	205
212	6336	6541	6745	6950	7155	7359	7563	7767	7972	8176	204
213	8380	8583	8787	8991	9194	9398	9601	9805	•••8	•211	203
214	330414	0017	0819	1022	1225	1427	1630	1832	2034	2236	202
215	2438	2640	2842	3044	3246	3447	3649	3850	4051	4253	202
216	4454	4655	4856	5057	5257	5458	5658	5859	6059	6260	201
217	6460	6660	6860	7060	7260	7459	7659	7858	8058	8257	200
218	8456	8656	8855	9054	9253	9451	9650	9849	••47	•246	199
219	340444	c642	0841	1039	1237	1435	1632	1830	2028	2225	198
N.	0	1	2	3	4	5	6	7	8	9	D.

N.	0	1	2	3	4	5	6	7	8	9	D.
220	342423	2620	2817	3014	3212	3409	3606	3802	3999	4196	197
221	4392	4589	4785	4981	5178	5374	5570	5766	5962	6157	196
222	6353	6549	6744	6939	7135	7330	7525	7720	7915	8110	195
223	8305	8500	8694	8889	9083	9278	9472	9666	9860	••54	194
224	350248	0442	0636	0829	1023	1216	1410	1603	1796	1989	193
225	2183	2375	2568	2761	2954	3147	3339	3532	3724	3916	193
226	4108	4301	4493	4685	4876	5068	5260	5452	5643	5834	192
227	6026	6217	6408	6599	6790	6981	7172	7363	7554	7744	191
228	7935	8125	8316	8506	8696	8886	9076	9266	9456	9646	190
229	9835	••25	•215	•404	•593	•783	•972	1161	1350	1539	189
230	361728	1917	2105	2294	2482	2671	2859	3048	3236	3424	188
231	3612	3800	3988	4176	4363	4551	4739	4926	5113	5301	188
232	5488	5675	5862	6049	6236	6423	6610	6796	6983	7169	187
233	7356	7542	7729	7915	8101	8287	8473	8659	8845	9030	186
234	9216	9401	9587	9772	9958	•143	•328	•513	•698	•883	185
235	371068	1253	1437	1622	1806	1991	2175	2360	2544	2728	184
236	2912	3096	3280	3464	3647	3831	4015	4198	4382	4565	184
237	4748	4932	5115	5298	5481	5664	5846	6029	6212	6394	183
238	6577	6759	6942	7124	7306	7488	7670	7852	8034	8216	182
239	8398	8580	8761	8943	9124	9306	9487	9668	9849	••30	181
240	380211	0392	0573	0754	0934	1115	1296	1476	1656	1837	181
241	2017	2197	2377	2557	2737	2917	3097	3277	3456	3636	180
242	3815	3995	4174	4353	4533	4712	4891	5070	5249	5428	179
243	5606	5785	5964	6142	6321	6499	6677	6856	7034	7212	178
244	7390	7568	7746	7923	8101	8279	8456	8634	8811	8989	178
245	9166	9343	9520	9698	9875	••51	•228	•405	•582	•759	177
246	390935	1112	1288	1464	1641	1817	1993	2169	2345	2521	176
247	2697	2873	3048	3224	3400	3575	3751	3926	4101	4277	176
248	4452	4627	4802	4977	5152	5326	5501	5676	5850	6025	175
249	6199	6374	6548	6722	6896	7071	7245	7419	7592	7766	174
250	397940	8114	8287	8461	8634	8808	8981	9154	9328	9501	173
251	9674	9847	••20	•192	•365	•538	•711	•883	1056	1228	173
252	401401	1573	1745	1917	2089	2261	2433	2605	2777	2949	172
253	3121	3292	3464	3635	3807	3978	4149	4320	4492	4663	171
254	4834	5005	5176	5346	5517	5688	5858	6029	6199	6370	171
255	6540	6710	6881	7051	7221	7391	7561	7731	7901	8070	170
256	8240	8410	8579	8749	8918	9087	9257	9426	9595	9764	169
257	9933	•102	•271	•440	•609	•777	•946	1114	1283	1451	169
258	411620	1788	1956	2124	2293	2461	2629	2796	2964	3132	168
259	3300	3467	3635	3803	3970	4137	4305	4472	4639	4806	167
260	414973	5140	5307	5474	5641	5808	5974	6141	6308	6474	167
261	6641	6807	6973	7139	7306	7472	7638	7804	7970	8135	166
262	8301	8467	8633	8798	8964	9129	9295	9460	9625	9791	165
263	9956	•121	•286	•451	•616	•781	•945	1110	1275	1439	165
264	421604	1788	1933	2097	2261	2426	2590	2754	2918	3082	164
265	3246	3410	3574	3737	3901	4065	4228	4392	4555	4718	164
266	4882	5045	5208	5371	5534	5697	5860	6023	6186	6349	163
267	6511	6674	6836	6999	7161	7324	7486	7648	7811	7973	162
268	8135	8297	8459	8621	8783	8944	9106	9268	9429	9591	162
269	7752	9914	••75	•236	•398	•559	•720	•881	1042	1203	161
270	431364	1525	1685	1846	2007	2167	2328	2488	2649	2809	161
271	2969	3130	3290	3450	3610	3770	3930	4090	4249	4409	160
272	4569	4729	4888	5048	5207	5367	5526	5685	5844	6004	159
273	6163	6322	6481	6640	6798	6957	7116	7275	7433	7592	159
274	7751	7909	8067	8226	8384	8542	8701	8859	9017	9175	158
275	9333	9491	9648	9806	9964	•122	•279	•437	•594	•752	158
276	440909	1066	1224	1381	1538	1695	1852	2009	2166	2323	157
277	2480	2637	2793	2950	3106	3263	3419	3576	3732	3889	157
278	4045	4201	4357	4513	4669	4825	4981	5137	5293	5449	156
279	5604	5760	5915	6071	6226	6382	6537	6692	6848	7003	155
N.	0	1	2	3	4	5	6	7	8	9	D.

N.	0	1	2	3	4	5	6	7	8	9	D.
280	447158	7313	7468	7623	7778	7933	8088	8242	8397	8552	155
281	8706	8861	9015	9170	9324	9478	9633	9787	9941	••95	154
282	450249	0403	0557	0711	0865	1018	1172	1326	1479	1633	154
283	1786	1940	2093	2247	2400	2553	2706	2859	3012	3165	153
284	3318	3471	3624	3777	3930	4082	4235	4387	4540	4692	153
285	4845	4997	5150	5302	5454	5606	5758	5910	6062	6214	152
286	6366	6518	6670	6821	6973	7125	7276	7428	7579	7731	152
287	7882	8033	8184	8336	8487	8638	8789	8940	9091	9242	151
288	9392	9543	9694	9845	9995	•146	•296	•447	•597	•748	151
289	460898	1048	1198	1348	1499	1649	1799	1948	2098	2248	150
290	462398	2548	2697	2847	2997	3146	3296	3445	3594	3744	150
291	3893	4042	4191	4340	4490	4639	4788	4936	5085	5234	149
292	5383	5532	5680	5829	5977	6126	6274	6423	6571	6719	149
293	6868	7016	7164	7312	7460	7608	7756	7904	8052	8200	148
294	8347	8495	8643	8790	8938	9085	9233	9380	9527	9675	148
295	9822	9969	•116	•263	•410	•557	•704	•851	•998	1145	147
296	471292	1438	1585	1732	1878	2025	2171	2318	2464	2610	146
297	2756	2903	3049	3195	3341	3487	3633	3779	3925	4071	146
298	4216	4362	4508	4653	4799	4944	5090	5235	5381	5526	146
299	5671	5816	5962	6107	6252	6397	6542	6687	6832	6976	145
300	477121	7266	7411	7555	7700	7844	7989	8133	8278	8422	145
301	8566	8711	8855	8999	9143	9287	9431	9575	9719	9863	144
302	480007	0151	0294	0438	0582	0725	0869	1012	1156	1299	144
303	1443	1586	1729	1872	2016	2159	2302	2445	2588	2731	143
304	2874	3016	3159	3302	3445	3587	3730	3872	4015	4157	143
305	4300	4442	4585	4727	4869	5011	5153	5295	5437	5579	142
306	5721	5863	6005	6147	6289	6430	6572	6714	6855	6997	142
307	7138	7280	7421	7563	7704	7845	7986	8127	8269	8410	141
308	8551	8692	8833	8974	9114	9255	9396	9537	9677	9818	141
309	9958	••99	•239	•380	•520	•661	•801	•941	1081	1222	140
310	491362	1502	1642	1782	1922	2062	2201	2341	2481	2621	140
311	2760	2900	3040	3179	3319	3458	3597	3737	3876	4015	139
312	4155	4294	4433	4572	4711	4850	4989	5128	5267	5406	139
313	5544	5683	5822	5960	6099	6238	6376	6515	6653	6791	139
314	6930	7068	7206	7344	7483	7621	7759	7897	8035	8173	138
315	8311	8448	8586	8724	8862	8999	9137	9275	9412	9550	138
316	9687	9824	9962	••99	•236	•374	•511	•648	•785	•922	137
317	501059	1196	1333	1470	1607	1744	1880	2017	2154	2291	137
318	2427	2564	2700	2837	2973	3109	3246	3382	3518	3655	136
319	3791	3927	4063	4199	4335	4471	4607	4743	4878	5014	136
320	505150	5286	5421	5557	5693	5828	5964	6099	6234	6370	136
321	6505	6640	6776	6911	7046	7181	7316	7451	7586	7721	135
322	7856	7991	8126	8260	8395	8530	8664	8799	8934	9068	135
323	9203	9337	9471	9606	9740	9874	•••9	•143	•277	•411	134
324	510545	0679	0813	0947	1081	1215	1349	1482	1616	1750	134
325	1883	2017	2151	2284	2418	2551	2684	2818	2951	3084	133
326	3218	3351	3484	3617	3750	3883	4016	4149	4282	4414	133
327	4548	4681	4813	4946	5079	5211	5344	5476	5609	5741	133
328	5874	6006	6139	6271	6403	6535	6668	6800	6932	7064	132
329	7196	7328	7460	7592	7724	7855	7987	8119	8251	8382	132
330	518514	8646	8777	8909	9040	9171	9303	9434	9566	9697	131
331	9828	9959	••90	•221	•353	•484	•615	•745	•876	1007	131
332	521138	1269	1400	1530	1661	1792	1922	2053	2183	2314	131
333	2444	2575	2705	2835	2966	3096	3226	3356	3486	3616	130
334	3746	3876	4006	4136	4266	4396	4526	4656	4785	4915	130
335	5045	5174	5304	5434	5563	5693	5822	5951	6081	0210	129
336	6339	6469	6598	6727	6856	6985	7114	7243	7372	7501	129
337	7630	7759	7888	8016	8145	8274	8402	8531	8660	8788	129
338	8917	9045	9174	9302	9430	9559	9687	9815	9943	••72	128
339	530200	0328	0456	0584	0712	0840	0968	1096	1223	1351	128
N.	0	1	2	3	4	5	6	7	8	9	D.

N.	0	1	2	3	4	5	6	7	8	9	D.
340	531479	1607	1734	1862	1990	2117	2245	2372	2500	2627	128
341	2754	2882	3009	3136	3264	3391	3518	3645	3772	3899	127
342	4026	4153	4280	4407	4534	4661	4787	4914	5041	5167	127
343	5294	5421	5547	5674	5800	5927	6053	6180	6306	6432	126
344	6558	6685	6811	6937	7063	7189	7315	7441	7567	7693	126
345	7819	7945	8071	8197	8322	8448	8574	8699	8825	8951	126
346	9076	9202	9327	9452	9578	9703	9829	9954	••79	•204	125
347	540329	0455	0580	0705	0830	0955	1080	1205	1330	1454	125
348	1579	1704	1829	1953	2078	2203	2327	2452	2576	2701	125
349	2825	2950	3074	3199	3323	3447	3571	3696	3820	3944	124
350	544068	4192	4316	4440	4564	4688	4812	4936	5060	5183	124
351	5307	5431	5555	5678	5802	5925	6049	6172	6296	6419	124
352	6543	6666	6789	6913	7036	7159	7282	7405	7529	7652	123
353	7775	7898	8021	8144	8267	8389	8512	8635	8758	8881	123
354	9003	9126	9249	9371	9494	9616	9739	9861	9984	•106	123
355	550228	0351	0473	0595	0717	0840	0962	1084	1206	1328	122
356	1450	1572	1694	1816	1938	2060	2181	2303	2425	2547	122
357	2668	2790	2911	3033	3155	3276	3398	3519	3640	3762	121
358	3883	4004	4126	4247	4368	4489	4610	4731	4852	4973	121
359	5094	5215	5336	5457	5578	5699	5820	5940	6061	6182	121
360	556303	6423	6544	6664	6785	6905	7026	7146	7267	7387	120
361	7507	7627	7748	7868	7988	8108	8228	8349	8469	8589	120
362	8709	8829	8948	9068	9188	9308	9428	9548	9667	9787	120
363	9907	••26	•146	•265	•385	•504	•624	•743	•863	•982	119
364	561101	1221	1340	1459	1578	1698	1817	1936	2055	2174	119
365	2293	2412	2531	2650	2769	2887	3006	3125	3244	3362	119
366	3481	3600	3718	3837	3955	4074	4192	4311	4429	4548	119
367	4666	4784	4903	5021	5139	5257	5376	5494	5612	5730	118
368	5848	5966	6084	6202	6320	6437	6555	6673	6791	6909	118
369	7026	7144	7262	7379	7497	7614	7732	7849	7967	8084	118
370	568202	8319	8436	8554	8671	8788	8905	9023	9140	9257	117
371	9374	9491	9608	9725	9842	9959	••76	•193	•309	•426	117
372	570543	0660	0776	0893	1010	1126	1243	1359	1476	1592	117
373	1709	1825	1942	2058	2174	2291	2407	2523	2639	2755	116
374	2872	2988	3104	3220	3336	3452	3568	3684	3800	3915	116
375	4031	4147	4263	4379	4494	4610	4726	4841	4957	5072	116
376	5188	5303	5419	5534	5650	5765	5880	5996	6111	6226	115
377	6341	6457	6572	6687	6802	6917	7032	7147	7262	7377	115
378	7492	7607	7722	7836	7951	8066	8181	8295	8410	8525	115
379	8639	8754	8868	8983	9097	9212	9326	9441	9555	9669	114
380	579784	9898	••12	•126	•241	•355	•469	•583	•697	•811	114
381	580925	1039	1153	1267	1381	1495	1608	1722	1836	1950	114
382	2063	2177	2291	2404	2518	2631	2745	2858	2972	3085	114
383	3199	3312	3426	3539	3652	3765	3879	3992	4105	4218	113
384	4331	4444	4557	4670	4783	4896	5009	5122	5235	5348	113
385	5461	5574	5686	5799	5912	6024	6137	6250	6362	6475	113
386	6587	6700	6812	6925	7037	7149	7262	7374	7486	7599	112
387	7711	7823	7935	8047	8160	8272	8384	8496	8608	8720	112
388	8832	8944	9056	9167	9279	9391	9503	9615	9726	9838	112
389	9950	••61	•173	•284	•396	•507	•619	•730	•842	•953	112
390	591065	1176	1287	1399	1510	1621	1732	1843	1955	2066	111
391	2177	2288	2399	2510	2621	2732	2843	2954	3064	3175	111
392	3286	3397	3508	3618	3729	3840	3950	4061	4171	4282	111
393	4393	4503	4614	4724	4834	4945	5055	5165	5276	5386	110
394	5496	5606	5717	5827	5937	6047	6157	6267	6377	6487	110
395	6597	6707	6817	6927	7037	7146	7256	7366	7476	7586	110
396	7695	7805	7914	8024	8134	8243	8353	8462	8572	8681	110
397	8791	8900	9009	9119	9228	9337	9446	9556	9665	9774	109
398	9883	9992	•101	•210	•319	•428	•537	•646	•755	•864	109
399	600973	1082	1191	1299	1408	1517	1625	1734	1843	1951	109
N.	0	1	2	3	4	5	6	7	8	9	D.

N.	0	1	2	3	4	5	6	7	8	9	D.
400	602060	2169	2277	2386	2494	2603	2711	2819	2928	3036	108
401	3144	3253	3361	3469	3577	3686	3794	3902	4010	4118	108
402	4226	4334	4442	4550	4658	4766	4874	4982	5089	5197	108
403	5305	5413	5521	5628	5736	5844	5951	6059	6166	6274	108
404	6381	6489	6596	6704	6811	6919	7026	7133	7241	7348	107
405	7455	7562	7669	7777	7884	7991	8098	8205	8312	8419	107
406	8526	8633	8740	8847	8954	9061	9167	9274	9381	9488	107
407	9594	9701	9808	9914	••21	•128	•234	•341	•447	•554	107
408	610660	0767	0873	0979	1086	1192	1298	1405	1511	1617	106
409	1723	1829	1936	2042	2148	2254	2360	2466	2572	2678	106
410	612784	2890	2996	3102	3207	3313	3419	3525	3630	3736	106
411	3842	3947	4053	4159	4264	4370	4475	4581	4686	4792	106
412	4897	5003	5108	5213	5319	5424	5529	5634	5740	5845	105
413	5950	6055	6160	6265	6370	6476	6581	6686	6790	6895	105
414	7000	7105	7210	7315	7420	7525	7629	7734	7839	7943	105
415	8048	8153	8257	8362	8466	8571	8676	8780	8884	8989	105
416	9093	9198	9302	9406	9511	9615	9719	9824	9928	••32	104
417	620136	0240	0344	0448	0552	0656	0760	0864	0968	1072	104
418	1176	1280	1384	1488	1592	1695	1799	1903	2007	2110	104
419	2214	2318	2421	2525	2628	2732	2835	2939	3042	3146	104
420	623249	3353	3456	3559	3663	3766	3869	3973	4076	4179	103
421	4282	4385	4488	4591	4695	4798	4901	5004	5107	5210	103
422	5312	5415	5518	5621	5724	5827	5929	6032	6135	6238	103
423	6340	6443	6546	6648	6751	6853	6956	7058	7161	7263	103
424	7366	7468	7571	7673	7775	7878	7980	8082	8185	8287	102
425	8389	8491	8593	8695	8797	8900	9002	9104	9206	9308	102
426	9410	9512	9613	9715	9817	9919	••21	•123	•224	•326	102
427	630428	0530	0631	0733	0835	0936	1038	1139	1241	1342	102
428	1444	1545	1647	1748	1849	1951	2052	2153	2255	2356	101
429	2457	2559	2660	2761	2862	2963	3064	3165	3266	3367	101
430	633468	3569	3670	3771	3872	3973	4074	4175	4276	4376	100
431	4477	4578	4679	4779	4880	4981	5081	5182	5283	5383	100
432	5484	5584	5685	5785	5886	5986	6087	6187	6287	6388	100
433	6488	6588	6688	6789	6889	6989	7089	7189	7290	7390	100
434	7490	7590	7690	7790	7890	7990	8090	8190	8290	8389	99
435	8489	8589	8689	8789	8888	8988	9088	9188	9287	9387	99
436	9486	9586	9686	9785	9885	9984	••84	•183	•283	•382	99
437	640481	0581	0680	0779	0879	0978	1077	1177	1276	1375	99
438	1474	1573	1672	1771	1871	1970	2069	2168	2267	2366	99
439	2465	2563	2662	2761	2860	2959	3058	3156	3255	3354	99
440	643453	3551	3650	3749	3847	3946	4044	4143	4242	4340	98
441	4439	4537	4636	4734	4832	4931	5029	5127	5226	5324	98
442	5422	5521	5619	5717	5815	5913	6011	6110	6208	6306	98
443	6404	6502	6600	6698	6796	6894	6992	7089	7187	7285	98
444	7383	7481	7579	7676	7774	7872	7969	8067	8165	8262	98
445	8360	8458	8555	8653	8750	8848	8945	9043	9140	9237	97
446	9335	9432	9530	9627	9724	9821	9919	••16	•113	•210	97
447	650308	0405	0502	0599	0696	0793	0890	0987	1084	1181	97
448	1278	1375	1472	1569	1666	1762	1859	1956	2053	2150	97
449	2246	2343	2440	2536	2633	2730	2826	2923	3019	3116	97
450	653213	3309	3405	3502	3598	3695	3791	3888	3984	4080	96
451	4177	4273	4369	4465	4562	4658	4754	4850	4946	5042	96
452	5138	5235	5331	5427	5523	5619	5715	5810	5906	6002	96
453	6098	6194	6290	6386	6482	6577	6673	6769	6864	6960	96
454	7056	7152	7247	7343	7438	7534	7629	7725	7820	7916	96
455	8011	8107	8202	8298	8393	8488	8584	8679	8774	8870	95
456	8965	9060	9155	9250	9346	9441	9536	9631	9726	9821	95
457	9916	••11	•106	•201	•296	•391	•486	•581	•676	•771	95
458	660865	0960	1055	1150	1245	1339	1434	1529	1623	1718	95
459	1813	1907	2002	2096	2191	2286	2380	2475	2569	2663	95
N.	0	1	2	3	4	5	6	7	8	9	D.

N.	0	1	2	3	4	5	6	7	8	9	D.
460	662758	2852	2947	3041	3135	3230	3324	3418	3512	3607	94
461	3701	3795	3889	3983	4078	4172	4266	4360	4454	4548	94
462	4642	4736	4830	4924	5018	5112	5206	5299	5393	5487	94
463	5581	5675	5769	5862	5956	6050	6143	6237	6331	6424	94
464	6518	6612	6705	6799	6892	6986	7079	7173	7266	7360	94
465	7453	7546	7640	7733	7826	7920	8013	8106	8199	8293	93
466	8386	8479	8572	8665	8759	8852	8945	9038	9131	9224	93
467	9317	9410	9503	9596	9689	9782	9875	9967	••60	•153	93
468	670246	0339	0431	0524	0617	0710	0802	0895	0988	1080	93
469	1173	1265	1358	1451	1543	1636	1728	1821	1913	2005	93
470	672098	2190	2283	2375	2467	2560	2652	2744	2836	2929	92
471	3021	3113	3205	3297	3390	3482	3574	3666	3758	3850	92
472	3942	4034	4126	4218	4310	4402	4494	4586	4677	4769	92
473	4861	4953	5045	5137	5228	5320	5412	5503	5595	5687	92
474	5778	5870	5962	6053	6145	6236	6328	6419	6511	6602	92
475	6694	6785	6876	6968	7059	7151	7242	7333	7424	7516	91
476	7607	7698	7789	7881	7972	8063	8154	8245	8336	8427	91
477	8518	8609	8700	8791	8882	8973	9064	9155	9246	9337	91
478	9428	9519	9610	9700	9791	9882	9973	••63	•154	•245	91
479	680336	0426	0517	0607	0698	0789	0879	0970	1060	1151	91
480	681241	1332	1422	1513	1603	1693	1784	1874	1964	2055	90
481	2145	2235	2326	2416	2506	2596	2686	2777	2867	2957	90
482	3047	3137	3227	3317	3407	3497	3587	3677	3767	3857	90
483	3947	4037	4127	4217	4307	4396	4486	4576	4666	4756	90
484	4845	4935	5025	5114	5204	5294	5383	5473	5563	5652	90
485	5742	5831	5921	6010	6100	6189	6279	6368	6458	6547	89
486	6636	6726	6815	6904	6994	7083	7172	7261	7351	7440	89
487	7529	7618	7707	7796	7886	7975	8064	8153	8242	8331	89
488	8420	8509	8598	8687	8776	8865	8953	9042	9131	9220	89
489	9309	9398	9486	9575	9664	9753	9841	9930	••19	•107	89
490	690196	0285	0373	0462	0550	0639	0728	0816	0905	0993	89
491	1081	1170	1258	1347	1435	1524	1612	1700	1789	1877	88
492	1965	2053	2142	2230	2318	2406	2494	2583	2671	2759	88
493	2847	2935	3023	3111	3199	3287	3375	3463	3551	3639	88
494	3727	3815	3903	3991	4078	4166	4254	4342	44[illegible]	4517	88
495	4605	4693	4781	4868	4956	5044	5131	5219	5307	5394	88
496	5482	5569	5657	5744	5832	5919	6007	6094	6182	6269	87
497	6356	6444	6531	6618	6706	6793	6880	6968	7055	7142	87
498	7229	7317	7404	7491	7578	7665	7752	7839	7926	8014	87
499	8101	8188	8275	8362	8449	8535	8622	8709	8796	8883	87
500	698970	9057	9144	9231	9317	9404	9491	9578	9664	9751	87
501	9838	9924	••11	••98	•184	•271	•358	•444	•531	•617	87
502	700704	0790	0877	0963	1050	1136	1222	1309	1395	1482	86
503	1568	1654	1741	1827	1913	1999	2086	2172	2258	2344	86
504	2431	2517	2603	2689	2775	2861	2947	3033	3119	3205	86
505	3291	3377	3463	3549	3635	3721	3807	3893	3979	4065	86
506	4151	4236	4322	4408	4494	4579	4665	4751	4837	4922	86
507	5008	5094	5179	5265	5350	5436	5522	5607	5693	5778	86
508	5864	5949	6035	6120	6206	6291	6376	6462	6547	6632	85
509	6718	6803	6888	6974	7059	7144	7229	7315	7400	7485	85
510	707570	7655	7740	7826	7911	7996	8081	8166	8251	8336	85
511	8421	8506	8591	8676	8761	8846	8931	9015	9100	9185	85
512	9270	9355	9440	9524	9609	9694	9779	9863	9948	••33	85
513	710117	0202	0287	0371	0456	0540	0625	0710	0794	0879	85
514	0963	1048	1132	1217	1301	1385	1470	1554	1639	1723	84
515	1807	1892	1976	2060	2144	2229	2313	2397	2481	2566	84
516	2650	2734	2818	2902	2986	3070	3154	3238	3323	3407	84
517	3491	3575	3659	3742	3826	3910	3994	4078	4162	4246	84
518	4330	4414	4497	4581	4665	4749	4833	4916	5000	5084	84
519	5167	5251	5335	5418	5502	5586	5669	5753	5836	5920	84
N.	0	1	2	3	4	5	6	7	8	9	D.

N.	0	1	2	3	4	5	6	7	8	9	D.
520	716003	6087	6170	6254	6337	6421	6504	6588	6671	6754	83
521	6838	6921	7004	7088	7171	7254	7338	7421	7504	7587	83
522	7671	7754	7837	7920	8003	8086	8169	8253	8336	8419	83
523	8502	8585	8668	8751	8834	8917	9000	9083	9165	9248	83
524	9331	9414	9497	9580	9663	9745	9828	9911	9994	••77	83
525	720159	0242	0325	0407	0490	0573	0655	0738	0821	0903	83
526	0986	1068	1151	1233	1316	1398	1481	1563	1646	1728	82
527	1811	1893	1975	2058	2140	2222	2305	2387	2469	2552	82
528	2634	2716	2798	2881	2963	3045	3127	3209	3291	3374	82
529	3456	3538	3620	3702	3784	3866	3948	4030	4112	4194	82
530	724276	4358	4440	4522	4604	4685	4767	4849	4931	5013	82
531	5095	5176	5258	5340	5422	5503	5585	5667	5748	5830	82
532	5912	5993	6075	6156	6238	6320	6401	6483	6564	6646	82
533	6727	6809	6890	6972	7053	7134	7216	7297	7379	7460	81
534	7541	7623	7704	7785	7866	7948	8029	8110	8191	8273	81
535	8354	8435	8516	8597	8678	8759	8841	8922	9003	9084	81
536	9165	9246	9327	9408	9489	9570	9651	9732	9813	9893	81
537	9974	••55	•136	•217	•298	•378	•459	•540	•621	•702	81
538	730782	0863	0944	1024	1105	1186	1266	1347	1428	1508	81
539	1589	1669	1750	1830	1911	1991	2072	2152	2233	2313	81
540	732394	2474	2555	2635	2715	2796	2876	2956	3037	3117	80
541	3197	3278	3358	3438	3518	3598	3679	3759	3839	3919	80
542	3999	4079	4160	4240	4320	4400	4480	4560	4640	4720	80
543	4800	4880	4960	5040	5120	5200	5279	5359	5439	5519	80
544	5599	5679	5759	5838	5918	5998	6078	6157	6237	6317	80
545	6397	6476	6556	6635	6715	6795	6874	6954	7034	7113	80
546	7193	7272	7352	7431	7511	7590	7670	7749	7829	7908	79
547	7987	8067	8146	8225	8305	8384	8463	8543	8622	8701	79
548	8781	8860	8939	9018	9097	9177	9256	9335	9414	9493	79
549	9572	9651	9731	9810	9889	9968	••47	•126	•205	•284	79
550	740363	0442	0521	0600	0678	0757	0836	0915	0994	1073	79
551	1152	1230	1309	1388	1467	1546	1624	1703	1782	1860	79
552	1939	2018	2096	2175	2254	2332	2411	2489	2568	2647	79
553	2725	2804	2882	2961	3039	3118	3196	3275	3353	3431	78
554	3510	3588	3667	3745	3823	3902	3980	4058	4136	4215	78
555	4293	4371	4449	4528	4606	4684	4762	4840	4919	4997	78
556	5075	5153	5231	5309	5387	5465	5543	5621	5699	5777	78
557	5855	5933	6011	6089	6167	6245	6323	6401	6479	6556	78
558	6634	6712	6790	6868	6945	7023	7101	7179	7256	7334	78
559	7412	7489	7567	7645	7722	7800	7878	7955	8033	8110	78
560	748188	8266	8343	8421	8498	8576	8653	8731	8808	8885	77
561	8963	9040	9118	9195	9272	9350	9427	9504	9582	9659	77
562	9736	9814	9891	9968	••45	•123	•200	•277	•354	•431	77
563	750508	0586	0663	0740	0817	0894	0971	1048	1125	1202	77
564	1279	1356	1433	1510	1587	1664	1741	1818	1895	1972	77
565	2048	2125	2202	2279	2356	2433	2509	2586	2663	2740	77
566	2816	2893	2970	3047	3123	3200	3277	3353	3430	3506	77
567	3583	3660	3736	3813	3889	3966	4042	4119	4195	4272	77
568	4348	4425	4501	4578	4654	4730	4807	4883	4960	5036	76
569	5112	5189	5265	5341	5417	5494	5570	5646	5722	5799	76
570	755875	5951	6027	6103	6180	6256	6332	6408	6484	6560	76
571	6636	6712	6788	6864	6940	7016	7092	7168	7244	7320	76
572	7396	7472	7548	7624	7700	7775	7851	7927	8003	8079	76
573	8155	8230	8306	8382	8458	8533	8609	8685	8761	8836	76
574	8912	8988	9063	9139	9214	9290	9366	9441	9517	9592	76
575	9668	9743	9819	9894	9970	••45	•121	•196	•272	•347	75
576	760422	0498	0573	0649	0724	0799	0875	0950	1025	1101	75
577	1176	1251	1326	1402	1477	1552	1627	1702	1778	1853	75
578	1928	2003	2078	2153	2228	2303	2378	2453	2529	2604	75
579	2679	2754	2829	2904	2978	3053	3128	3203	3278	3353	75
N.	0	1	2	3	4	5	6	7	8	9	D.

N.	0	1	2	3	4	5	6	7	8	9	D.
580	763428	3503	3578	3653	3727	3802	3877	3952	4027	4101	75
581	4176	4251	4326	4400	4475	4550	4624	4699	4774	4848	75
582	4923	4998	5072	5147	5221	5296	5370	5445	5520	5594	75
583	5669	5743	5818	5892	5966	6041	6115	6190	6264	6338	74
584	6413	6487	6562	6636	6710	6785	6859	6933	7007	7082	74
585	7156	7230	7304	7379	7453	7527	7601	7675	7749	7823	74
586	7898	7972	8046	8120	8194	8268	8342	8416	8490	8564	74
587	8638	8712	8786	8860	8934	9008	9082	9156	9230	9303	74
588	9377	9451	9525	9599	9673	9746	9820	9894	9968	••42	74
589	770115	0189	0263	0336	0410	0484	0557	0631	0705	0778	74
590	770852	0926	0999	1073	1146	1220	1293	1367	1440	1514	74
591	1587	1661	1734	1808	1881	1955	2028	2102	2175	2248	73
592	2322	2395	2468	2542	2615	2688	2762	2835	2908	2981	73
593	3055	3128	3201	3274	3348	3421	3494	3567	3640	3713	73
594	3786	3860	3933	4006	4079	4152	4225	4298	4371	4444	73
595	4517	4590	4663	4736	4809	4882	4955	5028	5100	5173	73
596	5246	5319	5392	5465	5538	5610	5683	5756	5829	5902	73
597	5974	6047	6120	6193	6265	6338	6411	6483	6556	6629	73
598	6701	6774	6846	6919	6992	7064	7137	7209	7282	7354	73
599	7427	7499	7572	7644	7717	7789	7862	7934	8006	8079	72
600	778151	8224	8296	8368	8441	8513	8585	8658	8730	8802	72
601	8874	8947	9019	9091	9163	9236	9308	9380	9452	9524	72
602	9596	9669	9741	9813	9885	9957	••29	•101	•173	•245	72
603	780317	0389	0461	0533	0605	0677	0749	0821	0893	0965	72
604	1037	1109	1181	1253	1324	1396	1468	1540	1612	1684	72
605	1755	1827	1899	1971	2042	2114	2186	2258	2329	2401	72
606	2473	2544	2616	2688	2759	2831	2902	2974	3046	3117	72
607	3189	3260	3332	3403	3475	3546	3618	3689	3761	3832	71
608	3904	3975	4046	4118	4189	4261	4332	4403	4475	4546	71
609	4617	4689	4760	4831	4902	4974	5045	5116	5187	5259	71
610	785330	5401	5472	5543	5615	5686	5757	5828	5899	5970	71
611	6041	6112	6183	6254	6325	6396	6467	6538	6609	6680	71
612	6751	6822	6893	6964	7035	7106	7177	7248	7319	7390	71
613	7460	7531	7602	7673	7744	7815	7885	7956	8027	8098	71
614	8168	8239	8310	8381	8451	8522	8593	8663	8734	8804	71
615	8875	8946	9016	9087	9157	9228	9299	9369	9440	9510	71
616	9581	9651	9722	9792	9863	9933	•••4	••74	•144	•215	70
617	790285	0356	0426	0496	0567	0637	0707	0778	0848	0918	70
618	0988	1059	1129	1199	1269	1340	1410	1480	1550	1620	70
619	1691	1761	1831	1901	1971	2041	2111	2181	2252	2322	70
620	792392	2462	2532	2602	2672	2742	2812	2882	2952	3022	70
621	3092	3162	3231	3301	3371	3441	3511	3581	3651	3721	70
622	3790	3860	3930	4000	4070	4139	4209	4279	4349	4418	70
623	4488	4558	4627	4697	4767	4836	4906	4976	5045	5115	70
624	5185	5254	5324	5393	5463	5532	5602	5672	5741	5811	70
625	5880	5949	6019	6088	6158	6227	6297	6366	6436	6505	69
626	6574	6644	6713	6782	6852	6921	6990	7060	7129	7198	69
627	7268	7337	7406	7475	7545	7614	7683	7752	7821	7890	69
628	7960	8029	8098	8167	8236	8305	8374	8443	8513	8582	69
629	8651	8720	8789	8858	8927	8996	9065	9134	9203	9272	69
630	799341	9409	9478	9547	9616	9685	9754	9823	9892	9961	69
631	800029	0098	0167	0236	0305	0373	0442	0511	0580	0648	69
632	0717	0786	0854	0923	0992	1061	1129	1198	1266	1335	69
633	1404	1472	1541	1609	1678	1747	1815	1884	1952	2021	69
634	2089	2158	2226	2295	2363	2432	2500	2568	2637	2705	69
635	2774	2842	2910	2979	3047	3116	3184	3252	3321	3389	68
636	3457	3525	3594	3662	3730	3798	3867	3935	4003	4071	68
637	4139	4208	4276	4344	4412	4480	4548	4616	4685	4753	68
638	4821	4889	4957	5025	5093	5161	5229	5297	5365	5433	68
639	5501	5569	5637	5705	5773	5841	5908	5976	6044	6112	68
N.	0	1	2	3	4	5	6	7	8	9	D.

N.	0	1	2	3	4	5	6	7	8	9	D.
640	806180	6248	6316	6384	6451	6519	6587	6655	6723	6790	68
641	6858	6926	6994	7061	7129	7197	7264	7332	7400	7467	68
642	7535	7603	7670	7738	7806	7873	7941	8008	8076	8143	68
643	8211	8279	8346	8414	8481	8549	8616	8684	8751	8818	67
644	8886	8953	9021	9088	9156	9223	9290	9358	9425	9492	67
645	9560	9627	9694	9762	9829	9896	9964	●●31	●●98	●165	67
646	810233	0300	0367	0434	0501	0569	0636	0703	0770	0837	67
647	0904	0971	1039	1106	1173	1240	1307	1374	1441	1508	67
648	1575	1642	1709	1776	1843	1910	1977	2044	2111	2178	67
649	2245	2312	2379	2445	2512	2579	2646	2713	2780	2847	67
650	812913	2980	3047	3114	3181	3247	3314	3381	3448	3514	67
651	3581	3648	3714	3781	3848	3914	3981	4048	4114	4181	67
652	4248	4314	4381	4447	4514	4581	4647	4714	4780	4847	67
653	4913	4980	5046	5113	5179	5246	5312	5378	5445	5511	66
654	5578	5644	5711	5777	5843	5910	5976	6042	6109	6175	66
655	6241	6308	6374	6440	6506	6573	6639	6705	6771	6838	66
656	6904	6970	7036	7102	7169	7235	7301	7367	7433	7499	66
657	7565	7631	7698	7764	7830	7896	7962	8028	8094	8160	66
658	8226	8292	8358	8424	8490	8556	8622	8688	8754	8820	66
659	8885	8951	9017	9083	9149	9215	9281	9346	9412	9478	66
660	819544	9610	9676	9741	9807	9873	9939	●●●4	●●70	●136	66
661	820201	0267	0333	0399	0464	0530	0595	0661	0727	0792	66
662	0858	0924	0989	1055	1120	1186	1251	1317	1382	1448	66
663	1514	1579	1645	1710	1775	1841	1906	1972	2037	2103	65
664	2168	2233	2299	2364	2430	2495	2560	2626	2691	2756	65
665	2822	2887	2952	3018	3083	3148	3213	3279	3344	3409	65
666	3474	3539	3605	3670	3735	3800	3865	3930	3996	4061	65
667	4126	4191	4256	4321	4386	4451	4516	4581	4646	4711	65
668	4776	4841	4906	4971	5036	5101	5166	5231	5296	5361	65
669	5426	5491	5556	5621	5686	5751	5815	5880	5945	6010	65
670	826075	6140	6204	6269	6334	6399	6464	6528	6593	6658	65
671	6723	6787	6852	6917	6981	7046	7111	7175	7240	7305	65
672	7369	7434	7499	7563	7628	7692	7757	7821	7886	7951	65
673	8015	8080	8144	8209	8273	8338	8402	8467	8531	8595	64
674	8660	8724	8789	8853	8918	8982	9046	9111	9175	9239	64
675	9304	9368	9432	9497	9561	9625	9690	9754	9818	9882	64
676	9947	●●11	●●75	●139	●204	●268	●332	●396	●460	●525	64
677	830589	0653	0717	0781	0845	0909	0973	1037	1102	1166	64
678	1230	1294	1358	1422	1486	1550	1614	1678	1742	1806	64
679	1870	1934	1998	2062	2126	2189	2253	2317	2381	2445	64
680	832509	2573	2637	2700	2764	2828	2892	2956	3020	3083	64
681	3147	3211	3275	3338	3402	3466	3530	3593	3657	3721	64
682	3784	3848	3912	3975	4039	4103	4166	4230	4294	4357	64
683	4421	4484	4548	4611	4675	4739	4802	4866	4929	4993	64
684	5056	5120	5183	5247	5310	5373	5437	5500	5564	5627	63
685	5691	5754	5817	5881	5944	6007	6071	6134	6197	6261	63
686	6324	6387	6451	6514	6577	6641	6704	6767	6830	6894	63
687	6957	7020	7083	7146	7210	7273	7336	7399	7462	7525	63
688	7588	7652	7715	7778	7841	7904	7967	8030	8093	8156	63
689	8219	8282	8345	8408	8471	8534	8597	8660	8723	8786	63
690	838849	8912	8975	9038	9101	9164	9227	9289	9352	9415	63
691	9478	9541	9604	9667	9729	9792	9855	9918	9981	●●43	63
692	840106	0169	0232	0294	0357	0420	0482	0545	0608	0671	63
693	0733	0796	0859	0921	0984	1046	1109	1172	1234	1297	63
694	1359	1422	1485	1547	1610	1672	1735	1797	1860	1922	63
695	1985	2047	2110	2172	2235	2297	2360	2422	2484	2547	62
696	2609	2672	2734	2796	2859	2921	2983	3046	3108	3170	62
697	3233	3295	3357	3420	3482	3544	3606	3669	3731	3793	62
698	3855	3918	3980	4042	4104	4166	4229	4291	4353	4415	62
699	4477	4539	4601	4664	4726	4788	4850	4912	4974	5036	62
N.	0	1	2	3	4	5	6	7	8	9	D.

N.	0	1	2	3	4	5	6	7	8	9	D.
700	845098	5160	5222	5284	5346	5408	5470	5532	5594	5656	62
701	5718	5780	5842	5904	5966	6028	6090	6151	6213	6275	62
702	6337	6399	6461	6523	6585	6646	6708	6770	6832	6894	62
703	6955	7017	7079	7141	7202	7264	7326	7388	7449	7511	62
704	7573	7634	7696	7758	7819	7881	7943	8004	8066	8128	62
705	8189	8251	8312	8374	8435	8497	8559	8620	8682	8743	62
706	8805	8866	8928	8989	9051	9112	9174	9235	9297	9358	61
707	9419	9481	9542	9604	9665	9726	9788	9849	9911	9972	61
708	850033	0095	0156	0217	0279	0340	0401	0462	0524	0585	61
709	0646	0707	0769	0830	0891	0952	1014	1075	1136	1197	61
710	851258	1320	1381	1442	1503	1564	1625	1686	1747	1809	61
711	1870	1931	1992	2053	2114	2175	2236	2297	2358	2419	61
712	2480	2541	2602	2663	2724	2785	2846	2907	2968	3029	61
713	3090	3150	3211	3272	3333	3394	3455	3516	3577	3637	61
714	3698	3759	3820	3881	3941	4002	4063	4124	4185	4245	61
715	4306	4367	4428	4488	4549	4610	4670	4731	4792	4852	61
716	4913	4974	5034	5095	5156	5216	5277	5337	5398	5459	61
717	5519	5580	5640	5701	5761	5822	5882	5943	6003	6064	61
718	6124	6185	6245	6306	6366	6427	6487	6548	6608	6668	60
719	6729	6789	6850	6910	6970	7031	7091	7152	7212	7272	60
720	857332	7393	7453	7513	7574	7634	7694	7755	7815	7875	60
721	7935	7995	8056	8116	8176	8236	8297	8357	8417	8477	60
722	8537	8597	8657	8718	8778	8838	8898	8958	9018	9078	60
723	9138	9198	9258	9318	9379	9439	9499	9559	9619	9679	60
724	9739	9799	9859	9918	9978	•038	•098	•158	•218	•278	60
725	860338	0398	0458	0518	0578	0637	0697	0757	0817	0877	60
726	0937	0996	1056	1116	1176	1236	1295	1355	1415	1475	60
727	1534	1594	1654	1714	1773	1883	1893	1952	2012	2072	60
728	2131	2191	2251	2310	2370	2430	2489	2549	2608	2668	60
729	2728	2787	2847	2906	2966	3025	3085	3144	3204	3263	60
730	863323	3382	3442	3501	3561	3620	3680	3739	3799	3858	59
731	3917	3977	4036	4096	4155	4214	4274	4333	4392	4452	59
732	4511	4570	4630	4689	4748	4808	4867	4926	4985	5045	59
733	5104	5163	5222	5282	5341	5400	5459	5519	5578	5637	59
734	5696	5755	5814	5874	5933	5992	6051	6110	6169	6228	59
735	6287	6346	6405	6465	6524	6583	6642	6701	6760	6819	59
736	6878	6937	6996	7055	7114	7173	7232	7291	7350	7409	59
737	7467	7526	7585	7644	7703	7762	7821	7880	7939	7998	59
738	8056	8115	8174	8233	8292	8350	8409	8468	8527	8586	59
739	8644	8703	8762	8821	8879	8938	8997	9056	9114	9173	59
740	869232	9290	9349	9408	9466	9525	9584	9642	9701	9760	59
741	9818	9877	9935	9994	•053	•111	•170	•228	•287	•345	59
742	870404	0462	0521	0579	0638	0696	0755	0813	0872	0930	58
743	0989	1047	1106	1164	1223	1281	1339	1398	1456	1515	58
744	1573	1631	1690	1748	1806	1865	1923	1981	2040	2098	58
745	2156	2215	2273	2331	2389	2448	2506	2564	2622	2681	58
746	2739	2797	2855	2913	2972	3030	3088	3146	3204	3262	58
747	3321	3379	3437	3495	3553	3611	3669	2727	3785	3844	58
748	3902	3960	4018	4076	4134	4192	4250	4308	4366	4424	58
749	4482	4540	4598	4656	4714	4772	4830	4888	4945	5003	58
750	875061	5119	5177	5235	5293	5351	5409	5466	5524	5582	58
751	5640	5698	5756	5813	5871	5929	5987	6045	6102	6160	58
752	6218	6276	6333	6391	6449	6507	6564	6622	6680	6737	58
753	6795	6853	6910	6968	7026	7083	7141	7199	7256	7314	58
754	7371	7429	7487	7544	7602	7659	7717	7774	7832	7889	58
755	7947	8004	8062	8119	8177	8234	8292	8349	8407	8464	57
756	8522	8579	8637	8694	8752	8809	8866	8924	8981	9039	57
757	9096	9153	9211	9268	9325	9383	9440	9497	9555	9612	57
758	9669	9726	9784	9841	9898	9956	•013	•070	•127	•185	57
759	880242	0299	0356	0413	0471	0528	0585	0642	0699	0756	57
N.	0	1	2	3	4	5	6	7	8	9	D.

N.	0	1	2	3	4	5	6	7	8	9	D.
760	880814	0871	0928	0985	1042	1099	1156	1213	1271	1328	57
761	1385	1442	1499	1556	1613	1670	1727	1784	1841	1898	57
762	1955	2012	2069	2126	2183	2240	2297	2354	2411	2468	57
763	2525	2581	2638	2695	2752	2809	2866	2923	2980	3037	57
764	3093	3150	3207	3264	3321	3377	3434	3491	3548	3605	57
765	3661	3718	3775	3832	3888	3945	4002	4059	4115	4172	57
766	4229	4285	4342	4399	4455	4512	4569	4625	4682	4739	57
767	4795	4852	4909	4965	5022	5078	5135	5192	5248	5305	57
768	5361	5418	5474	5531	5587	5644	5700	5757	5813	5870	57
769	5926	5983	6039	6096	6152	6209	6265	6321	6378	6434	56
770	886491	6547	6604	6660	6716	6773	6829	6885	6942	6998	56
771	7054	7111	7167	7223	7280	7336	7392	7449	7505	7561	56
772	7617	7674	7730	7786	7842	7898	7955	8011	8067	8123	56
773	8179	8236	8292	8348	8404	8460	8516	8573	8629	8685	56
774	8741	8797	8853	8909	8965	9021	9077	9134	9190	9246	56
775	9302	9358	9414	9470	9526	9582	9638	9694	9750	9806	56
776	9862	9918	9974	••30	••86	•141	•197	•253	•309	•365	56
777	890421	0477	0533	0589	0645	0700	0756	0812	0868	0924	56
778	0980	1035	1091	1147	1203	1259	1314	1370	1426	1482	56
779	1537	1593	1649	1705	1760	1816	1872	1928	1983	2039	56
780	892095	2150	2206	2262	2317	2373	2429	2484	2540	2595	56
781	2651	2707	2762	2818	2873	2929	2985	3040	3096	3151	56
782	3207	3262	3318	3373	3429	3484	3540	3595	3651	3706	56
783	3762	3817	3873	3928	3984	4039	4094	4150	4205	4261	55
784	4316	4371	4427	4482	4538	4593	4648	4704	4759	4814	55
785	4870	4925	4980	5036	5091	5146	5201	5257	5312	5367	55
786	5423	5478	5533	5588	5644	5699	5754	5809	5864	5920	55
787	5975	6030	6085	6140	6195	6251	6306	6361	6416	6471	55
788	6526	6581	6636	6692	6747	6802	6857	6912	6967	7022	55
789	7077	7132	7187	7242	7297	7352	7407	7462	7517	7572	55
790	897627	7682	7737	7792	7847	7902	7957	8012	8067	8122	55
791	8176	8231	8286	8341	8396	8451	8506	8561	8615	8670	55
792	8725	8780	8835	8890	8944	8999	9054	9109	9164	9218	55
793	9273	9328	9383	9437	9492	9547	9602	9656	9711	9766	55
794	9821	9875	9930	9985	••39	••94	•149	•203	•258	•312	55
795	900367	0422	0476	0531	0586	0640	0695	0749	0804	0859	55
796	0913	0968	1022	1077	1131	1186	1240	1295	1349	1404	55
797	1458	1513	1567	1622	1676	1731	1785	1840	1894	1948	54
798	2003	2057	2112	2166	2221	2275	2329	2384	2438	2492	54
799	2547	2601	2655	2710	2764	2818	2873	2927	2981	3036	54
800	903090	3144	3199	3253	3307	3361	3416	3470	3524	3578	54
801	3633	3687	3741	3795	3849	3904	3958	4012	4066	4120	54
802	4174	4229	4283	4337	4391	4445	4499	4553	4607	4661	54
803	4716	4770	4824	4878	4932	4986	5040	5094	5148	5202	54
804	5256	5310	5364	5418	5472	5526	5580	5634	5688	5742	54
805	5796	5850	5904	5958	6012	6066	6119	6173	6227	6281	54
806	6335	6389	6443	6497	6551	6604	6658	6712	6766	6820	54
807	6874	6927	6981	7035	7089	7143	7196	7250	7304	7358	54
808	7411	7465	7519	7573	7626	7680	7734	7787	7841	7895	54
809	7949	8002	8056	8110	8163	8217	8270	8324	8378	8431	54
810	908485	8539	8592	8646	8699	8753	8807	8860	8914	8967	54
811	9021	9074	9128	9181	9235	9289	9342	9396	9449	9503	54
812	9556	9610	9663	9716	9770	9823	9877	9930	9984	••37	53
813	910091	0144	0197	0251	0304	0358	0411	0464	0518	0571	53
814	0624	0678	0731	0784	0838	0891	0944	0998	1051	1104	53
815	1158	1211	1264	1317	1371	1424	1477	1530	1584	1637	53
816	1690	1743	1797	1850	1903	1956	2009	2063	2116	2169	53
817	2222	2275	2328	2381	2435	2488	2541	2594	2647	2700	53
818	2753	2806	2859	2913	2966	3019	3072	3125	3178	3231	53
819	3284	3337	3390	3443	3496	3549	3602	3655	3708	3761	53
N.	0	1	2	3	4	5	6	7	8	9	D.

N.	0	1	2	3	4	5	6	7	8	9	D.
820	913814	3867	3920	3973	4026	4079	4132	4184	4237	4290	53
821	4343	4396	4449	4502	4555	4608	4660	4713	4766	4819	53
822	4872	4925	4977	5030	5083	5136	5189	5241	5294	5347	53
823	5400	5453	5505	5558	5611	5664	5716	5769	5822	5875	53
824	5927	5980	6033	6085	6138	6191	6243	6296	6349	6401	53
825	6454	6507	6559	6612	6664	6717	6770	6822	6875	6927	53
826	6980	7033	7085	7138	7190	7243	7295	7348	7400	7453	53
827	7506	7558	7611	7663	7716	7768	7820	7873	7925	7978	52
828	8030	8083	8135	8188	8240	8293	8345	8397	8450	8502	52
829	8555	8607	8659	8712	8764	8816	8869	8921	8973	9026	52
830	919078	9130	9183	9235	9287	9340	9392	9444	9496	9549	52
831	9601	9653	9706	9758	9810	9862	9914	9967	••19	••71	52
832	920123	0176	0228	0280	0332	0384	0436	0489	0541	0593	52
833	0645	0697	0749	0801	0853	0906	0958	1010	1062	1114	52
834	1166	1218	1270	1322	1374	1426	1478	1530	1582	1634	52
835	1686	1738	1790	1842	1894	1946	1998	2050	2102	2154	52
836	2206	2258	2310	2362	2414	2466	2518	2570	2622	2674	52
837	2725	2777	2829	2881	2933	2985	3037	3089	3140	3192	52
838	3244	3296	3348	3399	3451	3503	3555	3607	3658	3710	52
839	3762	3814	3865	3917	3969	4021	4072	4124	4176	4228	52
840	924279	4331	4383	4434	4486	4538	4589	4641	4693	4744	52
841	4796	4848	4899	4951	5003	5054	5106	5157	5209	5261	52
842	5312	5364	5415	5467	5518	5570	5621	5673	5725	5776	52
843	5828	5879	5931	5982	6034	6085	6137	6188	6240	6291	51
844	6342	6394	6445	6497	6548	6600	6651	6702	6754	6805	51
845	6857	6908	6959	7011	7062	7114	7165	7216	7268	7319	51
846	7370	7422	7473	7524	7576	7627	7678	7730	7781	7832	51
847	7883	7935	7986	8037	8088	8140	9191	8242	8293	8345	51
848	8396	8447	8498	8549	8601	8652	8703	8754	8805	8857	51
849	8908	8959	9010	9061	9112	9163	9215	9266	9317	9368	51
850	929419	9470	9521	9572	9623	9674	9725	9776	9827	9879	51
851	9930	9981	••32	••83	•134	•185	•236	•287	•338	•389	51
852	930440	0491	0542	0592	0643	0694	0745	0796	0847	0898	51
853	0949	1000	1051	1102	1153	1204	1254	1305	1356	1407	51
854	1458	1509	1560	1610	1661	1712	1763	1814	1865	1915	51
855	1966	2017	2068	2118	2169	2220	2271	2322	2372	2423	51
856	2474	2524	2575	2626	2677	2727	2778	2829	2879	2930	51
857	2981	3031	3082	3133	3183	3234	3285	3335	3386	3437	51
858	3487	3538	3589	3639	3690	3740	3791	3841	3892	3943	51
859	3993	4044	4094	4145	4195	4246	4296	4347	4397	4448	51
860	934498	4549	4599	4650	4700	4751	4801	4852	4902	4953	50
861	5003	5054	5104	5154	5205	5255	5306	5356	5406	5457	50
862	5507	5558	5608	5658	5709	5759	5809	5860	5910	5960	50
863	6011	6061	6111	6162	6212	6262	6313	6363	6413	6463	50
864	6514	6564	6614	6665	6715	6765	6815	6865	6916	6966	50
865	7016	7066	7117	7167	7217	7267	7317	7367	7418	7468	50
866	7518	7568	7618	7668	7718	7769	7819	7869	7919	7969	50
867	8019	8069	8119	8169	8219	8269	8320	8370	8420	8470	50
868	8520	8570	8620	8670	8720	8770	8820	8870	8920	8970	50
869	9020	9070	9120	9170	9220	9270	9320	9369	9419	9469	50
870	939519	9569	9619	9669	9719	9769	9819	9869	9918	9968	50
871	940018	0068	0118	0168	0218	0267	0317	0367	0417	0467	50
872	0516	0566	0616	0666	0716	0765	0815	0865	0915	0964	50
873	1014	1064	1114	1163	1213	1263	1313	1362	1412	1462	50
874	1511	1561	1611	1660	1710	1760	1809	1859	1909	1958	50
875	2008	2058	2107	2157	2207	2256	2306	2355	2405	2455	50
876	2504	2554	2603	2653	2702	2752	2801	2851	2901	2950	50
877	3000	3040	3099	3148	3198	3247	3297	3346	3396	3445	59
878	3495	3544	3593	3643	3692	3742	3791	3841	3890	3939	59
879	3989	4038	4088	4137	4186	4236	4285	4335	4384	4433	59
N.	0	1	2	3	4	5	6	7	8	9	D.

N.	0	1	2	3	4	5	6	7	8	9	D.
880	944483	4532	4581	4631	4680	4729	4779	4828	4877	4927	49
881	4976	5025	5074	5124	5173	5222	5272	5321	5370	5419	49
882	5469	5518	5567	5616	5665	5715	5764	5813	5862	5912	49
883	5961	6010	6059	6108	6157	6207	6256	6305	6354	6403	49
884	6452	6501	6551	6600	6649	6698	6747	6796	6845	6894	49
885	6943	6992	7041	7090	7140	7189	7238	7287	7336	7385	49
886	7434	7483	7532	7581	7630	7679	7728	7777	7826	7875	49
887	7924	7973	8022	8070	8119	8168	8217	8266	8315	8364	49
888	8413	8462	8511	8560	8609	8657	8706	8755	8804	8853	49
889	8902	8951	8999	9048	9097	9146	9195	9244	9292	9341	49
890	949390	9439	9488	9536	9585	9634	9683	9731	9780	9829	49
891	9878	9926	9975	••24	••73	•121	•170	•219	•267	•316	49
892	950365	0414	0462	0511	0560	0608	0657	0706	0754	0803	49
893	0851	0900	0949	0997	1046	1095	1143	1192	1240	1289	49
894	1338	1386	1435	1483	1532	1580	1629	1677	1726	1775	49
895	1823	1872	1920	1969	2017	2066	2114	2163	2211	2260	48
896	2308	2356	2405	2453	2502	2550	2599	2647	2696	2744	48
897	2792	2841	2889	2938	2986	3034	3083	3131	3180	3228	48
898	3276	3325	3373	3421	3470	3518	3566	3615	3663	3711	48
899	3760	3808	3856	3905	3953	4001	4049	4098	4146	4194	48
900	954243	4291	4339	4387	4435	4484	4532	4580	4628	4677	48
901	4725	4773	4821	4869	4918	4966	5014	5062	5110	5158	48
902	5207	5255	5303	5351	5399	5447	5495	5543	5592	5640	48
903	5688	5736	5784	5832	5880	5928	5976	6024	6072	6120	48
904	6168	6216	6265	6313	6361	6409	6457	6505	6553	6601	48
905	6649	6697	6745	6793	6840	6888	6936	6984	7032	7080	48
906	7128	7176	7224	7272	7320	7368	7416	7464	7512	7559	48
907	7607	7655	7703	7751	7799	7847	7894	7942	7990	8038	48
908	8086	8134	8181	8229	8277	8325	8373	8421	8468	8516	48
909	8564	8612	8659	8707	8755	8803	8850	8898	8946	8994	48
910	959041	9089	9137	9185	9232	9280	9328	9375	9423	9471	48
911	9518	9566	9614	9661	9709	9757	9804	9852	9900	9947	48
912	9995	••42	••90	•138	•185	•233	•280	•328	•376	•423	48
913	960471	0518	0566	0613	0661	0709	0756	0804	0851	0899	48
914	0946	0994	1041	1089	1136	1184	1231	1279	1326	1374	47
915	1421	1469	1516	1563	1611	1658	1706	1753	1801	1848	47
916	1895	1943	1990	2038	2085	2132	2180	2227	2275	2322	47
917	2369	2417	2464	2511	2559	2606	2653	2701	2748	2795	47
918	2843	2890	2937	2985	3032	3079	3126	3174	3221	3268	47
919	3316	3363	3410	3457	3504	3552	3599	3646	3693	3741	47
920	963788	3835	3882	3929	3977	4024	4071	4118	4165	4212	47
921	4260	4307	4354	4401	4448	4495	4542	4590	4637	4684	47
922	4731	4778	4825	4872	4919	4966	5013	5061	5108	5155	47
923	5202	5249	5296	5343	5390	5437	5484	5531	5578	5625	47
924	5672	5719	5766	5813	5860	5907	5954	6001	6048	6095	47
925	6142	6189	6236	6283	6329	6376	6423	6470	6517	6564	47
926	6611	6658	6705	6752	6799	6845	6892	6939	6986	7033	47
927	7080	7127	7173	7220	7267	7314	7361	7408	7454	7501	47
928	7548	7595	7642	7688	7735	7782	7829	7875	7922	7969	47
929	8016	8062	8109	8156	8203	8249	8296	8343	8390	8436	47
930	968483	8530	8576	8623	8670	8716	8763	8810	8856	8903	47
931	8950	8996	9043	9090	9136	9183	9229	9276	9323	9369	47
932	9416	9463	9509	9556	9602	9649	9695	9742	9789	9835	47
933	9882	9928	9975	••21	••68	•114	•161	•207	•254	•300	47
934	970347	0393	0440	0486	0533	0579	0626	0672	0719	0765	46
935	0812	0858	0904	0951	0997	1044	1090	1137	1183	1229	46
936	1276	1322	1369	1415	1461	1508	1554	1601	1647	1693	46
937	1740	1786	1832	1879	1925	1971	2018	2064	2110	2157	46
938	2203	2249	2295	2342	2388	2434	2481	2527	2573	2619	46
939	2666	2712	2758	2804	2851	2897	2943	2989	3035	3082	46
N.	0	1	2	3	4	5	6	7	8	9	D.

N.	0	1	2	3	4	5	6	7	8	9	D.
940	973128	3174	3220	3266	3313	3359	3405	3451	3497	3543	46
941	3590	3636	3682	3728	3774	3820	3866	3913	3959	4005	46
942	4051	4097	4143	4189	4235	4281	4327	4374	4420	4466	46
943	4512	4558	4604	4650	4696	4742	4788	4834	4880	4926	46
944	4972	5018	5064	5110	5156	5202	5248	5294	5340	5386	46
945	5432	5478	5524	5570	5616	5662	5707	5753	5799	5845	46
946	5891	5937	5983	6029	6075	6121	6167	6212	6258	6304	46
947	6350	6396	6442	6488	6533	6579	6625	6671	6717	6763	46
948	6808	6854	6900	6946	6992	7037	7083	7129	7175	7220	46
949	7266	7312	7358	7403	7449	7495	7541	7586	7632	7678	46
950	977724	7769	7815	7861	7906	7952	7998	8043	8089	8135	46
951	8181	8226	8272	8317	8363	8409	8454	8500	8546	8591	46
952	8637	8683	8728	8774	8819	8865	8911	8956	9002	9047	46
953	9093	9138	9184	9230	9275	9321	9366	9412	9457	9503	46
954	9548	9594	9639	9685	9730	9776	9821	9867	9912	9958	46
955	980003	0049	0094	0140	0185	0231	0276	0322	0367	0412	45
956	0458	0503	0549	0594	0640	0685	0730	0776	0821	0867	45
957	0912	0957	1003	1048	1093	1139	1184	1229	1275	1320	45
958	1366	1411	1456	1501	1547	1592	1637	1683	1728	1773	45
959	1819	1864	1909	1954	2000	2045	2090	2135	2181	2226	45
960	982271	2316	2362	2407	2452	2497	2543	2588	2633	2678	45
961	2723	2769	2814	2859	2904	2949	2994	3040	3085	3130	45
962	3175	3220	3265	3310	3356	3401	3446	3491	3536	3581	45
963	3626	3671	3716	3762	3807	3852	3897	3942	3987	4032	45
964	4077	4122	4167	4212	4257	4302	4347	4392	4437	4482	45
965	4527	4572	4617	4662	4707	4752	4797	4842	4887	4932	45
966	4977	5022	5067	5112	5157	5202	5247	5292	5337	5382	45
967	5426	5471	5516	5561	5606	5651	5696	5741	5786	5830	45
968	5875	5920	5965	6010	6055	6100	6144	6189	6234	6279	45
969	6324	6369	6413	6458	6503	6548	6593	6637	6682	6727	45
970	986772	6817	6861	6906	6951	6996	7040	7085	7130	7175	45
971	7219	7264	7309	7353	7398	7443	7488	7532	7577	7622	45
972	7666	7711	7756	7800	7845	7890	7934	7979	8024	8068	45
973	8113	8157	8202	8247	8291	8336	8381	8425	8470	8514	45
974	8559	8604	8648	8693	8737	8782	8826	8871	8916	8960	45
975	9005	9049	9094	9138	9183	9227	9272	9316	9361	9405	45
976	9450	9494	9539	9583	9628	9672	9717	9761	9806	9850	44
977	9895	9939	9983	••28	••72	•117	•161	•206	•250	•294	44
978	990339	0383	0428	0472	0516	0561	0605	0650	0694	0738	44
979	0783	0827	0871	0916	0960	1004	1049	1093	1137	1182	44
980	991226	1270	1315	1359	1403	1448	1492	1536	1580	1625	44
981	1669	1713	1758	1802	1846	1890	1935	1979	2023	2067	44
982	2111	2156	2200	2244	2288	2333	2377	2421	2465	2509	44
983	2554	2598	2642	2686	2730	2774	2819	2863	2907	2951	44
984	2995	3039	3083	3127	3172	3216	3260	3304	3348	3392	44
985	3436	3480	3524	3568	3613	3657	3701	3745	3789	3833	44
986	3877	3921	3965	4009	4053	4097	4141	4185	4229	4273	44
987	4317	4361	4405	4449	4493	4537	4581	4625	4669	4713	44
988	4757	4801	4845	4889	4933	4977	5021	5065	5108	5152	44
989	5196	5240	5284	5328	5372	5416	5460	5504	5547	5591	44
990	995635	5679	5723	5767	5811	5854	5898	5942	5986	6030	44
991	6074	6117	6161	6205	6249	6293	6337	6380	6424	6468	44
992	6512	6555	6599	6643	6687	6731	6774	6818	6862	6906	44
993	6949	6993	7037	7080	7124	7168	7212	7255	7299	7343	44
994	7386	7430	7474	7517	7561	7605	7648	7692	7736	7779	44
995	7823	7867	7910	7954	7998	8041	8085	8129	8172	8216	44
996	8259	8303	8347	8390	8434	8477	8521	8564	8608	8652	44
997	8695	8739	8782	8826	8869	8913	8956	9000	9043	9087	44
998	9131	9174	9218	9261	9305	9348	9392	9435	9479	9522	44
999	9565	9609	9652	9696	9739	9783	9826	9870	9913	9957	43
N.	0	1	2	3	4	5	6	7	8	9	D.

A TABLE

OF

LOGARITHMIC

SINES AND TANGENTS

FOR EVERY

DEGREE AND MINUTE

OF THE QUADRANT.

REMARK. The minutes in the left-hand column of each page, increasing downwards, belong to the degrees at the top; and those increasing upwards, in the right-hand column, belong to the degrees below.

M.	Sine	D.	Cosine	D.	Tang.	D.	Cotang.	
0	0·000000		10·000000		0·000000		Infinite.	60
1	6·463726	5017·17	000000	·00	6·463726	5017·17	13·536274	59
2	764756	2934·85	000000	·00	764756	2934·83	235244	58
3	940847	2082·31	000000	·00	940847	2082·31	059153	57
4	7·065786	1615·17	000000	·00	7·065786	1615·17	12·934214	56
5	162696	1319·68	000000	·00	162696	1319·69	837304	55
6	241877	1115·75	9·999999	·01	241878	1115·78	758122	54
7	308824	966·53	999999	·01	308825	996·53	691175	53
8	366816	852·54	999999	·01	366817	852·54	633183	52
9	417968	762·63	999999	·01	417970	762·63	582030	51
10	463725	689·88	999998	·01	463727	689·88	536273	50
11	7·505118	629·81	9·999998	·01	7·505120	629·81	12·494880	49
12	542906	579·36	999997	·01	542909	579·33	457091	48
13	577668	536·41	999997	·01	577672	536·42	422328	47
14	609853	499·38	999996	·01	609857	499·39	390143	46
15	639816	467·14	999996	·01	639820	467·15	360180	45
16	667845	438·81	999995	·01	667849	438·82	332151	44
17	694173	413·72	999995	·01	694179	413·73	305821	43
18	718997	391·35	999994	·01	719004	391·36	280997	42
19	742477	371·27	999993	·01	742484	371·28	257516	41
20	764754	353·15	999993	·01	764761	351·36	235239	40
21	7·785943	336·72	9·999992	·01	7·785951	336·73	12·214049	39
22	806146	321·75	999991	·01	806155	321·76	193845	38
23	825451	308·05	999990	·01	825460	308·06	174540	37
24	843934	295·47	999989	·02	843944	295·49	156056	36
25	861662	283·88	999988	·02	861674	283·90	138326	35
26	878695	273·17	999988	·02	878708	273·18	121292	34
27	895085	263·23	999987	·02	895099	263·25	104901	33
28	910879	253·99	999986	·02	910894	254·01	089106	32
29	926119	245·38	999985	·02	926134	245·40	073866	31
30	940842	237·33	999983	·02	940858	237·35	059142	30
31	7·955082	229·80	9·999982	·02	7·955100	229·81	12·044900	29
32	968870	222·73	999981	·02	968889	222·75	031111	28
33	982233	216·08	999980	·02	982253	216·10	017747	27
34	995198	209·81	999979	·02	995219	209·83	004781	26
35	8·007787	203·90	999977	·02	8·007809	203·92	11·992191	25
36	020021	198·31	999976	·02	020045	198·33	979955	24
37	031919	193·02	999975	·02	031945	193·05	968055	23
38	043501	188·01	999973	·02	043527	188·03	956473	22
39	054781	183·25	999972	·02	054809	183·27	945191	21
40	065776	178·72	999971	·02	065806	178·74	934194	20
41	8·076500	174·41	9·999969	·02	8·076531	174·44	11·923469	19
42	086965	170·31	999968	·02	086997	170·34	913003	18
43	097183	166·39	999966	·02	097217	166·42	902783	17
44	107167	162·65	999964	·03	107202	162·68	892797	16
45	116926	159·08	999963	·03	116963	159·10	883037	15
46	126471	155·66	999961	·03	126510	155·68	873490	14
47	135810	152·38	999959	·03	135851	152·41	864149	13
48	144953	149·24	999958	·03	144996	149·27	855004	12
49	153907	146·22	999956	·03	153952	146·27	846048	11
50	162681	143·33	999954	·03	162727	143·36	837273	10
51	8·171280	140·54	9·999952	·03	8·171328	140·57	11·828672	9
52	179713	137·86	999950	·03	179763	137·90	820237	8
53	187985	135·29	999948	·03	188036	135·32	811964	7
54	196102	132·80	999946	·03	196156	132·84	803844	6
55	204070	130·41	999944	·03	204126	130·44	795874	5
56	211895	128·10	999942	·04	211953	128·14	788047	4
57	219581	125·87	999940	·04	219641	125·90	780359	3
58	227134	123·72	999938	·04	227195	123·76	772805	2
59	234557	121·64	999936	·04	234621	121·68	765379	1
60	241855	119·63	999934	·04	241921	119·67	758079	0
	Cosine	D.	Sine	89°	Cotang.	D.	Tang.	M.

M.	Sine	D.	Cosine	D.	Tang.	D.	Cotang.	
0	8·241855	119·63	9·999934	·04	8·241921	119·67	11·758079	60
1	249033	117·68	999932	·04	249102	117·72	750898	59
2	256094	115·80	999929	·04	256165	115·84	743835	58
3	263042	113·98	999927	·04	263115	114·02	736885	57
4	269881	112·21	999925	·04	269956	112·25	730044	56
5	276614	110·50	999922	·04	276691	110·54	723309	55
6	283243	108·83	999920	·04	283323	108·87	716677	54
7	289773	107·21	999918	·04	289856	107·26	710144	53
8	296207	105·65	999915	·04	296292	105·70	703708	52
9	302546	104·13	999913	·04	302634	104·18	697366	51
10	308794	102·66	999910	·04	308884	102·70	691116	50
11	8·314904	101·22	9·999907	·04	8·315046	101·26	11·684954	49
12	321027	99·82	999905	·04	321122	99·87	678878	48
13	327016	98·47	999902	·04	327114	98·51	672886	47
14	332924	97·14	999899	·05	333025	97·19	666975	46
15	338753	95·86	999897	·05	338856	95·90	661144	45
16	344504	94·60	999894	·05	344610	94·65	655390	44
17	350181	93·38	999891	·05	350289	93·43	649711	43
18	355783	92·19	999888	·05	355895	92·24	644105	42
19	361315	91·03	999885	·05	361430	91·08	638570	41
20	366777	89·90	999882	·05	366895	89·95	633105	40
21	8·372171	88·80	9·999879	·05	8·372292	88·85	11·627708	39
22	377499	87·72	999876	·05	377622	87·77	622378	38
23	382762	86·67	999873	·05	382889	86·72	617111	37
24	387962	85·64	999870	·05	388092	85·70	611908	36
25	393101	84·64	999867	·05	393234	84·70	606766	35
26	398179	83·66	999864	·05	398315	83·71	601685	34
27	403199	82·71	999861	·05	403338	82·76	596662	33
28	408161	81·77	999858	·05	408304	81·82	591696	32
29	413068	80·86	999854	·05	413213	80·91	586787	31
30	417919	79·96	999851	·06	418068	80·02	581932	30
31	8·422717	79·09	9·999848	·06	8·422869	79·14	11·577131	29
32	427462	78·23	999844	·06	427618	78·30	572382	28
33	432156	77·40	999841	·06	432315	77·45	567685	27
34	436800	76·57	999838	·06	436962	76·63	563038	26
35		75·77	999834	·06	441560	75·83	558440	25
36		74·99	999831	·06	446110	75·05	553890	24
37		74·22	999827	·06	450613	74·28	549387	23
38	454893	73·46	999823	·06	455070	73·52	544930	22
39	459301	72·73	999820	·06	459481	72·79	540519	21
40	463665	72·00	999816	·06	463849	72·06	536151	20
41	8·467985	71·29	9·999812	·06	8·468172	71·35	11·531828	19
42	472263	70·60	999809	·06	472454	70·66	527546	18
43	476498	69·91	999805	·06	476693	69·98	523307	17
44	480693	69·24	999801	·06	480892	69·31	519108	16
45	484848	68·59	999797	·07	485050	68·65	514950	15
46	488963	67·94	999793	·07	489170	68·02	510830	14
47	493040	67·31	999790	·07	493250	67·38	506750	13
48	497078	66·69	999786	·07	497293	66·76	502707	12
49	501080	66·08	999782	·07	501298	66·15	498702	11
50	505045	65·48	999778	·07	505267	65·55	494733	10
51	8·508974	64·89	9·999774	·07	8·509200	64·96	11·490800	9
52	512867	64·31	999769	·07	513098	64·39	486902	8
53	516726	63·75	999765	·07	516961	63·82	483039	7
54	520551	63·19	999761	·07	520790	63·26	479210	6
55	524343	62·64	999757	·07	524586	62·72	475414	5
56	528102	62·11	999753	·07	528349	62·18	471651	4
57	531828	61·58	999748	·07	532080	61·65	467920	3
58	535523	61·06	999744	·07	535779	61·13	464221	2
59	539186	60·55	999740	·07	539447	60·62	460553	1
60	542819	60·04	999735	·07	543084	60·12	456916	0
	Cosine	D.	Sine	88°	Cotang.	D.	Tang	

M.	Sine	D.	Cosine	D.	Tang.	D.	Cotang.	
0	8·542819	60·04	9·999735	·07	8·543084	60·12	11·456916	60
1	546422	59·55	999731	·07	546691	59·62	453309	59
2	549995	59·06	999726	·07	550268	59·14	449732	58
3	553539	58·58	999722	·08	553817	58·66	446183	57
4	557054	58·11	999717	·08	557336	58·19	442664	56
5	560540	57·65	999713	·08	560828	57·73	439172	55
6	563999	57·19	999708	·08	564291	57·27	435709	54
7	567431	56·74	999704	·08	567727	56·82	432273	53
8	570836	56·30	999699	·08	571137	56·38	428863	52
9	574214	55·87	999694	·08	574520	55·95	425480	51
10	577566	55·44	999689	·08	577877	55·52	422123	50
11	8·580892	55·02	9·999685	·08	8·581208	55·10	11·418792	49
12	584193	54·60	999680	·08	584514	54·68	415486	48
13	587469	54·19	999675	·08	587795	54·27	412205	47
14	590721	53·79	999670	·08	591051	53·87	408949	46
15	593948	53·39	999665	·08	594283	53·47	405717	45
16	597152	53·00	999660	·08	597492	53·08	402508	44
17	600332	52·61	999655	·08	600677	52·70	399323	43
18	603489	52·23	999650	·08	603839	52·32	396161	42
19	606623	51·86	999645	·09	606978	51·94	393022	41
20	609734	51·49	999640	·09	610094	51·58	389906	40
21	8·612823	51·12	9·999635	·09	8·613189	51·21	11·386811	39
22	615891	50·76	999629	·09	616262	50·85	383738	38
23	618937	50·41	999624	·09	619313	50·50	380687	37
24	621962	50·06	999619	·09	622343	50·15	377657	36
25	624965	49·72	999614	·09	625352	49·81	374648	35
26	627948	49·38	999608	·09	628340	49·47	371660	34
27	630911	49·04	999603	·09	631308	49·13	368692	33
28	633854	48·71	999597	·09	634256	48·80	365744	32
29	636776	48·39	999592	·09	637184	48·48	362816	31
30	639680	48·06	999586	·09	640093	48·16	359907	30
31	8·642563	47·75	9·999581	·09	8·642982	47·84	11·357018	29
32	645428	47·43	999575	·09	645853	47·53	354147	28
33	648274	47·12	999570	·09	648704	47·22	351296	27
34	651102	46·82	999564	·09	651537	46·91	348463	26
35	653911	46·52	999558	·10	654352	46·61	345648	25
36	656702	46·22	999553	·10	657149	46·31	342851	24
37	659475	45·92	999547	·10	659928	46·02	340072	23
38	662230	45·63	999541	·10	662689	45·73	337311	22
39	664968	45·35	999535	·10	665433	45·44	334567	21
40	667689	45·06	999529	·10	668160	45·26	331840	20
41	8·670393	44·79	9·999524	·10	8·670870	44·88	11·329130	19
42	673080	44·51	999518	·10	673563	44·61	326437	18
43	675751	44·24	999512	·10	676239	44·34	323761	17
44	678405	43·97	999506	·10	678900	44·17	321100	16
45	681043	43·70	999500	·10	681544	43·80	318456	15
46	683665	43·44	999493	·10	684172	43·54	315828	14
47	686272	43·18	999487	·10	686784	43·28	313216	13
48	688863	42·92	999481	·10	689381	43·03	310619	12
49	691438	42·67	999475	·10	691963	42·77	308037	11
50	693998	42·42	999469	·10	694529	42·52	305471	10
51	8·696543	42·17	9·999463	·11	8·697081	42·28	11·302919	9
52	699073	41·92	999456	·11	699617	42·03	300383	8
53	701589	41·68	999450	·11	702139	41·79	297861	7
54	704090	41·44	999443	·11	704646	41·55	295354	6
55	706577	41·21	999437	·11	707140	41·32	292860	5
56	709049	40·97	999431	·11	709618	41·08	290382	4
57	711507	40·74	999424	·11	712083	40·85	287917	3
58	713952	40·51	999418	·11	714534	40·62	285465	2
59	716383	40·29	999411	·11	716972	40·40	283028	1
60	718800	40·06	999404	·11	719396	40·17	280604	0
	Cosine	D.	Sine	87°	Cotang.	D.	Tang.	M.

M.	Sine	D.	Cosine	D.	Tang.	D.	Cotang.	
0	8·718800	40·06	9·999404	·11	8·719396	40·17	11·280604	60
1	721204	39·84	999398	·11	721806	39·95	278194	59
2	723595	39·62	999391	·11	724204	39·74	275796	58
3	725972	39·41	999384	·11	726588	39·52	273412	57
4	728337	39·19	999378	·11	728959	39·30	271041	56
5	730688	38·98	999371	·11	731317	39·09	268683	55
6	733027	38·77	999364	·12	733663	38·89	266337	54
7	735354	38·57	999357	·12	735996	38·68	264004	53
8	737667	38·36	999350	·12	738317	38·48	261683	52
9	739969	38·16	999343	·12	740626	38·27	259374	51
10	742259	37·96	999336	·12	742922	38·07	257078	50
11	8·744536	37·76	9·999329	·12	8·745207	37·87	11·254793	49
12	746802	37·56	999322	·12	747479	37·68	252521	48
13	749055	37·37	999315	·12	749740	37·49	250260	47
14	751297	37·17	999308	·12	751989	37·29	248011	46
15	753528	36·98	999301	·12	754227	37·10	245773	45
16	755747	36·79	999294	·12	756453	36·92	243547	44
17	757955	36·61	999286	·12	758668	36·73	241332	43
18	760151	36·42	999279	·12	760872	36·55	239128	42
19	762337	36·24	999272	·12	763065	36·36	236935	41
20	764511	36·06	999265	·12	765246	36·18	234754	40
21	8·766675	35·88	9·999257	·12	8·767417	36·00	11·232583	39
22	768828	35·70	999250	·13	769578	35·83	230422	38
23	770970	35·53	999242	·13	771727	35·65	228273	37
24	773101	35·35	999235	·13	773866	35·48	226134	36
25	775223	35·18	999227	·13	775995	35·31	224005	35
26	777333	35·01	999220	·13	778114	35·14	221886	34
27	779434	34·84	999212	·13	780222	34·97	219778	33
28	781524	34·67	999205	·13	782320	34·80	217680	32
29	783605	34·51	999197	·13	784408	34·64	215592	31
30	785675	34·31	999189	·13	786486	34·47	213514	30
31	8·787736	34·18	9·999181	·13	8·788554	34·31	11·211446	29
32	789787	34·02	999174	·13	790613	34·15	209387	28
33	791828	33·86	999166	·13	792662	33·99	207338	27
34	793859	33·70	999158	·13	794701	33·83	205299	26
35	795881	33·54	999150	·13	796731	33·68	203269	25
36	797894	33·39	999142	·13	798752	33·52	201248	24
37	799897	33·23	999134	·13	800763	33·37	199237	23
38	801892	33·08	999126	·13	802765	33·22	197235	22
39	803876	32·93	999118	·13	804758	33·07	195242	21
40	805852	32·78	999110	·13	806742	32·92	193258	20
41	8·807819	32·63	9·999102	·13	8·808717	32·78	11·191283	19
42	809777	32·49	999094	·14	810683	32·62	189317	18
43	811726	32·34	999086	·14	812641	32·48	187359	17
44	813667	32·19	999077	·14	814589	32·33	185411	16
45	815599	32·05	999069	·14	816529	32·19	183471	15
46	817522	31·91	999061	·14	818461	32·05	181539	14
47	819436	31·77	999053	·14	820384	31·91	179616	13
48	821343	31·63	999044	·14	822298	31·77	177702	12
49	823240	31·49	999036	·14	824205	31·63	175795	11
50	825130	31·35	999027	·14	826103	31·50	173897	10
51	8·827011	31·22	9·999019	·14	8·827992	31·36	11·172008	9
52	828884	31·08	999010	·14	829874	31·23	170126	8
53	830749	30·95	999002	·14	831748	31·10	168252	7
54	832607	30·82	998993	·14	833613	30·96	166387	6
55	834456	30·69	998984	·14	835471	30·83	164529	5
56	836297	30·56	998976	·14	837321	30·70	162679	4
57	838130	30·43	998967	·15	839163	30·57	160837	3
58	839956	30·30	998958	·15	840998	30·45	159002	2
59	841774	30·17	998950	·15	842825	30·32	157175	1
60	843585	30·00	998941	·15	844644	30·19	155356	0
	Cosine	D.	Sine	86°	Cotang.	D.	Tang.	M.

M.	Sine	D.	Cosine	D.	Tang.	D.	Cotang.	
0	8·843585	30·05	9·998941	·15	8·844644	30·19	11·155356	60
1	845387	29·92	998932	·15	846455	30·07	153545	59
2	847183	29·80	998923	·15	848260	29·95	151740	58
3	848971	29·67	998914	·15	850057	29·82	149943	57
4	850751	29·55	998905	·15	851846	29·70	148154	56
5	852525	29·43	998896	·15	853628	29·58	146372	55
6	854291	29·31	998887	·15	855403	29·46	144597	54
7	856049	29·19	998878	·15	857171	29·35	142829	53
8	857801	29·07	998869	·15	858932	29·23	141068	52
9	859546	28·96	998860	·15	860686	29·11	139314	51
10	861283	28·84	998851	·15	862433	29·00	137567	50
11	8·863014	28·73	9·998841	·15	8·864173	28·88	11·135827	49
12	864738	28·61	998832	·15	865906	28·77	134094	48
13	866455	28·50	998823	·16	867632	28·66	132368	47
14	868165	28·39	998813	·16	869351	28·54	130649	46
15	869868	28·28	998804	·16	871064	28·43	128936	45
16	871565	28·17	998795	·16	872770	28·32	127230	44
17	873255	28·06	998785	·16	874469	28·21	125531	43
18	874938	27·95	998776	·16	876162	28·11	123838	42
19	876615	27·86	998766	·16	877849	28·00	122151	41
20	878285	27·73	998757	·16	879529	27·89	120471	40
21	8·879949	27·63	9·998747	·16	8·881202	27·79	11·118798	39
22	881607	27·52	998738	·16	882869	27·68	117131	38
23	883258	27·42	998728	·16	884530	27·58	115470	37
24	884903	27·31	998718	·16	886185	27·47	113815	36
25	886542	27·21	998708	·16	887833	27·37	112167	35
26	888174	27·11	998699	·16	889476	27·27	110524	34
27	889801	27·00	998689	·16	891112	27·17	108888	33
28	891421	26·90	998679	·16	892742	27·07	107258	32
29	893035	26·80	998669	·17	894366	26·97	105634	31
30	894643	26·70	998659	·17	895984	26·87	104016	30
31	8·896246	26·60	9·998649	·17	8·897596	26·77	11·102404	29
32	897842	26·51	998639	·17	899203	26·67	100797	28
33	899432	26·41	9986	·17	900803	26·58	099197	27
34	901017	26·31	9986	·17	902398	26·48	097602	26
35	902596	26·22	99	·17	903987	26·38	096013	25
36	904169	26·12	99	·17	905570	26·29	094430	24
37	905736	26·03	99	·17	907147	26·20	092853	23
38	907297	25·93	99	·17	908719	26·10	091281	22
39	908853	25·84	99	·17	910285	26·01	089715	21
40	910404	25·75	99	·17	911846	25·92	088154	20
41	8·911949	25·66	9·99	·17	8·913401	25·83	11·086599	19
42	913488	25·56	99	·17	914951	25·74	085049	18
43	915022	25·47	99	·17	916495	25·65	083505	17
44	916550	25·38	99	·18	918034	25·56	081966	16
45	918073	25·29	99	·18	919568	25·47	080432	15
46	919591	25·20	99	·18	921096	25·38	078904	14
47	921103	25·12	99	·18	922619	25·30	077381	13
48	922610	25·03	99	·18	924136	25·21	075864	12
49	924112	24·94	99	·18	925649	25·12	074351	11
50	925609	24·86	99	·18	927156	25·03	072844	10
51	8·927100	24·77	9·99	·18	8·928658	24·95	11·071342	9
52	928587	24·69	998431	·18	930155	24·86	069845	8
53	930068	24·60	998421	·18	931647	24·78	068353	7
54	931544	24 52	998410	·18	933134	24·70	066866	6
55	933015	24·43	998399	·18	934616	24·61	065384	5
56	934481	24·35	998388	·18	936093	24·53	063907	4
57	935942	24·27	998377	·18	937565	24·45	062435	3
58	937398	24·19	998366	·18	939032	24·37	060968	2
59	938850	24·11	998355	·18	940494	24·30	059506	1
60	940296	24·03	998344	·18	941952	24·21	058048	0
	Cosine	D.	Sine	85°	Cotang.	D.	Tang.	M.

M.	Sine	D.	Cosine	D.	Tang.	D.	Cotang.	
0	8·940296	24·03	9·998344	·19	8·941952	24·21	11·058048	60
1	941738	23·94	998333	·19	943404	24·13	056596	59
2	943174	23·87	998322	·19	944852	24·05	055148	58
3	944606	23·79	998311	·19	946295	23·97	053705	57
4	946034	23·71	998300	·19	947734	23·90	052266	56
5	947456	23·63	998289	·19	949168	23·82	050832	55
6	948874	23·55	998277	·19	950597	23·74	049403	54
7	950287	23·48	998266	·19	952021	23·66	047979	53
8	951696	23·40	998255	·19	953441	23·60	046559	52
9	953100	23·32	998243	·19	954856	23·51	045144	51
10	954499	23·25	998232	·19	956267	23·44	043733	50
11	8·955894	23·17	9·998220	·19	8·957674	23·37	11·042326	49
12	957284	23·10	998209	·19	959075	23·29	040925	48
13	958670	23·02	998197	·19	960473	23·23	039527	47
14	960052	22·95	998186	·19	961866	23·14	038134	46
15	961429	22·88	998174	·19	963255	23·07	036745	45
16	962801	22·80	998163	·19	964639	23·00	035361	44
17	964170	22·73	998151	·19	966019	22·93	033981	43
18	965534	22·66	998139	·20	967394	22·86	032606	42
19	966893	22·59	998128	·20	968766	22·79	031234	41
20	968249	22·52	998116	·20	970133	22·71	029867	40
21	8·969600	22·44	9·998104	·20	8·971496	22·65	11·028504	39
22	970947	22·38	998092	·20	972855	22·57	027145	38
23	972289	22·31	998080	·20	974209	22·51	025791	37
24	973628	22·24	998068	·20	975560	22·44	024440	36
25	974962	22·17	998056	·20	976906	22·37	023094	35
26	976293	22·10	998044	·20	978248	22·30	021752	34
27	977619	22·03	998032	·20	979586	22·23	020414	33
28	978941	21·97	998020	·20	980921	22·17	019079	32
29	980259	21·90	998008	·20	982251	22·10	017749	31
30	981573	21·83	997996	·20	983577	22·04	016423	30
31	8·982883	21·77	9·997985	·20	8·984899	21·97	11·015101	29
32	984189	21·70	997972	·20	986217	21·91	013783	28
33	985491	21·63	997959	·20	987532	21·84	012468	27
34	986789	21·57	997947	·20	988842	21·78	011158	26
35	988083	21·50	997935	·21	990149	21·71	009851	25
36	989374	21·44	997922	·21	991451	21·65	008549	24
37	990660	21·38	997910	·21	992750	21·58	007250	23
38	991943	21·31	997897	·21	994045	21·52	005955	22
39	993222	21·25	997885	·21	995337	21·46	004663	21
40	994497	21·19	997872	·21	996624	21·40	003376	20
41	8·995768	21·12	9·997860	·21	8·997908	21·34	11·002092	19
42	997036	21·06	997847	·21	999188	21·27	000812	18
43	998299	21·00	997835	·21	9·000465	21·21	10·999535	17
44	999560	20·94	997822	·21	001738	21·15	998262	16
45	9·000816	20·87	997809	·21	003007	21·09	996993	15
46	002069	20·82	997797	·21	004272	21·03	995728	14
47	003318	20·76	997784	·21	005534	20·97	994466	13
48	004563	20·70	997771	·21	006792	20·91	993208	12
49	005805	20·64	997758	·21	008047	20·85	991953	11
50	007044	20·58	997745	·21	009298	20·80	990702	10
51	9·008278	20·52	9·997732	·21	9·010546	20·74	10·989454	9
52	009510	20·46	997719	·21	011790	20·68	988210	8
53	010737	20·40	997706	·21	013031	20·62	986969	7
54	011962	20·34	997693	·22	014268	20·56	985732	6
55	013182	20·29	997680	·22	015502	20·51	984498	5
56	014400	20·23	997667	·22	016732	20·45	983268	4
57	015613	20·17	997654	·22	017959	20·40	982041	3
58	016824	20·12	997641	·22	019183	20·33	980817	2
59	018031	20·06	997628	·22	020403	20·28	979597	1
60	019235	20·00	997614	·22	021620	20·23	978380	0
	Cosine	D.	Sine	84°	Cotang.	D.	Tang.	M.

M.	Sine	D.	Cosine	D.	Tang.	D.	Cotang.	
0	9·019235	20·00	9·997614	·22	9·021620	20·23	10·978380	60
1	020435	19·95	997601	·22	022834	20·17	977166	59
2	021632	19·89	997588	·22	024044	20·11	975956	58
3	022825	19·84	997574	·22	025251	20·06	974749	57
4	024016	19·78	997561	·22	026455	20·00	973545	56
5	025203	19·73	997547	·22	027655	19·95	972345	55
6	026386	19·67	997534	·23	028852	19·90	971148	54
7	027567	19·62	997520	·23	030046	19·85	969954	53
8	028744	19·57	997507	·23	031237	19·79	968763	52
9	029918	19·51	997493	·23	032425	19·74	967575	51
10	031089	19·47	997480	·23	033609	19·69	966391	50
11	9·032257	19·41	9·997466	·23	9·034791	19·64	10·965209	49
12	033421	19·36	997452	·23	035969	19·58	964031	48
13	034582	19·30	997439	·23	037144	19·53	962856	47
14	035741	19·25	997425	·23	038316	19·48	961684	46
15	036896	19·20	997411	·23	039485	19·43	960515	45
16	038048	19·15	997397	·23	040651	19·38	959349	44
17	039197	19·10	997383	·23	041813	19·33	958187	43
18	040342	19·05	997369	·23	042973	19·28	957027	42
19	041485	18·99	997355	·23	044130	19·23	955870	41
20	042625	18·94	997341	·23	045284	19·18	954716	40
21	9·043762	18·89	9·997327	·24	9·046434	19·13	10·953566	39
22	044895	18·84	997313	·24	047582	19·08	952418	38
23	046026	18·79	697299	·24	048727	19·03	951273	37
24	047154	18·75	997285	·24	049869	18·98	950131	36
25	048279	18·70	997271	·24	051008	18·93	948992	35
26	049400	18·65	997257	·24	052144	18·89	947856	34
27	050519	18·60	997242	·24	053277	18·84	946723	33
28	051635	18·55	997228	·24	054407	18·79	945593	32
29	052749	18·50	997214	·24	055535	18·74	944465	31
30	053859	18·45	997199	·24	056659	18·70	943341	30
31	9·054966	18·41	9·997185	·24	9·057781	18·65	10·942219	29
32	056071	18·36	997170	·24	058900	18·60	941100	28
33	057172	18·31	997156	·24	060016	18·55	939984	27
34	058271	18·27	997141	·24	061130	18·51	938870	26
35	059367	18·22	997127	·24	062240	18·46	937760	25
36	060460	18·17	997112	·24	063348	18·42	936652	24
37	061551	18·13	997098	·24	064453	18·37	935547	23
38	062639	18·08	997083	·25	065556	18·33	934444	22
39	063724	18·04	997068	·25	066655	18·28	933345	21
40	064806	17·99	997053	·25	067752	18·24	932248	20
41	9·065885	17·94	9·997039	·25	9·068846	18·19	10·931154	19
42	066962	17·90	997024	·25	069938	18·15	930062	18
43	068036	17·86	997009	·25	071027	18·10	928973	17
44	069107	17·81	996994	·25	072113	18·06	927887	16
45	070176	17·77	996979	·25	073197	18·02	926803	15
46	071242	17·72	996964	·25	074278	17·97	925722	14
47	072306	17·68	996949	·25	075356	17·93	924644	13
48	073366	17·63	996934	·25	076432	17·89	923568	12
49	074424	17·59	996919	·25	077505	17·84	922495	11
50	075480	17·55	996904	·25	078576	17·80	921424	10
51	9·076533	17·50	9·996889	·25	9·079644	17·76	10·920356	9
52	077583	17·46	996874	·25	080710	17·72	919290	8
53	078631	17·42	996858	·25	081773	17·67	918227	7
54	079676	17·38	996843	·25	082833	17·63	917167	6
55	080719	17·33	996828	·25	083891	17·59	916109	5
56	081759	17·29	996812	·26	084947	17·55	915053	4
57	082797	17·25	996797	·26	086000	17·51	914000	3
58	083832	17·21	996782	·26	087050	17·47	912950	2
59	084864	17·17	996766	·26	088098	17·43	911902	1
60	085894	17·13	996751	·26	089144	17·38	910856	0
	Cosine	D.	Sine	83°	Cotang.	D.	Tang.	M.

M.	Sine	D.	Cosine	D.	Tang.	D.	Cotang.	
0	9·085894	17·13	9·996751	·26	9·089144	17·38	10·910856	60
1	086922	17·09	996735	·26	090187	17·34	909813	59
2	087947	17·04	996720	·26	091228	17·30	908772	58
3	088970	17·00	996704	·26	092266	17·27	907734	57
4	089990	16·96	996688	·26	093302	17·22	906698	56
5	091008	16·92	996673	·26	094336	17·19	905664	55
6	092024	16·88	996657	·26	095367	17·15	904633	54
7	093037	16·84	996641	·26	096395	17·11	903605	53
8	094047	16·80	996625	·26	097422	17·07	902578	52
9	095056	16·76	996610	·26	098446	17·03	901554	51
10	096062	16·73	996594	·26	099468	16·99	900532	50
11	9·097065	16·68	9·996578	·27	9·100487	16·95	10·899513	49
12	098066	16·65	996562	·27	101504	16·91	898496	48
13	099065	16·61	996546	·27	102519	16·87	897481	47
14	100062	16·57	996530	·27	103532	16·84	896468	46
15	101056	16·53	996514	·27	104542	16·80	895458	45
16	102048	16·49	996498	·27	105550	16·76	894450	44
17	103037	16·45	996482	·27	106556	16·72	893444	43
18	104025	16·41	996465	·27	107559	16·69	892441	42
19	105010	16·38	996449	·27	108560	16·65	891440	41
20	105992	16·34	996433	·27	109559	16·61	890441	40
21	9·106973	16·30	9·996417	·27	9·110556	16·58	10·889444	39
22	107951	16·27	996400	·27	111551	16·54	888449	38
23	108927	16·23	996384	·27	112543	16·50	887457	37
24	109901	16·19	996368	·27	113533	16·46	886467	36
25	110873	16·16	996351	·27	114521	16·43	885479	35
26	111842	16·12	996335	·27	115507	16·39	884493	34
27	112809	16·08	996318	·27	116491	16·36	883509	33
28	113774	16·05	996302	·28	117472	16·32	882528	32
29	114737	16·01	996285	·28	118452	16·29	881548	31
30	115698	15·97	996269	·28	119429	16·25	880571	30
31	9·116656	15·94	9·996252	·28	9·120404	16·22	10·879596	29
32	117613	15·90	996235	·28	121377	16·18	878623	28
33	118567	15·87	996219	·28	122348	16·15	877652	27
34	119519	15·83	996202	·28	123317	16·11	876683	26
35	120469	15·80	996185	·28	124284	16·07	875716	25
36	121417	15·76	996168	·28	125249	16·04	874751	24
37	122362	15·73	996151	·28	126211	16·01	873789	23
38	123306	15·69	996134	·28	127172	15·97	872828	22
39	124248	15·66	996117	·28	128130	15·94	871870	21
40	125187	15·62	996100	·28	129087	15·91	870913	20
41	9·126125	15·59	9·996083	·29	9·130041	15·87	10·869959	19
42	127060	15·56	996066	·29	130994	15·84	869006	18
43	127993	15·52	996049	·29	131944	15·81	868056	17
44	128925	15·49	996032	·29	132893	15·77	867107	16
45	129854	15·45	996015	·29	133839	15·74	866161	15
46	130781	15·42	995998	·29	134784	15·71	865216	14
47	131706	15·39	995980	·29	135726	15·67	864274	13
48	132630	15·35	995963	·29	136667	15·64	863333	12
49	133551	15·32	995946	·29	137605	15·61	862395	11
50	134470	15·29	995928	·29	138542	15·58	861458	10
51	9·135387	15·25	9·995911	·29	9·139476	15·55	10·860524	9
52	136303	15·22	995894	·29	140409	15·51	859591	8
53	137216	15·19	995876	·29	141340	15·48	858660	7
54	138128	15·16	995859	·29	142269	15·45	857731	6
55	139037	15·12	995841	·29	143196	15·42	856804	5
56	139944	15·09	995823	·29	144121	15·39	855879	4
57	140850	15·06	995806	·29	145044	15·35	854956	3
58	141754	15·03	995788	·29	145966	15·32	854034	2
59	142655	15·00	995771	·29	146885	15·29	853115	1
60	143555	14·96	995753	·29	147803	15·26	852197	0
	Cosine	D.	Sine	82°	Cotang.	D.	Tang.	M.

M.	Sine	D.	Cosine	D.	Tang.	D.	Cotang.	
0	9.143555	14.96	9.995753	.30	9.147803	15.26	10.852197	60
1	144453	14.93	995735	.30	148718	15.23	851282	59
2	145349	14.90	995717	.30	149632	15.20	850368	58
3	146243	14.87	995699	.30	150544	15.17	849456	57
4	147136	14.84	995681	.30	151454	15.14	848546	56
5	148026	14.81	995664	.30	152363	15.11	847637	55
6	148915	14.78	995646	.30	153269	15.08	846731	54
7	149802	14.75	995628	.30	154174	15.05	845826	53
8	150686	14.72	995610	.30	155077	15.02	844923	52
9	151569	14.69	995591	.30	155978	14.99	844022	51
10	152451	14.66	995573	.30	156877	14.96	843123	50
11	9.153330	14.63	9.995555	.30	9.157775	14.93	10.842225	49
12	154208	14.60	995537	.30	158671	14.90	841329	48
13	155083	14.57	995519	.30	159565	14.87	840435	47
14	155957	14.54	995501	.31	160457	14.84	839543	46
15	156830	14.51	995482	.31	161347	14.81	838653	45
16	157700	14.48	995464	.31	162236	14.79	837764	44
17	158569	14.45	995446	.31	163123	14.76	836877	43
18	159435	14.42	995427	.31	164008	14.73	835992	42
19	160301	14.39	995409	.31	164892	14.70	835108	41
20	161164	14.36	995390	.31	165774	14.67	834226	40
21	9.162025	14.33	9.995372	.31	9.166654	14.64	10.833346	39
22	162885	14.30	995353	.31	167532	14.61	832468	38
23	163743	14.27	995334	.31	168409	14.58	831591	37
24	164600	14.24	995316	.31	169284	14.55	830716	36
25	165454	14.22	995297	.31	170157	14.53	829843	35
26	166307	14.19	995278	.31	171029	14.50	828971	34
27	167159	14.16	995260	.31	171899	14.47	828101	33
28	168008	14.13	995241	.32	172767	14.44	827233	32
29	168856	14.10	995222	.32	173634	14.42	826366	31
30	169702	14.07	995203	.32	174499	14.39	825501	30
31	9.170547	14.05	9.995184	.32	9.175362	14.36	10.824638	29
32	171389	14.02	995165	.32	176224	14.33	823776	28
33	172230	13.99	995146	.32	177084	14.31	822916	27
34	173070	13.96	995127	.32	177942	14.28	822058	26
35	173908	13.94	995108	.32	178799	14.25	821201	25
36	174744	13.91	995089	.32	179655	14.23	820345	24
37	175578	13.88	995070	.32	180508	14.20	819492	23
38	176411	13.86	995051	.32	181360	14.17	818640	22
39	177242	13.83	995032	.32	182211	14.15	817789	21
40	178072	13.80	995013	.32	183059	14.12	816941	20
41	9.178900	13.77	9.994993	.32	9.183907	14.09	10.816093	19
42	179726	13.74	994974	.32	184752	14.07	815248	18
43	180551	13.72	994955	.32	185597	14.04	814403	17
44	181374	13.69	994935	.32	186439	14.02	813561	16
45	182196	13.66	994916	.33	187280	13.99	812720	15
46	183016	13.64	994896	.33	188120	13.96	811880	14
47	183834	13.61	994877	.33	188958	13.93	811042	13
48	184651	13.59	994857	.33	189794	13.91	810206	12
49	185466	13.56	994838	.33	190629	13.89	809371	11
50	186280	13.53	994818	.33	191462	13.86	808538	10
51	9.187092	13.51	9.994798	.33	9.192294	13.84	10.807706	9
52	187903	13.48	994779	.33	193124	13.81	806876	8
53	188712	13.46	994759	.33	193953	13.79	806047	7
54	189519	13.43	994739	.33	194780	13.76	805220	6
55	190325	13.41	994719	.33	195606	13.74	804394	5
56	191130	13.38	994700	.33	196430	13.71	803570	4
57	191933	13.36	994680	.33	197253	13.69	802747	3
58	192734	13.33	994660	.33	198074	13.66	801926	2
59	193534	13.30	994640	.33	198894	13.64	801106	1
60	194332	13.28	994620	.33	199713	13.61	800287	0
	Cosine	D.	Sine	81°	Cotang.	D.	Tang.	M.

M.	Sine	D.	Cosine	D.	Tang.	D.	Cotang.	
0	9·194332	13·28	9·994620	·33	9·199713	13·61	10·800287	60
1	195129	13·26	994600	·33	200529	13·59	799471	59
2	195925	13·23	994580	·33	201345	13·56	798655	58
3	196719	13·21	994560	·34	202159	13·54	797841	57
4	197511	13·18	994540	·34	202971	13·52	797029	56
5	198302	13·16	994519	·34	203782	13·49	796218	55
6	199091	13·13	994499	·34	204592	13·47	795408	54
7	199879	13·11	994479	·34	205400	13·45	794600	53
8	200666	13·08	994459	·34	206207	13·42	793793	52
9	201451	13·06	994438	·34	207013	13·40	792987	51
10	202234	13·04	994418	·34	207817	13·38	792183	50
11	9·203017	13·01	9·994397	·34	9·208619	13·35	10·791381	49
12	203797	12·99	994377	·34	209420	13·33	790580	48
13	204577	12·96	994357	·34	210220	13·31	789780	47
14	205354	12·94	994336	·34	211018	13·28	788982	46
15	206131	12·92	994316	·34	211815	13·26	788185	45
16	206906	12·89	994295	·34	212611	13·24	787389	44
17	207679	12·87	994274	·35	213405	13·21	786595	43
18	208452	12·85	994254	·35	214198	13·19	785802	42
19	209222	12·82	994233	·35	214989	13·17	785011	41
20	209992	12·80	994212	·35	215780	13·15	784220	40
21	9·210760	12·78	9·994191	·35	9·216568	13·12	10·783432	39
22	211526	12·75	994171	·35	217356	13·10	782644	38
23	212291	12·73	994150	·35	218142	13·08	781858	37
24	213055	12·71	994129	·35	218926	13·05	781074	36
25	213818	12·68	994108	·35	219710	13·03	780290	35
26	214579	12·66	994087	·35	220492	13·01	779508	34
27	215338	12·64	994066	·35	221272	12·99	778728	33
28	216097	12·61	994045	·35	222052	12·97	777948	32
29	216854	12·59	994024	·35	222830	12·94	777170	31
30	217609	12·57	994003	·35	223606	12·92	776394	30
31	9·218363	12·55	9·993981	·35	9·224382	12·90	10·775618	29
32	219116	12·53	993960	·35	225156	12·88	774844	28
33	219868	12·50	993939	·35	225929	12·86	774071	27
34	220618	12·48	993918	·35	226700	12·84	773300	26
35	221367	12·46	993896	·36	227471	12·81	772529	25
36	222115	12·44	993875	·36	228239	12·79	771761	24
37	222861	12·42	993854	·36	229007	12·77	770993	23
38	223606	12·39	993832	·36	229773	12·75	770227	22
39	224349	12·37	993811	·36	230539	12·73	769461	21
40	225092	12·35	993789	·36	231302	12·71	768698	20
41	9·225833	12·33	9·993768	·36	9·232065	12·69	10·767935	19
42	226573	12·31	993746	·36	232826	12·67	767174	18
43	227311	12·28	993725	·36	233586	12·65	766414	17
44	228048	12·26	993703	·36	234345	12·62	765655	16
45	228784	12·24	993681	·36	235103	12·60	764897	15
46	229518	12·22	993660	·36	235859	12·58	764141	14
47	230252	12·20	993638	·36	236614	12·56	763386	13
48	230984	12·18	993616	·36	237368	12·54	762632	12
49	231714	12·16	993594	·37	238120	12·52	761880	11
50	232444	12·14	993572	·37	238872	12·50	761128	10
51	9·233172	12·12	9·993550	·37	9·239622	12·48	10·760378	9
52	233899	12·09	993528	·37	240371	12·46	759629	8
53	234625	12·07	993506	·37	241118	12·44	758882	7
54	235349	12·05	993484	·37	241865	12·42	758135	6
55	236073	12·03	993462	·37	242610	12·40	757390	5
56	236795	12·01	993440	·37	243354	12·38	756646	4
57	237515	11·99	993418	·37	244097	12·36	755903	3
58	238235	11·97	993396	·37	244839	12·34	755161	2
59	238953	11·95	993374	·37	245579	12·32	754421	1
60	239670	11·93	993351	·37	246319	12·30	753681	0
	Cosine	D.	Sine	80°	Cotang.	D.	Tang.	

M.	Sine	D.	Cosine	D.	Tang.	D.	Cotang.	
0	9·239670	11·93	9·993351	·37	9·246319	12·30	10·753681	60
1	240386	11·91	993329	·37	247057	12·28	752943	59
2	241101	11·89	993307	·37	247794	12·26	752206	58
3	241814	11·87	993285	·37	248530	12·24	751470	57
4	242526	11·85	993262	·37	249264	12·22	750736	56
5	243237	11·83	993240	·37	249998	12·20	750002	55
6	243947	11·81	993217	·38	250730	12·18	749270	54
7	244656	11·79	993195	·38	251461	12·17	748539	53
8	245363	11·77	993172	·38	252191	12·15	747809	52
9	246069	11·75	993149	·38	252920	12·13	747080	51
10	246775	11·73	993127	·38	253648	12·11	746352	50
11	9·247478	11·71	9·993104	·38	9·254374	12·09	10·745626	49
12	248181	11·69	993081	·38	255100	12·07	744900	48
13	248883	11·67	993059	·38	255824	12·05	744176	47
14	249583	11·65	993036	·38	256547	12·03	743453	46
15	250282	11·63	993013	·38	257269	12·01	742731	45
16	250980	11·61	992990	·38	257990	12·00	742010	44
17	251677	11·59	992967	·38	258710	11·98	741290	43
18	252373	11·58	992944	·38	259429	11·96	740571	42
19	253067	11·56	992921	·38	260146	11·94	739854	41
20	253761	11·54	992898	·38	260863	11·92	739137	40
21	9·254453	11·52	9·992875	·38	9·261578	11·90	10·738422	39
22	255144	11·50	992852	·38	262292	11·89	737708	38
23	255834	11·48	992829	·39	263005	11·87	736995	37
24	256523	11·46	992806	·39	263717	11·85	736283	36
25	257211	11·44	992783	·39	264428	11·83	735572	35
26	257898	11·42	992759	·39	265138	11·81	734862	34
27	258583	11·41	992736	·39	265847	11·79	734153	33
28	259268	11·39	992713	·39	266555	11·78	733445	32
29	259951	11·37	992690	·39	267261	11·76	732739	31
30	260633	11·35	992666	·39	267967	11·74	732033	30
31	9·261314	11·33	9·992643	·39	9·268671	11·72	10·731329	29
32	261994	11·31	992619	·39	269375	11·70	730625	28
33	262673	11·30	992596	·39	270077	11·69	729923	27
34	263351	11·28	992572	·39	270779	11·67	729221	26
35	264027	11·26	992549	·39	271479	11·65	728521	25
36	264703	11·24	992525	·39	272178	11·64	727822	24
37	265377	11·22	992501	·39	272876	11·62	727124	23
38	266051	11·20	992478	·40	273573	11·60	726427	22
39	266723	11·19	992454	·40	274269	11·58	725731	21
40	267395	11·17	992430	·40	274964	11·57	725036	20
41	9·268065	11·15	9·992406	·40	9·275658	11·55	10·724342	19
42	268734	11·13	992382	·40	276351	11·53	723649	18
43	269402	11·11	992359	·40	277043	11·51	722957	17
44	270069	11·10	992335	·40	277734	11·50	722266	16
45	270735	11·08	992311	·40	278424	11·48	721576	15
46	271400	11·06	992287	·40	279113	11·47	720887	14
47	272064	11·05	992263	·40	279801	11·45	720199	13
48	272726	11·03	992239	·40	280488	11·43	719512	12
49	273388	11·01	992214	·40	281174	11·41	718826	11
50	274049	10·99	992190	·40	281858	11·40	718142	10
51	9·274708	10·98	9·992166	·40	9·282542	11·38	10·717458	9
52	275367	10·96	992142	·40	283225	11·36	716775	8
53	276024	10·94	992117	·41	283907	11·35	716093	7
54	276681	10·92	992093	·41	284588	11·33	715412	6
55	277337	10·91	992069	·41	285268	11·31	714732	5
56	277991	10·89	992044	·41	285947	11·30	714053	4
57	278644	10·87	992020	·41	286624	11·28	713376	3
58	279297	10·86	991996	·41	287301	11·26	712699	2
59	279948	10·84	991971	·41	287977	11·25	712023	1
60	280599	10·82	991947	·41	288652	11·23	711348	0
	Cosine	D.	Sine	79°	Cotang.	D.	Tang.	M.

M.	Sine	D.	Cosine	D.	Tang.	D.	Cotang.	
0	9·280599	10·82	9·991947	·41	9·288652	11·23	10·711348	60
1	281248	10·81	991922	·41	289326	11·22	710674	59
2	281897	10·79	991897	·41	289999	11·20	710001	58
3	282544	10·77	991873	·41	290671	11·18	709329	57
4	283190	10·76	991848	·41	291342	11·17	708658	56
5	283836	10·74	991823	·41	292013	11·15	707987	55
6	284480	10·72	991799	·41	292682	11·14	707318	54
7	285124	10·71	991774	·42	293350	11·12	706650	53
8	285766	10·69	991749	·42	294017	11·11	705983	52
9	286408	10·67	991724	·42	294684	11·09	705316	51
10	287048	10·66	991699	·42	295349	11·07	704651	50
11	9·287687	10·64	9·991674	·42	9·296013	11·06	10·703987	49
12	288326	10·63	991649	·42	296677	11·04	703323	48
13	288964	10·61	991624	·42	297339	11·03	702661	47
14	289600	10·59	991599	·42	298001	11·01	701999	46
15	290236	10·58	991574	·42	298662	11·00	701338	45
16	290870	10·56	991549	·42	299322	10·98	700678	44
17	291504	10·54	991524	·42	299980	10·96	700020	43
18	292137	10·53	991498	·42	300638	10·95	699362	42
19	292768	10·51	991473	·42	301295	10·93	698705	41
20	293399	10·50	991448	·42	301951	10·92	698049	40
21	9·294029	10·48	9·991422	·42	9·302607	10·90	10·697393	39
22	294658	10·46	991397	·42	303261	10·89	696739	38
23	295286	10·45	991372	·43	303914	10·87	696086	37
24	295913	10·43	991346	·43	304567	10·86	695433	36
25	296539	10·42	991321	·43	305218	10·84	694782	35
26	297164	10·40	991295	·43	305869	10·83	694131	34
27	297788	10·39	991270	·43	306519	10·81	693481	33
28	298412	10·37	991244	·43	307168	10·80	692832	32
29	299034	10·36	991218	·43	307815	10·78	692185	31
30	299655	10·34	991193	·43	308463	10·77	691537	30
31	9·300276	10·32	9·991167	·43	9·309109	10·75	10·690891	29
32	300895	10·31	991141	·43	309754	10·74	690246	28
33	301514	10·29	991115	·43	310398	10·73	689602	27
34	302132	10·28	991090	·43	311042	10·71	688958	26
35	302748	10·26	991064	·43	311685	10·70	688315	25
36	303364	10·25	991038	·43	312327	10·68	687673	24
37	303979	10·23	991012	·43	312967	10·67	687033	23
38	304593	10·22	990986	·43	313608	10·65	686392	22
39	305207	10·20	990960	·43	314247	10·64	685753	21
40	305819	10·19	990934	·44	314885	10·62	685115	20
41	9·306430	10·17	9·990908	·44	9·315523	10·61	10·684477	19
42	307041	10·16	990882	·44	316159	10·60	683841	18
43	307650	10·14	990855	·44	316795	10·58	683205	17
44	308259	10·13	990829	·44	317430	10·57	682570	16
45	308867	10·11	990803	·44	318064	10·55	681936	15
46	309474	10·10	990777	·44	318697	10·54	681303	14
47	310080	10·08	990750	·44	319329	10·53	680671	13
48	310685	10·07	990724	·44	319961	10·51	680039	12
49	311289	10·05	990697	·44	320592	10·50	679408	11
50	311893	10·04	990671	·44	321222	10·48	678778	10
51	9·312495	10·03	9·990644	·44	9·321851	10·47	10·678149	9
52	313097	10·01	990618	·44	322479	10·45	677521	8
53	313698	10·00	990591	·44	323106	10·44	676894	7
54	314297	9·98	990565	·44	323733	10·43	676267	6
55	314897	9·97	990538	·44	324358	10·41	675642	5
56	315495	9·96	990511	·45	324983	10·40	675017	4
57	316092	9·94	990485	·45	325607	10·39	674393	3
58	316689	9·93	990458	·45	326231	10·37	673769	2
59	317284	9·91	990431	·45	326853	10·36	673147	1
60	317879	9·90	990404	·45	327475	10·35	672525	0
	Cosine	D.	Sine	78°	Cotang.	D.	Tang.	M.

M.	Sine	D.	Cosine	D.	Tang.	D.	Cotang.	
0	9·317879	9·90	9·990404	·45	9·327474	10·35	10·672526	60
1	318473	9·88	990378	·45	328095	10·33	671905	59
2	319066	9·87	990351	·45	328715	10·32	671285	58
3	319658	9·86	990324	·45	329334	10·30	670666	57
4	320249	9·84	990297	·45	329953	10·29	670047	56
5	320840	9·83	990270	·45	330570	10·28	669430	55
6	321430	9·82	990243	·45	331187	10·26	668813	54
7	322019	9·80	990215	·45	331803	10·25	668197	53
8	322607	9·79	990188	·45	332418	10·24	667582	52
9	323194	9·77	990161	·45	333033	10·23	666967	51
10	323780	9·76	990134	·45	333646	10·21	666354	50
11	9·324366	9·75	9·990107	·46	9·334259	10·20	10·665741	49
12	324950	9·73	990079	·46	334871	10·19	665129	48
13	325534	9·72	990052	·46	335482	10·17	664518	47
14	326117	9·70	990025	·46	336093	10·16	663907	46
15	326700	9·69	989997	·46	336702	10·15	663298	45
16	327281	9·68	989970	·46	337311	10·13	662689	44
17	327862	9·66	989942	·46	337919	10·12	662081	43
18	328442	9·65	989915	·46	338527	10·11	661473	42
19	329021	9·64	989887	·46	339133	10·10	660867	41
20	329599	9·62	989860	·46	339739	10·08	660261	40
21	9·330176	9·61	9·989832	·46	9·340344	10·07	10·659656	39
22	330753	9·60	989804	·46	340948	10·06	659052	38
23	331329	9·58	989777	·46	341552	10·04	658448	37
24	331903	9·57	989749	·47	342155	10·03	657845	36
25	332478	9·56	989721	·47	342757	10·02	657243	35
26	333051	9·54	989693	·47	343358	10·00	656642	34
27	333624	9·53	989665	·47	343958	9·99	656042	33
28	334195	9·52	989637	·47	344558	9·98	655442	32
29	334766	9·50	989609	·47	345157	9·97	654843	31
30	335337	9·49	989582	·47	345755	9·96	654245	30
31	9·335906	9·48	9·989553	·47	9·346353	9·94	10·653647	29
32	336475	9·46	989525	·47	346949	9·93	653051	28
33	337043	9·45	989497	·47	347545	9·92	652455	27
34	337610	9·44	989469	·47	348141	9·91	651859	26
35	338176	9·43	989441	·47	348735	9·90	651265	25
36	338742	9·41	989413	·47	349329	9·88	650671	24
37	339306	9·40	989384	·47	349922	9·87	650078	23
38	339871	9·39	989356	·47	350514	9·86	649486	22
39	340434	9·37	989328	·47	351106	9·85	648894	21
40	340996	9·36	989300	·47	351697	9·83	648303	20
41	9·341558	9·35	9·989271	·47	9·352287	9·82	10·647713	19
42	342119	9·34	989243	·47	352876	9·81	647124	18
43	342679	9·32	989214	·47	353465	9·80	646535	17
44	343239	9·31	989186	·47	354053	9·79	645947	16
45	343797	9·30	989157	·47	354640	9·77	645360	15
46	344355	9·29	989128	·48	355227	9·76	644773	14
47	344912	9·27	989100	·48	355813	9·75	644187	13
48	345469	9·26	989071	·48	356398	9·74	643602	12
49	346024	9·25	989042	·48	356982	9·73	643018	11
50	346579	9·24	989014	·48	357566	9·71	642434	10
51	9·347134	9·22	9·988985	·48	9·358149	9·70	10·641851	9
52	347687	9·21	988956	·48	358731	9·69	641269	8
53	348240	9·20	988927	·48	359313	9·68	640687	7
54	348792	9·19	988898	·48	359893	9·67	640107	6
55	349343	9·17	988869	·48	360474	9·66	639526	5
56	349893	9·16	988840	·48	361053	9·65	638947	4
57	350443	9·15	988811	·49	361632	9·63	638368	3
58	350992	9·14	988782	·49	362210	9·62	637790	2
59	351540	9·13	988753	·49	362787	9·61	637213	1
60	352088	9·11	988724	·49	363364	9·60	636636	0
	Cosine	D.	Sine	77°	Cotang.	D.	Tang.	M.

M.	Sine	D.	Cosine	D.	Tang.	D.	Cotang.	
0	9·352088	9·11	9·988724	·49	9·363364	9·60	10·636636	60
1	352635	9·10	988695	·49	363940	9·59	636060	59
2	353181	9·09	988666	·49	364515	9·58	635485	58
3	353726	9·08	988636	·49	365090	9·57	634910	57
4	354271	9·07	988607	·49	365664	9·55	634336	56
5	354815	9·05	988578	·49	366237	9·54	633763	55
6	355358	9·04	988548	·49	366810	9·53	633190	54
7	355901	9·03	988519	·49	367382	9·52	632618	53
8	356443	9·02	988489	·49	367953	9·51	632047	52
9	356984	9·01	988460	·49	368524	9·50	631476	51
10	357524	8·99	988430	·49	369094	9·49	630906	50
11	9·358064	8·98	9·988401	·49	9·369663	9·48	10·630337	49
12	358603	8·97	988371	·49	370232	9·46	629768	48
13	359141	8·96	988342	·49	370799	9·45	629201	47
14	359678	8·95	988312	·50	371367	9·44	628633	46
15	360215	8·93	988282	·50	371933	9·43	628067	45
16	360752	8·92	988252	·50	372499	9·42	627501	44
17	361287	8·91	988223	·50	373064	9·41	626936	43
18	361822	8·90	988193	·50	373629	9·40	626371	42
19	362356	8·89	988163	·50	374193	9·39	625807	41
20	362889	8·88	988133	·50	374756	9·38	625244	40
21	9·363422	8·87	9·988103	·50	9·375319	9·37	10·624681	39
22	363954	8·85	988073	·50	375881	9·35	624119	38
23	364485	8·84	988043	·50	376442	9·34	623558	37
24	365016	8·83	988013	·50	377003	9·33	622997	36
25	365546	8·82	987983	·50	377563	9·32	622437	35
26	366075	8·81	987953	·50	378122	9·31	621878	34
27	366604	8·80	987922	·50	378681	9·30	621319	33
28	367131	8·79	987892	·50	379239	9·29	620761	32
29	367659	8·77	987862	·50	379797	9·28	620203	31
30	368185	8·76	987832	·51	380354	9·27	619646	30
31	9·368711	8·75	9·987801	·51	9·380910	9·26	10·619090	29
32	369236	8·74	987771	·51	381466	9·25	618534	28
33	369761	8·73	987740	·51	382020	9·24	617980	27
34	370285	8·72	987710	·51	382575	9·23	617425	26
35	370808	8·71	987679	·51	383129	9·22	616871	25
36	371330	8·70	987649	·51	383682	9·21	616318	24
37	371852	8·69	987618	·51	384234	9·20	615766	23
38	372373	8·67	987588	·51	384786	9·19	615214	22
39	372894	8·66	987557	·51	385337	9·18	614663	21
40	373414	8·65	987526	·51	385888	9·17	614112	20
41	9·373933	8·64	9·987496	·51	9·386438	9·15	10·613562	19
42	374452	8·63	987465	·51	386987	9·14	613013	18
43	374970	8·62	987434	·51	387536	9·13	612464	17
44	375487	8·61	987403	·52	388084	9·12	611916	16
45	376003	8·60	687372	·52	388631	9·11	611369	15
46	376519	8·59	987341	·52	389178	9·10	610822	14
47	377035	8·58	987310	·52	389724	9·09	610276	13
48	377549	8·57	987279	·52	390270	9·08	609730	12
49	378063	8·56	987248	·52	390815	9·07	609185	11
50	378577	8·54	987217	·52	391360	9·06	608640	10
51	9·379089	8·53	9·987186	·52	9·391903	9·05	10·608097	9
52	379601	8·52	987155	·52	392447	9·04	607553	8
53	380113	8·51	987124	·52	392989	9·03	607011	7
54	380624	8·50	987092	·52	393531	9·02	606469	6
55	381134	8·49	987061	·52	394073	9·01	605927	5
56	381643	8·48	987030	·52	394614	9·00	605386	4
57	382152	8·47	986998	·52	395154	8·99	604846	3
58	382661	8·46	986967	·52	395694	8·98	604306	2
59	383168	8·45	986936	·52	396233	8·97	603767	1
60	383675	8·44	986904	·52	396771	8·96	603229	0
	Cosine	D.	Sine	76°	Cotang.	D.	Tang.	M.

M.	Sine	D.	Cosine	D.	Tang.	D.	Cotang.	
0	9·383675	8·44	9·986904	·52	9·396771	8·96	10·603229	60
1	384182	8·43	986873	·53	397309	8·96	602691	59
2	384687	8·42	986841	·53	397846	8·95	602154	58
3	385192	8·41	986809	·53	398383	8·94	601617	57
4	385697	8·40	986778	·53	398919	8·93	601081	56
5	386201	8·39	986746	·53	399455	8·92	600545	55
6	386704	8·38	986714	·53	399990	8·91	600010	54
7	387207	8·37	986683	·53	400524	8·90	599476	53
8	387709	8·36	986651	·53	401058	8·89	598942	52
9	388210	8·35	986619	·53	401591	8·88	598409	51
10	388711	8·34	986587	·53	402124	8·87	597876	50
11	9·389211	8·33	9·986555	·53	9·402656	8·86	10·597344	49
12	389711	8·32	986523	·53	403187	8·85	596813	48
13	390210	8·31	986491	·53	403718	8·84	596282	47
14	390708	8·30	986459	·53	404249	8·83	595751	46
15	391206	8·28	986427	·53	404778	8·82	595222	45
16	391703	8·27	986395	·53	405308	8·81	594692	44
17	392199	8·26	986363	·54	405836	8·80	594164	43
18	392695	8·25	986331	·54	406364	8·79	593636	42
19	393191	8·24	986299	·54	406892	8·78	593108	41
20	393685	8·23	986266	·54	407419	8·77	592581	40
21	9·394179	8·22	9·986234	·54	9·407945	8·76	10·592055	39
22	394673	8·21	986202	·54	408471	8·75	591529	38
23	395166	8·20	986169	·54	408997	8·74	591003	37
24	395658	8·19	986137	·54	409521	8·74	590479	36
25	396150	8·18	986104	·54	410045	8·73	589955	35
26	396641	8·17	986072	·54	410569	8·72	589431	34
27	397132	8·17	986039	·54	411092	8·71	588908	33
28	397621	8·16	986007	·54	411615	8·70	588385	32
29	398111	8·15	985974	·54	412137	8·69	587863	31
30	398600	8·14	985942	·54	412658	8·68	587342	30
31	9·399088	8·13	9·985909	·55	9·413179	8·67	10·586821	29
32	399575	8·12	985876	·55	413699	8·66	586301	28
33	400062	8·11	985843	·55	414219	8· 5	585781	27
34	400549	8·10	985811	·55	414738	8· 4	585262	26
35	401035	8·09	985778	·55	415257	8· 4	584743	25
36	401520	8·08	985745	·55	415775	8· 3	584225	24
37	402005	8·07	985712	·55	416293	8· 2	583707	23
38	402489	8·06	985679	·55	416810	8·61	583190	22
39	402972	8·05	985646	·55	417326	8·60	582674	21
40	403455	8·04	985613	·55	417842	8·59	582158	20
41	9·403938	8·03	9·985580	·55	9·418358	8·58	10·58·642	19
42	404420	8·02	985547	·55	418873	8·57	581127	18
43	404901	8·01	985514	·55	419387	8·56	580613	17
44	405382	8·00	985480	·55	419901	8·55	580099	16
45	405862	7·99	985447	·55	420415	8·55	579585	15
46	406341	7· 8	985414	·56	420927	8·54	579073	14
47	406820	7· 7	985380	·56	421440	8·53	578560	13
48	407299	7· 6	985347	·56	421952	8·52	578048	12
49	407777	7· 5	985314	·56	422463	8·51	577537	11
50	408254	7· 4	985280	·56	422974	8·50	577026	10
51	9·408731	7· 4	9·985247	·56	9·423484	8·49	10·576516	9
52	409207	7· 3	985213	·56	423993	8·48	576007	8
53	409682	7·92	985180	·56	424503	8·48	575497	7
54	410157	7·91	985146	·56	425011	8·47	574989	6
55	410632	7·90	985113	·56	425519	8·46	574481	5
56	411106	7·89	985079	·56	426027	8·45	573973	4
57	411579	7·88	985045	·56	426534	8·44	573466	3
58	412052	7·87	985011	·56	427041	8·43	572959	2
59	412524	7·86	984978	·56	427547	8·43	572453	1
60	412996	7·85	984944	·56	428052	8·42	571948	0
	Cosine	D.	Sine	75°	Cotang.	D.	Tang.	M.

M.	Sine	D.	Cosine	D.	Tang.	D.	Cotang.	
0	9·412996	7·85	9·984944	·57	9·428052	8·42	10·571948	60
1	413467	7·84	984910	·57	428557	8·41	571443	59
2	413938	7·83	984876	·57	429062	8·40	570938	58
3	414408	7·83	984842	·57	429566	8·39	570434	57
4	414878	7·82	984808	·57	430070	8·38	569930	56
5	415347	7·81	984774	·57	430573	8·38	569427	55
6	415815	7·80	984740	·57	431075	8·37	568925	54
7	416283	7·79	984706	·57	431577	8·36	568423	53
8	416751	7·78	984672	·57	432079	8·35	567921	52
9	417217	7·77	984637	·57	432580	8·34	567420	51
10	417684	7·76	984603	·57	433080	8·33	566920	50
11	9·418150	7·75	9·984569	·57	9·433580	8·32	10·566420	49
12	418615	7·74	984535	·57	434080	8·32	565920	48
13	419079	7·73	984500	·57	434579	8·31	565421	47
14	419544	7·73	984466	·57	435078	8·30	564922	46
15	420007	7·72	984432	·58	435576	8·29	564424	45
16	420470	7·71	984397	·58	436073	8·28	563927	44
17	420933	7·70	984363	·58	436570	8·28	563430	43
18	421395	7·69	984328	·58	437067	8·27	562933	42
19	421857	7·68	984294	·58	437563	8·26	562437	41
20	422318	7·67	984259	·58	438059	8·25	561941	40
21	9·422778	7·67	9·984224	·58	9·438554	8·24	10·561446	39
22	423238	7·66	984190	·58	439048	8·23	560952	38
23	423697	7·65	984155	·58	439543	8·23	560457	37
24	424156	7·64	984120	·58	440036	8·22	559964	36
25	424615	7·63	984085	·58	440529	8·21	559471	35
26	425073	7·62	984050	·58	441022	8·20	558978	34
27	425530	7·61	984015	·58	441514	8·19	558486	33
28	425987	7·60	983981	·58	442006	8·19	557994	32
29	426443	7·60	983946	·58	442497	8·18	557503	31
30	426899	7·59	983911	·58	442988	8·17	557012	30
31	9·427354	7·58	9·983875	·58	9·443479	8·16	10·556521	29
32	427809	7·57	983840	·59	443968	8·16	556032	28
33	428263	7·56	983805	·59	444458	8·15	555542	27
34	428717	7·55	983770	·59	444947	8·14	555053	26
35	429170	7·54	983735	·59	445435	8·13	554565	25
36	429623	7·53	983700	·59	445923	8·12	554077	24
37	430075	7·52	983664	·59	446411	8·12	553589	23
38	430527	7·52	983629	·59	446898	8·11	553102	22
39	430978	7·51	983594	·59	447384	8·10	552616	21
40	431429	7·50	983558	·59	447870	8·09	552130	20
41	9·431879	7·49	9·983523	·59	9·448356	8·09	10·551644	19
42	432329	7·49	983487	·59	448841	8·08	551159	18
43	432778	7·48	983452	·59	449326	8·07	550674	17
44	433226	7·47	983416	·59	449810	8·06	550190	16
45	433675	7·46	983381	·59	450294	8·06	549706	15
46	434122	7·45	983345	·59	450777	8·05	549223	14
47	434569	7·44	983309	·59	451260	8·04	548740	13
48	435016	7·44	983273	·60	451743	8·03	548257	12
49	435462	7·43	983238	·60	452225	8·02	547775	11
50	435908	7·42	983202	·60	452706	8·02	547294	10
51	9·436353	7·41	9·983166	·60	9·453187	8·01	10·546813	9
52	436798	7·40	983130	·60	453668	8·00	546332	8
53	437242	7·40	983094	·60	454148	7·99	545852	7
54	437686	7·39	983058	·60	454628	7·99	545372	6
55	438129	7·38	983022	·60	455107	7·98	544893	5
56	438572	7·37	982986	·60	455586	7·97	544414	4
57	439014	7·36	982950	·60	456064	7·96	543936	3
58	439456	7·36	982914	·60	456542	7·96	543458	2
59	439897	7·35	982878	·60	457019	7·95	542981	1
60	440338	7·34	982842	·60	457496	7·94	542504	0
	Cosine	D.	Sine	74°	Cotang.	D.	Tang.	M.

M.	Sine	D.	Cosine	D.	Tang.	D.	Cotang.	
0	9·440338	7·32	9·982842	·60	9·457496	7·94	10·542504	60
1	440778	7·33	982805	·60	457973	7·93	542027	59
2	441218	7·32	982769	·61	458449	7·93	541551	58
3	441658	7·31	982733	·61	458925	7·92	541075	57
4	442096	7·31	982696	·61	459400	7·91	540600	56
5	442535	7·30	982660	·61	459875	7·90	540125	55
6	442973	7·29	982624	·61	460349	7·90	539651	54
7	443410	7·28	982587	·61	460823	7·89	539177	53
8	443847	7·27	982551	·61	461297	7·88	538703	52
9	444284	7·27	982514	·61	461770	7·88	538230	51
10	444720	7·26	982477	·61	462242	7·87	537758	50
11	9·445155	7·25	9·982441	·61	9·462714	7·86	10·537286	49
12	445590	7·24	982404	·61	463186	7·85	536814	48
13	446025	7·23	982367	·61	463658	7·85	536342	47
14	446459	7·23	982331	·61	464129	7·84	535871	46
15	446893	7·22	982294	·61	464599	7·83	535401	45
16	447326	7·21	982257	·61	465069	7·83	534931	44
17	447759	7·20	982220	·62	465539	7·82	534461	43
18	448191	7·20	982183	·62	466008	7·81	533992	42
19	448623	7·19	982146	·62	466476	7·80	533524	41
20	449054	7·18	982109	·62	466945	7·80	533055	40
21	9·449485	7·17	9·982072	·62	9·467413	7·79	10·532587	39
22	449915	7·16	982035	·62	467880	7·78	532120	38
23	450345	7·16	981998	·62	468347	7·78	531653	37
24	450775	7·15	981961	·62	468814	7·77	531186	36
25	451204	7·14	981924	·62	469280	7·76	530720	35
26	451632	7·13	981886	·62	469746	7·75	530254	34
27	452060	7·13	981849	·62	470211	7·75	529789	33
28	452488	7·12	981812	·62	470676	7·74	529324	32
29	452915	7·11	981774	·62	471141	7·73	528859	31
30	453342	7·10	981737	·62	471605	7·73	528395	30
31	9·453768	7·10	9·981699	·63	9·472068	7·72	10·527932	29
32	454194	7·09	981662	·63	472532	7·71	527468	28
33	454619	7·08	981625	·63	472995	7·71	527005	27
34	455044	7·07	981587	·63	473457	7·70	526543	26
35	455469	7·07	981549	·63	473919	7·69	526081	25
36	455893	7·06	981512	·63	474381	7·69	525619	24
37	456316	7·05	981474	·63	474842	7·68	525158	23
38	456739	7·04	981436	·63	475303	7·67	524697	22
39	457162	7·04	981399	·63	475763	7·67	524237	21
40	457584	7·03	981361	·63	476223	7·66	523777	20
41	9·458006	7·02	9·981323	·63	9·476683	7·65	10·523317	19
42	458427	7·01	981285	·63	477142	7·65	522858	18
43	458848	7·01	981247	·63	477601	7·64	522399	17
44	459268	7·00	981209	·63	478059	7·63	521941	16
45	459688	6·99	981171	·63	478517	7·63	521483	15
46	460108	6·98	981133	·64	478975	7·62	521025	14
47	460527	6·98	981095	·64	479432	7·61	520568	13
48	460946	6·97	981057	·64	479889	7·61	520111	12
49	461364	6·96	981019	·64	480345	7·60	519655	11
50	461782	6·95	980981	·64	480801	7·59	519199	10
51	9·462199	6·95	9·980942	·64	9·481257	7·59	10·518743	9
52	462616	6·94	980904	·64	481712	7·58	518288	8
53	463032	6·93	980866	·64	482167	7·57	517833	7
54	463448	6·93	980827	·64	482621	7·57	517379	6
55	463864	6·92	980789	·64	483075	7·56	516925	5
56	464279	6·91	980750	·64	483529	7·55	516471	4
57	464694	6·90	980712	·64	483982	7·55	516018	3
58	465108	6·90	980673	·64	484435	7·54	515565	2
59	465522	6·89	980635	·64	484887	7·53	515113	1
60	465935	6·88	980596	·64	485339	7·53	514661	0
	Cosine	D.	Sine	73°	Cotang.	D.	Tang.	M.

M.	Sine	D.	Cosine	D.	Tang.	D.	Cotang.	
0	9·465935	6·88	9·980596	·64	9·485339	7·55	10·514661	60
1	466348	6·88	980558	·64	485791	7·52	514209	59
2	466761	6·87	980519	·65	486242	7·51	513758	58
3	467173	6·86	980480	·65	486693	7·51	513307	57
4	467585	6·85	980442	·65	487143	7·50	512857	56
5	467996	6·85	980403	·65	487593	7·49	512407	55
6	468407	6·84	980364	·65	488043	7·49	511957	54
7	468817	6·83	980325	·65	488492	7·48	511508	53
8	469227	6·83	980286	·65	488941	7·47	511059	52
9	469637	6·82	980247	·65	489390	7·47	510610	51
10	470046	6·81	980208	·65	489838	7·46	510162	50
11	9·470455	6·80	9·980169	·65	9·490286	7·46	10·509714	49
12	470863	6·80	980130	·65	490733	7·45	509267	48
13	471271	6·79	980091	·65	491180	7·44	508820	47
14	471679	6·78	980052	·65	491627	7·44	508373	46
15	472086	6·78	980012	·65	492073	7·43	507927	45
16	472492	6·77	979973	·65	492519	7·43	507481	44
17	472898	6·76	979934	·66	492965	7·42	507035	43
18	473304	6·76	979895	·66	493410	7·41	506590	42
19	473710	6·75	979855	·66	493854	7·40	506146	41
20	474115	6·74	979816	·66	494299	7·40	505701	40
21	9·474519	6·74	9·979776	·66	9·494743	7·40	10·505257	39
22	474923	6·73	979737	·66	495186	7·39	504814	38
23	475327	6·72	979697	·66	495630	7·38	504370	37
24	475730	6·72	979658	·66	496073	7·37	503927	36
25	476133	6·71	979618	·66	496515	7·37	503485	35
26	476536	6·70	979579	·66	496957	7·36	503043	34
27	476938	6·69	979539	·66	497399	7·36	502601	33
28	477340	6·69	979499	·66	497841	7·35	502159	32
29	477741	6·68	979459	·66	498282	7·34	501718	31
30	478142	6·67	979420	·66	498722	7·34	501278	30
31	9·478542	6·67	9·979380	·66	9·499163	7·33	10·500837	29
32	478942	6·66	979340	·66	499603	7·33	500397	28
33	479342	6·65	979300	·67	500042	7·32	499958	27
34	479741	6·65	979260	·67	500481	7·31	499519	26
35	480140	6·64	979220	·67	500920	7·31	499080	25
36	480539	6·63	979180	·67	501359	7·30	498641	24
37	480937	6·63	979140	·67	501797	7·30	498203	23
38	481334	6·62	979100	·67	502235	7·29	497765	22
39	481731	6·61	979059	·67	502672	7·28	497328	21
40	482128	6·61	979019	·67	503109	7·28	496891	20
41	9·482525	6·60	9·978979	·67	9·503546	7·27	10·496454	19
42	482921	6·59	978939	·67	503982	7·27	496018	18
43	483316	6·59	978898	·67	504418	7·26	495582	17
44	483712	6·58	978858	·67	504854	7·25	495146	16
45	484107	6·57	978817	·67	505289	7·25	494711	15
46	484501	6·57	978777	·67	505724	7·24	494276	14
47	484895	6·56	978736	·67	506159	7·24	493841	13
48	485289	6·55	978696	·68	506593	7·23	493407	12
49	485682	6·55	978655	·68	507027	7·22	492973	11
50	486075	6·54	978615	·68	507460	7·22	492540	10
51	9·486467	6·53	9·978574	·68	9·507893	7·21	10·492107	9
52	486860	6·53	978533	·68	508326	7·21	491674	8
53	487251	6·52	978493	·68	508759	7·20	491241	7
54	487643	6·51	978452	·68	509191	7·19	490809	6
55	488034	6·51	978411	·68	509622	7·19	490378	5
56	488424	6·50	978370	·68	510054	7·18	489946	4
57	488814	6·50	978329	·68	510485	7·18	489515	3
58	489204	6·49	978288	·68	510916	7·17	489084	2
59	489593	6·48	978247	·68	511346	7·16	488654	1
60	489982	6·48	978206	·68	511776	7·16	488224	0
	Cosine	D.	Sine	72°	Cotang.	D.	Tang.	M.

M.	Sine	D.	Cosine	D.	Tang.	D.	Cotang.	
0	9·489982	6·48	9·978206	·68	9·511776	7·16	10·488224	60
1	490371	6·48	978165	·68	512206	7·16	487794	59
2	490759	6·47	978124	·68	512635	7·15	487365	58
3	491147	6·46	978083	·69	513064	7·14	486936	57
4	491535	6·46	978042	·69	513493	7·14	486507	56
5	491922	6·45	978001	·69	513921	7·13	486079	55
6	492308	6·44	977959	·69	514349	7·13	485651	54
7	492695	6·44	977918	·69	514777	7·12	485223	53
8	493081	6·43	977877	·69	515204	7·12	484796	52
9	493466	6·42	977835	·69	515631	7·11	484369	51
10	493851	6·42	977794	·69	516057	7·10	483943	50
11	9·494236	6·41	9·977752	·69	9·516484	7·10	10·483516	49
12	494621	6·41	977711	·69	516910	7·09	483090	48
13	495005	6·40	977669	·69	517335	7·09	482665	47
14	495388	6·39	977628	·69	517761	7·08	482239	46
15	495772	6·39	977586	·69	518185	7·08	481815	45
16	496154	6·38	977544	·70	518610	7·07	481390	44
17	496537	6·37	977503	·70	519034	7·06	480966	43
18	496919	6·37	977461	·70	519458	7·06	480542	42
19	497301	6·36	977419	·70	519882	7·05	480118	41
20	497682	6·36	977377	·70	520305	7·05	479695	40
21	9·498064	6·35	9·977335	·70	9·520728	7·04	10·479272	39
22	498444	6·34	977293	·70	521151	7·03	478849	38
23	498825	6·34	977251	·70	521573	7·03	478427	37
24	499204	6·33	977209	·70	521995	7·03	478005	36
25	499584	6·32	977167	·70	522417	7·02	477583	35
26	499963	6·32	977125	·70	522838	7·02	477162	34
27	500342	6·31	977083	·70	523259	7·01	476741	33
28	500721	6·31	977041	·70	523680	7·01	476320	32
29	501099	6·30	976999	·70	524100	7·00	475900	31
30	501476	6·29	976957	·70	524520	6·99	475480	30
31	9·501854	6·29	9·976914	·70	9·524939	6·99	10·475061	29
32	502231	6·28	976872	·71	525359	6·98	474641	28
33	502607	6·28	976830	·71	525778	6·98	474222	27
34	502984	6·27	976787	·71	526197	6·97	473803	26
35	503360	6·26	976745	·71	526615	6·97	473385	25
36	503735	6·26	976702	·71	527033	6·96	472967	24
37	504110	6·25	976660	·71	527451	6·96	472549	23
38	504485	6·25	976617	·71	527868	6·95	472132	22
39	504860	6·24	976574	·71	528285	6·95	471715	21
40	505234	6·23	976532	·71	528702	6·94	471298	20
41	9·505608	6·23	9·976489	·71	9·529119	6·93	10·470881	19
42	505981	6·22	976446	·71	529535	6·93	470465	18
43	506354	6·22	976404	·71	529950	6·93	470050	17
44	506727	6·21	976361	·71	530366	6·92	469634	16
45	507099	6·20	676318	·71	530781	6·91	469219	15
46	507471	6·20	976275	·71	531196	6·91	468804	14
47	507843	6·19	976232	·72	531611	6·90	468389	13
48	508214	6·19	976189	·72	532025	6·90	467975	12
49	508585	6·18	976146	·72	532439	6·89	467561	11
50	508956	6·18	976103	·72	532853	6·89	467147	10
51	9·509326	6·17	9·976060	·72	9·533266	6·88	10·466734	9
52	509696	6·16	976017	·72	533679	6·88	466321	8
53	510065	6·16	975974	·72	534092	6·87	465908	7
54	510434	6·15	975930	·72	534504	6·87	465496	6
55	510803	6·15	975887	·72	534916	6·86	465084	5
56	511172	6·14	975844	·72	535328	6·86	464672	4
57	511540	6·13	975800	·72	535739	6·85	464261	3
58	511907	6·13	975757	·72	536150	6·85	463850	2
59	512275	6·12	975714	·72	536561	6·84	463439	1
60	512642	6·12	975670	·72	536972	6·84	463028	0
	Cosine	D.	Sine	71°	Cotang.	D.	Tang.	M.

M.	Sine	D.	Cosine	D.	Tang.	D.	Cotang.	
0	9·512642	6·12	9·975670	·73	9·536972	6·84	10·463028	60
1	513009	6·11	975627	·73	537382	6·83	462618	59
2	513375	6·11	975583	·73	537792	6·83	462208	58
3	513741	6·10	975539	·73	538202	6·82	461798	57
4	514107	6·09	975496	·73	538611	6·82	461389	56
5	514472	6·09	975452	·73	539020	6·81	460980	55
6	514837	6·08	975408	·73	539429	6·81	460571	54
7	515202	6·08	975365	·73	539837	6·80	460163	53
8	515566	6·07	975321	·73	540245	6·80	459755	52
9	515930	6·07	975277	·73	540653	6·79	459347	51
10	516294	6·06	975233	·73	541061	6·79	458939	50
11	9·516657	6·05	9·975189	·73	9·541468	6·78	10·458532	49
12	517020	6·05	975145	·73	541875	6·78	458125	48
13	517382	6·04	975101	·73	542281	6·77	457719	47
14	517745	6·04	975057	·73	542688	6·77	457312	46
15	518107	6·03	975013	·73	543094	6·76	456906	45
16	518468	6·03	974969	·74	543499	6·76	456501	44
17	518829	6·02	974925	·74	543905	6·75	456095	43
18	519190	6·01	974880	·74	544310	6·75	455690	42
19	519551	6·01	974836	·74	544715	6·74	455285	41
20	519911	6·00	974792	·74	545119	6·74	454881	40
21	9·520271	6·00	9·974748	·74	9·545524	6·73	10·454476	39
22	520631	5·99	974703	·74	545928	6·73	454072	38
23	520990	5·99	974659	·74	546331	6·72	453669	37
24	521349	5·98	974614	·74	546735	6·72	453265	36
25	521707	5·98	974570	·74	547138	6·71	452862	35
26	522066	5·97	974525	·74	547540	6·71	452460	34
27	522424	5·96	974481	·74	547943	6·70	452057	33
28	522781	5·96	974436	·74	548345	6·70	451655	32
29	523138	5·95	974391	·74	548747	6·69	451253	31
30	523495	5·95	974347	·75	549149	6·69	450851	30
31	9·523852	5·94	9·974302	·75	9·549550	6·68	10·450450	29
32	524208	5·94	974257	·75	549951	6·68	450049	28
33	524564	5·93	974212	·75	550352	6·67	449648	27
34	524920	5·93	974167	·75	550752	6·67	449248	26
35	525275	5·92	974122	·75	551152	6·66	448848	25
36	525630	5·91	974077	·75	551552	6·66	448448	24
37	525984	5·91	974032	·75	551952	6·65	448048	23
38	526339	5·90	973987	·75	552351	6·65	447649	22
39	526693	5·90	973942	·75	552750	6·65	447250	21
40	527046	5·89	973897	·75	553149	6·64	446851	20
41	9·527400	5·89	9·973852	·75	9·553548	6·64	10·446452	19
42	527753	5·88	973807	·75	553946	6·63	446054	18
43	528105	5·88	973761	·75	554344	6·63	445656	17
44	528458	5·87	973716	·76	554741	6·62	445259	16
45	528810	5·87	973671	·76	555139	6·62	444861	15
46	529161	5·86	973625	·76	555536	6·61	444464	14
47	529513	5·86	973580	·76	555933	6·61	444067	13
48	529864	5·85	973535	·76	556329	6·60	443671	12
49	530215	5·85	973489	·76	556725	6·60	443275	11
50	530565	5·84	973444	·76	557121	6·59	442879	10
51	9·530915	5·84	9·973398	·76	9·557517	6·59	10·442483	9
52	531265	5·83	973352	·76	557913	6·59	442087	8
53	531614	5·82	973307	·76	558308	6·58	441692	7
54	531963	5·82	973261	·76	558702	6·58	441298	6
55	532312	5·81	973215	·76	559097	6·57	440903	5
56	532661	5·81	973169	·76	559491	6·57	440509	4
57	533009	5·80	973124	·76	559885	6·56	440115	3
58	533357	5·80	973078	·76	560279	6·56	439721	2
59	533704	5·79	973032	·77	560673	6·55	439327	1
60	534052	5·78	972986	·77	561066	6·55	438934	0
	Cosine	D.	Sine	70°	Cotang.	D.	Tang.	M.

M.	Sine	D.	Cosine	D.	Tang.	D.	Cotang.	
0	9·534052	5·78	9·972986	·77	9·561066	6·55	10·438934	60
1	534399	5·77	972940	·77	561459	6·54	438541	59
2	534745	5·77	972894	·77	561851	6·54	438149	58
3	535092	5·77	972848	·77	562244	6·53	437756	57
4	535438	5·76	972802	·77	562636	6·53	437364	56
5	535783	5·76	972755	·77	563028	6·53	436972	55
6	536129	5·75	972709	·77	563419	6·52	436581	54
7	536474	5·74	972663	·77	563811	6·52	436189	53
8	536818	5·74	972617	·77	564202	6·51	435798	52
9	537163	5·73	972570	·77	564592	6·51	435408	51
10	537507	5·73	972524	·77	564983	6·50	435017	50
11	9·537851	5·72	9·972478	·77	9·565373	6·50	10·434627	49
12	538194	5·72	972431	·78	565763	6·49	434237	48
13	538538	5·71	972385	·78	566153	6·49	433847	47
14	538880	5·71	972338	·78	566542	6·49	433458	46
15	539223	5·70	972291	·78	566932	6·48	433068	45
16	539565	5·70	972245	·78	567320	6·48	432680	44
17	539907	5·69	972198	·78	567709	6·47	432291	43
18	540249	5·69	972151	·78	568098	6·47	431902	42
19	540590	5·68	972105	·78	568486	6·46	431514	41
20	540931	5·68	972058	·78	568873	6·46	431127	40
21	9·541272	5·67	9·972011	·78	9·569261	6·45	10·430739	39
22	541613	5·67	971964	·78	569648	6·45	430352	38
23	541953	5·66	971917	·78	570035	6·45	429965	37
24	542293	5·66	971870	·78	570422	6·44	429578	36
25	542632	5·65	971823	·78	570809	6·44	429191	35
26	542971	5·65	971776	·78	571195	6·43	428805	34
27	543310	5·64	971729	·79	571581	6·43	428419	33
28	543649	5·64	971682	·79	571967	6·42	428033	32
29	543987	5·63	971635	·79	572352	6·42	427648	31
30	544325	5·63	971588	·79	572738	6·42	427262	30
31	9·544663	5·62	9·971540	·79	9·573123	6·41	10·426877	29
32	545000	5·62	971493	·79	573507	6·41	426493	28
33	545338	5·61	971446	·79	573892	6·40	426108	27
34	545674	5·61	971398	·79	574276	6·40	425724	26
35	546011	5·60	971351	·79	574660	6·39	425340	25
36	546347	5·60	971303	·79	575044	6·39	424956	24
37	546683	5·59	971256	·79	575427	6·39	424573	23
38	547019	5·59	971208	·79	575810	6·38	424190	22
39	547354	5·58	971161	·79	576193	6·38	423807	21
40	547689	5·58	971113	·79	576576	6·37	423424	20
41	9·548024	5·57	9·971066	·80	9·576958	6·37	10·423041	19
42	548359	5·57	971018	·80	577341	6·36	422659	18
43	548693	5·56	970970	·80	577723	6·36	422277	17
44	549027	5·56	970922	·80	578104	6·36	421896	16
45	549360	5·55	970874	·80	578486	6·35	421514	15
46	549693	5·55	970827	·80	578867	6·35	421133	14
47	550026	5·54	970779	·80	579248	6·34	420752	13
48	550359	5·54	970731	·80	579629	6·34	420371	12
49	550692	5·53	970683	·80	580009	6·34	419991	11
50	551024	5·53	970635	·80	580389	6·33	419611	10
51	9·551356	5·52	9·970586	·80	9·580769	6·33	10·419231	9
52	551687	5·52	970538	·80	581149	6·32	418851	8
53	552018	5·52	970490	·80	581528	6·32	418472	7
54	552349	5·51	970442	·80	581907	6·32	418093	6
55	552680	5·51	970394	·80	582286	6·31	417714	5
56	553010	5·50	970345	·81	582665	6·31	417335	4
57	553341	5·50	970297	·81	583043	6·30	416957	3
58	553670	5·49	970249	·81	583422	6·30	416578	2
59	554000	5·49	970200	·81	583800	6·29	416200	1
60	554329	5·48	970152	·81	584177	6·29	415823	0
	Cosine	D.	Sine	69°	Cotang.	D.	Tang.	M.

M.	Sine	D.	Cosine	D.	Tang.	D.	Cotang.	
0	9·554329	5·48	9·970152	·81	9·584177	6·29	10·415823	60
1	554658	5·48	970103	·81	584555	6·29	415445	59
2	554987	5·47	970055	·81	584932	6·28	415068	58
3	555315	5·47	970006	·81	585309	6·28	414691	57
4	555643	5·46	969957	·81	585686	6·27	414314	56
5	555971	5·46	969909	·81	586062	6·27	413938	55
6	556299	5·45	969860	·81	586439	6·27	413561	54
7	556626	5·45	969811	·81	586815	6·26	413185	53
8	556953	5·44	969762	·81	587190	6·26	412810	52
9	557280	5·44	969714	·81	587566	6·25	412434	51
10	557606	5·43	969665	·81	587941	6·25	412059	50
11	9·557932	5·43	9·969616	·82	9·588316	6·25	10·411684	49
12	558258	5·43	969567	·82	588691	6·24	411309	48
13	558584	5·42	969518	·82	589066	6·24	410934	47
14	558909	5·42	969469	·82	589440	6·23	410560	46
15	559234	5·41	969420	·82	589814	6·23	410186	45
16	559558	5·41	969370	·82	590188	6·23	409812	44
17	559883	5·40	969321	·82	590562	6·22	409438	43
18	560207	5·40	969272	·82	590935	6·22	409065	42
19	560531	5·39	969223	·82	591308	6·22	408692	41
20	560855	5·39	969173	·82	591681	6·21	408319	40
21	9·561178	5·38	9·969124	·82	9·592054	6·21	10·407946	39
22	561501	5·38	969075	·82	592426	6·20	407574	38
23	561824	5·37	969025	·82	592798	6·20	407202	37
24	562146	5·37	968976	·82	593170	6·19	406829	36
25	562468	5·36	968926	·83	593542	6·19	406458	35
26	562790	5·36	968877	·83	593914	6·18	406086	34
27	563112	5·36	968827	·83	594285	6·18	405715	33
28	563433	5·35	968777	·83	594656	6·18	405344	32
29	563755	5·35	968728	·83	595027	6·17	404973	31
30	564075	5·34	968678	·83	595398	6·17	404602	30
31	9·564396	5·34	9·968628	·83	9·595768	6·17	10·404232	29
32	564716	5·33	968578	·83	596138	6·16	403862	28
33	565036	5·33	968528	·83	596508	6·16	403492	27
34	565356	5·32	968479	·83	596878	6·16	403122	26
35	565676	5·32	968429	·83	597247	6·15	402753	25
36	565995	5·31	968379	·83	597616	6·15	402384	24
37	566314	5·31	968329	·83	597985	6·15	402015	23
38	566632	5·31	968278	·83	598354	6·14	401646	22
39	566951	5·30	968228	·84	598722	6·14	401278	21
40	567269	5·30	968178	·84	599091	6·13	400909	20
41	9·567587	5·29	9·968128	·84	9·599459	6·13	10·400541	19
42	567904	5·29	968078	·84	599827	6·13	400173	18
43	568222	5·28	968027	·84	600194	6·12	399806	17
44	568539	5·28	967977	·84	600562	6·12	399438	16
45	568856	5·28	967927	·84	600929	6·11	399071	15
46	569172	5·27	967876	·84	601296	6·11	398704	14
47	569488	5·27	967826	·84	601662	6·11	398338	13
48	569804	5·26	967775	·84	602029	6·10	397971	12
49	570120	5·26	967725	·84	602395	6·10	397605	11
50	570435	5·25	967674	·84	602761	6·10	397239	10
51	9·570751	5·25	9·967624	·84	9·603127	6·09	10·396873	9
52	571066	5·24	967573	·84	603493	6·09	396507	8
53	571380	5·24	967522	·85	603858	6·09	396142	7
54	571695	5·23	967471	·85	604223	6·08	395777	6
55	572009	5·23	967421	·85	604588	6·08	395412	5
56	572323	5·23	967370	·85	604953	6·07	395047	4
57	572636	5·22	967319	·85	605317	6·07	394683	3
58	572950	5·22	967268	·85	605682	6·07	394318	2
59	573263	5·21	967217	·85	606046	6·06	393954	1
60	573575	5·21	967166	·85	606410	6·06	393590	0
	Cosine	D.	Sine	68°	Cotang.	D.	Tang.	M.

M.	Sine	D.	Cosine	D.	Tang.	D.	Cotang.	
0	9·573575	5·21	9·967166	·85	9·606410	6·06	10·393590	60
1	573888	5·20	967115	·85	606773	6·06	393227	59
2	574200	5·20	967064	·85	607137	6·05	392863	58
3	574512	5·19	967013	·85	607500	6·05	392500	57
4	574824	5·19	966961	·85	607863	6·04	392137	56
5	575136	5·19	966910	·85	608225	6·04	391775	55
6	575447	5·18	966859	·85	608588	6·04	391412	54
7	575758	5·18	966808	·85	608950	6·03	391050	53
8	576069	5·17	966756	·86	609312	6·03	390688	52
9	576379	5·17	966705	·86	609674	6·03	390326	51
10	576689	5·16	966653	·86	610036	6·02	389964	50
11	9·576999	5·16	9·966602	·86	9·610397	6·02	10·389603	49
12	577309	5·16	966550	·86	610759	6·02	389241	48
13	577618	5·15	966499	·86	611120	6·01	388880	47
14	577927	5·15	966447	·86	611480	6·01	388520	46
15	578236	5·14	966395	·86	611841	6·01	388159	45
16	578545	5·14	966344	·86	612201	6·00	387799	44
17	578853	5·13	966292	·86	612561	6·00	387439	43
18	579162	5·13	966240	·86	612921	6·00	387079	42
19	579470	5·13	966188	·86	613281	5·99	386719	41
20	579777	5·12	966136	·86	613641	5·99	386359	40
21	9·580085	5·12	9·966085	·87	9·614000	5·98	10·386000	39
22	580392	5·11	966033	·87	614359	5·98	385641	38
23	580699	5·11	965981	·87	614718	5·98	385282	37
24	581005	5·11	965928	·87	615077	5·97	384923	36
25	581312	5·10	965876	·87	615435	5·97	384565	35
26	581618	5·10	965824	·87	615793	5·97	384207	34
27	581924	·	965772	·87	616151	5·96	383849	33
28	582229	·	965720	·87	616509	5·96	383491	32
29	582535	·	965668	·87	616867	5·96	383133	31
30	582840	·	965615	·87	617224	5·95	382776	30
31	9·583145		9·965563	·87	9·617582	5·95	10·382418	29
32	583449		965511	·87	617939	5·95	382061	28
33	583754	7	965458	·87	618295	5·94	381705	27
34	584058		965406	·87	618652	5·94	381348	26
35	584361		965353	·88	619008	5·94	380992	25
36	584665		965301	·88	619364	5·93	380636	24
37	584968	·	965248	·88	619721	5·93	380279	23
38	585272	·	965195	·88	620076	5·93	379924	22
39	585574	5 09	965143	·88	620432	5·92	379568	21
40	585877	5·[illegible]	965090	·88	620787	5·92	379213	20
41	9·586179	5·03	9·965037	·88	9·621142	5·92	10·378858	19
42	586482	5·03	964984	·88	621497	5·91	378503	18
43	586783	5·03	964931	·88	621852	5·91	378148	17
44	587085	5·02	964879	·88	622207	5·90	377793	16
45	587386	5·02	964826	·88	622561	5·90	377439	15
46	587688	5·01	964773	·88	622915	5·90	377085	14
47	587989	5·01	964720	·88	623269	5·89	376731	13
48	588289	5·01	964666	·89	623623	5·89	376377	12
49	588590	5·00	964613	·89	623976	5·89	376024	11
50	588890	5·00	964560	·89	624330	5·88	375670	10
51	9·589190	4·99	9·964507	·89	9·624683	5·88	10·375317	9
52	589489	4·99	964454	·89	625036	5·88	374964	8
53	589789	4·99	964400	·89	625388	5·87	374612	7
54	590088	4·98	964347	·89	625741	5·87	374259	6
55	590387	4·98	964294	·89	626093	5·87	373907	5
56	590686	4·97	964240	·89	626445	5·86	373555	4
57	590984	4·97	964187	·89	626797	5·86	373203	3
58	591282	4·97	964133	·89	627149	5·86	372851	2
59	591580	4·96	964080	·89	627501	5·85	372499	1
60	591878	4·96	964026	·89	627852	5·85	372148	0
	Cosine	D.	Sine	67°	Cotang.	D.	Tang.	M.

M.	Sine	D.	Cosine	D.	Tang.	D.	Cotang.	
0	9·591878	4·96	9·964026	·89	9·627852	5·85	10·372148	60
1	592176	4·95	963972	·89	628203	5·85	371797	59
2	592473	4·95	963919	·89	628554	5·85	371446	58
3	592770	4·95	963865	·90	628905	5·84	371095	57
4	593067	4·94	963811	·90	629255	5·84	370745	56
5	593363	4·94	963757	·90	629606	5·83	370394	55
6	593659	4·93	963704	·90	629956	5·83	370044	54
7	593955	4·93	963650	·90	630306	5·83	369694	53
8	594251	4·93	963596	·90	630656	5·83	369344	52
9	594547	4· 2	963542	·90	631005	5·82	368995	51
10	594842	4· 2	963488	·90	631355	5·82	368645	50
11	9·595137	4· 1	9·963434	·90	9·631704	5·82	10·368296	49
12	595432	4·91	963379	·90	632053	5·81	367947	48
13	595727	4·91	963325	·90	632401	5·81	367599	47
14	596021	4·90	963271	·90	632750	5·81	367250	46
15	596315	4·90	963217	·90	633098	5·80	366902	45
16	596609	4·89	963163	·90	633447	5·80	366553	44
17	596903	4·89	963108	·91	633795	5·80	366205	43
18	597196	4·89	963054	·91	634143	5·79	365857	42
19	597490	4·88	962999	·91	634490	5·79	365510	41
20	597783	4·88	962945	·91	634838	5·79	365162	40
21	9·598075	4·87	9·962890	·91	9·635185	5·78	10·364815	39
22	598368	4·87	962836	·91	635532	5·78	364468	38
23	598660	4·87	962781	·91	635879	5·78	364121	37
24	598952	4·86	962727	·91	636226	5·77	363774	36
25	599244	4·86	962672	·91	636572	5·77	363428	35
26	599536	4·85	962617	·91	636919	5·77	363081	34
27	599827	4·85	962562	·91	637265	5·77	362735	33
28	600118	4·85	962508	·91	637611	5·76	362389	32
29	600409	4·84	962453	·91	637956	5·76	362044	31
30	600700	4·84	962398	·92	638302	5·76	361698	30
31	9·600990	4·84	9·962343	·92	9·638647	5·75	10·361353	29
32	601280	4·83	962288	·92	638992	5·75	361008	28
33	601570	4·83	962233	·92	639337	5·75	360663	27
34	601860	4·82	962178	·92	639682	5·74	360318	26
35	602150	4·82	962123	·92	640027	5·74	359973	25
36	602439	4·82	962067	·92	640371	5·74	359629	24
37	602728	4·81	962012	·92	640716	5·73	359284	23
38	603017	4·81	961957	·92	641060	5·73	358940	22
39	603305	4·81	961902	·92	641404	5·73	358596	21
40	603594	4·80	961846	·92	641747	5·72	358253	20
41	9·603882	4·80	9·961791	·92	9·642091	5·72	10·357909	19
42	604170	4·79	961735	·92	642434	5·72	357566	18
43	604457	4·79	961680	·92	642777	5·72	357223	17
44	604745	4·79	961624	·93	643120	5·71	356880	16
45	605032	4·78	961569	·93	643463	5·71	356537	15
46	605319	4·78	961513	·93	643806	5·71	356194	14
47	605606	4·78	961458	·93	644148	5·70	355852	13
48	605892	4·77	961402	·93	644490	5·70	355510	12
49	606179	4·77	961346	·93	644832	5·70	355168	11
50	606465	4·76	961290	·93	645174	5·69	354826	10
51	9·606751	4·76	9·961235	·93	9·645516	5·69	10·354484	9
52	607036	4·76	961179	·93	645857	5·69	354143	8
53	607322	4·75	961123	·93	646199	5·69	353801	7
54	607607	4·75	961067	·93	646540	5·68	353460	6
55	607892	4·74	961011	·93	646881	5·68	353119	5
56	608177	4·74	960955	·93	647222	5·68	352778	4
57	608461	4·74	960899	·93	647562	5·67	352438	3
58	608745	4·73	960843	·94	647903	5·67	352097	2
59	609029	4·73	960786	·94	648243	5·67	351757	1
60	609313	4·73	960730	·94	648583	5·66	351417	0
	Cosine	D.	Sine	66°	Cotang.	D.	Tang.	M.

M.	Sine	D.	Cosine	D.	Tang.	D.	Cotang.	
0	9·609313	4·73	9·960730	·94	9·648583	5·66	10·351417	60
1	609597	4·72	960674	·94	648923	5·66	351077	59
2	609880	4·72	960618	·94	649263	5·66	350737	58
3	610164	4·72	960561	·94	649602	5·66	350398	57
4	610447	4·71	960505	·94	649942	5·65	350058	56
5	610729	4·71	960448	·94	650281	5·65	349719	55
6	611012	4·70	960392	·94	650620	5·65	349380	54
7	611294	4·70	960335	·94	650959	5·64	349041	53
8	611576	4·70	960279	·94	651297	5·64	348703	52
9	611858	4·69	960222	·94	651636	5·64	348364	51
10	612140	4·69	960165	·94	651974	5·63	348026	50
11	9·612421	4·69	9·960109	·95	9·652312	5·63	10·347688	49
12	612702	4·68	960052	·95	652650	5·63	347350	48
13	612983	4·68	959995	·95	652988	5·63	347012	47
14	613264	4·67	959938	·95	653326	5·62	346674	46
15	613545	4·67	959882	·95	653663	5·62	346337	45
16	613825	4·67	959825	·95	654000	5·62	346000	44
17	614105	4·66	959768	·95	654337	5·61	345663	43
18	614385	4·66	959711	·95	654674	5·61	345326	42
19	614665	4·66	959654	·95	655011	5·61	344989	41
20	614944	4·65	959596	·95	655348	5·61	344652	40
21	9·615223	4·65	9·959539	·95	9·655684	5·60	10·344316	39
22	615502	4·65	959482	·95	656020	5·60	343980	38
23	615781	4·64	959425	·95	656356	5·60	343644	37
24	616060	4·64	959368	·95	656692	5·59	343308	36
25	616338	4·64	959310	·96	657028	5·59	342972	35
26	616616	4·63	959253	·96	657364	5·59	342636	34
27	616894	4·63	959195	·96	657699	5·59	342301	33
28	617172	4·62	959138	·96	658034	5·58	341966	32
29	617450	4·62	959081	·96	658369	5·58	341631	31
30	617727	4·62	959023	·96	658704	5·58	341296	30
31	9·618004	4·61	9·958965	·96	9·659039	5·58	10·340961	29
32	618281	4·61	958908	·96	659373	5·57	340627	28
33	618558	4·61	958850	·96	659708	5·57	340292	27
34	618834	4·60	958792	·96	660042	5·57	339958	26
35	619110	4·60	958734	·96	660376	5·57	339624	25
36	619386	4·60	958677	·96	660710	5·56	339290	24
37	619662	4·59	958619	·96	661043	5·56	338957	23
38	619938	4·59	958561	·96	661377	5·56	338623	22
39	620213	4·59	958503	·97	661710	5·55	338290	21
40	620488	4·58	958445	·97	662043	5·55	337957	20
41	9·620763	4·58	9·958387	·97	9·662376	5·55	10·337624	19
42	621038	4·57	958329	·97	662709	5·54	337291	18
43	621313	4·57	958271	·97	663042	5·54	336958	17
44	621587	4·57	958213	·97	663375	5·54	336625	16
45	621861	4·56	958154	·97	663707	5·54	336293	15
46	622135	4·56	958096	·97	664039	5·53	335961	14
47	622409	4·56	958038	·97	664371	5·53	335629	13
48	622682	4·55	957979	·97	664703	5·53	335297	12
49	622956	4·55	957921	·97	665035	5·53	334965	11
50	623229	4·55	957863	·97	665366	5·52	334634	10
51	9·623502	4·54	9·957804	·97	9·665697	5·52	10·334303	9
52	623774	4·54	957746	·98	666029	5·52	333971	8
53	624047	4·54	957687	·98	666360	5·51	333640	7
54	624319	4·53	957628	·98	666691	5·51	333309	6
55	624591	4·53	957570	·98	667021	5·51	332979	5
56	624863	4·53	957511	·98	667352	5·51	332648	4
57	625135	4·52	957452	·98	667682	5·50	332318	3
58	625406	4·52	957393	·98	668013	5·50	331987	2
59	625677	4·52	957335	·98	668343	5·50	331657	1
60	625948	4·51	957276	·98	668672	5·50	331328	0
	Cosine	D.	Sine	65°	Cotang.	D.	Tang.	M.

M.	Sine	D.	Cosine	D.	Tang.	D.	Cotang.	
0	9.625948	4.51	9.957276	.98	9.668673	5.50	10.331327	60
1	626219	4.51	957217	.98	669002	5.49	330998	59
2	626490	4.51	957158	.98	669332	5.49	330668	58
3	626760	4.50	957099	.98	669661	5.49	330339	57
4	627030	4.50	957040	.98	669991	5.48	330009	56
5	627300	4.50	956981	.98	670320	5.48	329680	55
6	627570	4.49	956921	.99	670649	5.48	329351	54
7	627840	4.49	956862	.99	670977	5.48	329023	53
8	628109	4.49	956803	.99	671306	5.47	328694	52
9	628378	4.48	956744	.99	671634	5.47	328366	51
10	628647	4.48	956684	.99	671963	5.47	328037	50
11	9.628916	4.47	9.956625	.99	9.672291	5.47	10.327709	49
12	629185	4.47	956566	.99	672619	5.46	327381	48
13	629453	4.47	956506	.99	672947	5.46	327053	47
14	629721	4.46	956447	.99	673274	5.46	326726	46
15	629989	4.46	956387	.99	673602	5.46	326398	45
16	630257	4.46	956327	.99	673929	5.45	326071	44
17	630524	4.46	956268	.99	674257	5.45	325743	43
18	630792	4.45	956208	1.00	674584	5.45	325416	42
19	631059	4.45	956148	1.00	674910	5.44	325090	41
20	631326	4.45	956089	1.00	675237	5.44	324763	40
21	9.631593	4.44	9.956029	1.00	9.675564	5.44	10.324436	39
22	631859	4.44	955969	1.00	675890	5.44	324110	38
23	632125	4.44	955909	1.00	676216	5.43	323784	37
24	632392	4.43	955849	1.00	676543	5.43	323457	36
25	632658	4.43	955789	1.00	676869	5.43	323131	35
26	632923	4.43	955729	1.00	677194	5.43	322806	34
27	633189	4.42	955669	1.00	677520	5.42	322480	33
28	633454	4.42	955609	1.00	677846	5.42	322154	32
29	633719	4.42	955548	1.00	678171	5.42	321829	31
30	633984	4.41	955488	1.00	678496	5.42	321504	30
31	9.634249	4.41	9.955428	1.01	9.678821	5.41	10.321179	29
32	634514	4.40	955368	1.01	679146	5.41	320854	28
33	634778	4.40	955307	1.01	679471	5.41	320529	27
34	635042	4.40	955247	1.01	679795	5.41	320205	26
35	635306	4.39	955186	1.01	680120	5.40	319880	25
36	635570	4.39	955126	1.01	680444	5.40	319556	24
37	635834	4.39	955065	1.01	680768	5.40	319232	23
38	636097	4.38	955005	1.01	681092	5.40	318908	22
39	636360	4.38	954944	1.01	681416	5.39	318584	21
40	636623	4.38	954883	1.01	681740	5.39	318260	20
41	9.636886	4.37	9.954823	1.01	9.682063	5.39	10.317937	19
42	637148	4.37	954762	1.01	682387	5.39	317613	18
43	637411	4.37	954701	1.01	682710	5.38	317290	17
44	637673	4.37	954640	1.01	683033	5.38	316967	16
45	637935	4.36	954579	1.01	683356	5.38	316644	15
46	638197	4.36	954518	1.02	683679	5.38	316321	14
47	638458	4.36	954457	1.02	684001	5.37	315999	13
48	638720	4.35	954396	1.02	684324	5.37	315676	12
49	638981	4.35	954335	1.02	684646	5.37	315354	11
50	639242	4.35	954274	1.02	684968	5.37	315032	10
51	9.639503	4.34	9.954213	1.02	9.685290	5.36	10.314710	9
52	639764	4.34	954152	1.02	685612	5.36	314388	8
53	640024	4.34	954090	1.02	685934	5.36	314066	7
54	640284	4.33	954029	1.02	686255	5.36	313745	6
55	640544	4.33	953968	1.02	686577	5.35	313423	5
56	640804	4.33	953906	1.02	686898	5.35	313102	4
57	641064	4.32	953845	1.02	687219	5.35	312781	3
58	641324	4.32	953783	1.02	687540	5.35	312460	2
59	641584	4.32	953722	1.03	687861	5.34	312139	1
60	641842	4.31	953660	1.03	688182	5.34	311818	0
	Cosine	D.	Sine	64°	Cotang.	D.	Tang.	M.

M.	Sine	D.	Cosine	D.	Tang.	D.	Cotang.	
0	9·641842	4·31	9·953660	1·03	9·688182	5·34	10·311818	60
1	642101	4·31	953599	1·03	688502	5·34	311498	59
2	642360	4·31	953537	1·03	688823	5·34	311177	58
3	642618	4·30	953475	1·03	689143	5·33	310857	57
4	642877	4·30	953413	1·03	689463	5·33	310537	56
5	643135	4·30	953352	1·03	689783	5·33	310217	55
6	643393	4·30	953290	1·03	690103	5·33	309897	54
7	643650	4·29	953228	1·03	690423	5·33	309577	53
8	643908	4·29	953166	1·03	690742	5·32	309258	52
9	644165	4·29	953104	1·03	691062	5·32	308938	51
10	644423	4·28	953042	1·03	691381	5·32	308619	50
11	9·644680	4·28	9·952980	1·04	9·691700	5·31	10·308300	49
12	644936	4·28	952918	1·04	692019	5·31	307981	48
13	645193	4·27	952855	1·04	692338	5·31	307662	47
14	645450	4·27	952793	1·04	692656	5·31	307344	46
15	645706	4·27	952731	1·04	692975	5·31	307025	45
16	645962	4·26	952669	1·04	693293	5·30	306707	44
17	646218	4·26	952606	1·04	693612	5·30	306388	43
18	646474	4·26	952544	1·04	693930	5·30	306070	42
19	646729	4·25	952481	1·04	694248	5·30	305752	41
20	646984	4·25	952419	1·04	694566	5·29	305434	40
21	9·647240	4·25	9·952356	1·04	9·694883	5·29	10·305117	39
22	647494	4·24	952294	1·04	695201	5·29	304799	38
23	647749	4·24	952231	1·04	695518	5·29	304482	37
24	648004	4·24	952168	1·05	695836	5·29	304164	36
25	648258	4·24	952106	1·05	696153	5·28	303847	35
26	648512	4·23	952043	1·05	696470	5·28	303530	34
27	648766	4·23	951980	1·05	696787	5·28	303213	33
28	649020	4·23	951917	1·05	697103	5·28	302897	32
29	649274	4·22	951854	1·05	697420	5·27	302580	31
30	649527	4·22	951791	1·05	697736	5·27	302264	30
31	9·649781	4·22	9·951728	1·05	9·698053	5·27	10·301947	29
32	650034	4·22	951665	1·05	698369	5·27	301631	28
33	650287	4·21	951602	1·05	698685	5·26	301315	27
34	650539	4·21	951539	1·05	699001	5·26	300999	26
35	650792	4·21	951476	1·05	699316	5·26	300684	25
36	651044	4·20	951412	1·05	699632	5·26	300368	24
37	651297	4·20	951349	1·06	699947	5·26	300053	23
38	651549	4·20	951286	1·06	700263	5·25	299737	22
39	651800	4·19	951222	1·06	700578	5·25	299422	21
40	652052	4·19	951159	1·06	700893	5·25	299107	20
41	9·652304	4·19	9·951096	1·06	9·701208	5·24	10·298792	19
42	652555	4·18	951032	1·06	701523	5·24	298477	18
43	652806	4·18	950968	1·06	701837	5·24	298163	17
44	653057	4·18	950905	1·06	702152	5·24	297848	16
45	653308	4·18	950841	1·06	702466	5·24	297534	15
46	653558	4·17	950778	1·06	702780	5·23	297220	14
47	653808	4·17	950714	1·06	703095	5·23	296905	13
48	654059	4·17	950650	1·06	703409	5·23	296591	12
49	654309	4·16	950586	1·06	703723	5·23	296277	11
50	654558	4·16	950522	1·07	704036	5·22	295964	10
51	9·654808	4·16	9·950458	1·07	9·704350	5·22	10·295650	9
52	655058	4·16	950394	1·07	704663	5·22	295337	8
53	655307	4·15	950330	1·07	704977	5·22	295023	7
54	655556	4·15	950266	1·07	705290	5·22	294710	6
55	655805	4·15	950202	1·07	705603	5·21	294397	5
56	656054	4·14	950138	1·07	705916	5·21	294084	4
57	656302	4·14	950074	1·07	706228	5·21	293772	3
58	656551	4·14	950010	1·07	706541	5·21	293459	2
59	656799	4·13	949945	1·07	706854	5·21	293146	1
60	657047	4·13	949881	1·07	707166	5·20	292834	0
	Cosine	D.	Sine	63°	Cotang.	D.	Tang.	M.

M.	Sine	D.	Cosine	D.	Tang.	D.	Cotang.	
0	9·657047	4·13	9·949881	1·07	9·707166	5·20	10·292834	60
1	657295	4·13	949816	1·07	707478	5·20	292522	59
2	657542	4·12	949752	1·07	707790	5·20	292210	58
3	657790	4·12	949688	1·08	708102	5·20	291898	57
4	658037	4·12	949623	1·08	708414	5·19	291586	56
5	658284	4·12	949558	1·08	708726	5·19	291274	55
6	658531	4·11	949494	1·08	709037	5·19	290963	54
7	658778	4·11	949429	1·08	709349	5·19	290651	53
8	659025	4·11	949364	1·08	709660	5·19	290340	52
9	659271	4·10	949300	1·08	709971	5·18	290029	51
10	659517	4·10	949235	1·08	710282	5·18	289718	50
11	9·659763	4·10	9·949170	1·08	9·710593	5·18	10·289407	49
12	660009	4·09	949105	1·08	710904	5·18	289096	48
13	660255	4·09	949040	1·08	711215	5·18	288785	47
14	660501	4·09	948975	1·08	711525	5·17	288475	46
15	660746	4·09	948910	1·08	711836	5·17	288164	45
16	660991	4·08	948845	1·08	712146	5·17	287854	44
17	661236	4·08	948780	1·09	712456	5·17	287544	43
18	661481	4·08	948715	1·09	712766	5·16	287234	42
19	661726	4·07	948650	1·09	713076	5·16	286924	41
20	661970	4·07	948584	1·09	713386	5·16	286614	40
21	9·662214	4·07	9·948519	1·09	9·713696	5·16	10·286304	39
22	662459	4·07	948454	1·09	714005	5·16	285995	38
23	662703	4·06	948388	1·09	714314	5·15	285686	37
24	662946	4·06	948323	1·09	714624	5·15	285376	36
25	663190	4·06	948257	1·09	714933	5·15	285067	35
26	663433	4·05	948192	1·09	715242	5·15	284758	34
27	663677	4·05	948126	1·09	715551	5·14	284449	33
28	663920	4·05	948060	1·09	715860	5·14	284140	32
29	664163	4·05	947995	1·10	716168	5·14	283832	31
30	664406	4·04	947929	1·10	716477	5·14	283523	30
31	9·664648	4·04	9·947863	1·10	9·716785	5·14	10·283215	29
32	664891	4·04	947797	1·10	717093	5·13	282907	28
33	665133	4·03	947731	1·10	717401	5·13	282599	27
34	665375	4·03	947665	1·10	717709	5·13	282291	26
35	665617	4·03	947600	1·10	718017	5·13	281983	25
36	665859	4·02	947533	1·10	718325	5·13	281670	24
37	666100	4·02	947467	1·10	718633	5·12	281367	23
38	666342	4·02	947401	1·10	718940	5·12	281060	22
39	666583	4·02	947335	1·10	719248	5·12	280752	21
40	666824	4·01	947269	1·10	719555	5·12	280445	20
41	9·667065	4·01	9·947203	1·10	9·719862	5·12	10·280138	19
42	667305	4·01	947136	1·11	720169	5·11	279831	18
43	667546	4·01	947070	1·11	720476	5·11	279524	17
44	667786	4·00	947004	1·11	720783	5·11	279217	16
45	668027	4·00	946937	1·11	721089	5·11	278911	15
46	668267	4·00	946871	1·11	721396	5·11	278604	14
47	668506	3·99	946804	1·11	721702	5·10	278298	13
48	668746	3·99	946738	1·11	722009	5·10	277991	12
49	668986	3·99	946671	1·11	722315	5·10	277685	11
50	669225	3·99	946604	1·11	722621	5·10	277379	10
51	9·669464	3·98	9·946538	1·11	9·722927	5·10	10·277073	9
52	669703	3·98	946471	1·11	723232	5·09	276768	8
53	669942	3·98	946404	1·11	723538	5·09	276462	7
54	670181	3·97	946337	1·11	723844	5·09	276156	6
55	670419	3·97	946270	1·12	724149	5·09	275851	5
56	670658	3·97	946203	1·12	724454	5·09	275546	4
57	670896	3·97	946136	1·12	724759	5·08	275241	3
58	671134	3·96	946069	1·12	725065	5·08	274935	2
59	671372	3·96	946002	1·12	725369	5·08	274631	1
60	671609	3·96	945935	1·12	725674	5·08	274326	0
	Cosine	D.	Sine	62°	Cotang.	D.	Tang.	M.

M.	Sine	D.	Cosine	D.	Tang.	D.	Cotang.	
0	9·671609	3·96	9·945935	1·12	9·725674	5·08	10·274326	60
1	671847	3·95	945868	1·12	725979	5·08	274021	59
2	672084	3·95	945800	1·12	726284	5·07	273716	58
3	672321	3·95	945733	1·12	726588	5·07	273412	57
4	672558	3·95	945666	1·12	726892	5·07	273108	56
5	672795	3·94	945598	1·12	727197	5·07	272803	55
6	673032	3·94	945531	1·12	727501	5·07	272499	54
7	673268	3·94	945464	1·13	727805	5·06	272195	53
8	673505	3·94	945396	1·13	728109	5·06	271891	52
9	673741	3·93	945328	1·13	728412	5·06	271588	51
10	673977	3·93	945261	1·13	728716	5·06	271284	50
11	9·674213	3·93	9·945193	1·13	9·729020	5·06	10·270980	49
12	674448	3·92	945125	1·13	729323	5·05	270677	48
13	674684	3·92	945058	1·13	729626	5·05	270374	47
14	674919	3·92	944990	1·13	729929	5·05	270071	46
15	675155	3·92	944922	1·13	730233	5·05	269767	45
16	675390	3·91	944854	1·13	730535	5·05	269465	44
17	675624	3·91	944786	1·13	730838	5·04	269162	43
18	675859	3·91	944718	1·13	731141	5·04	268859	42
19	676094	3·91	944650	1·13	731444	5·04	268556	41
20	676328	3·90	944582	1·14	731746	5·04	268254	40
21	9·676562	3·90	9·944514	1·14	9·732048	5·04	10·267952	39
22	676796	3·90	944446	1·14	732351	5·03	267649	38
23	677030	3·90	944377	1·14	732653	5·03	267347	37
24	677264	3·89	944309	1·14	732955	5·03	267045	36
25	677498	3·89	944241	1·14	733257	5·03	266743	35
26	677731	3·89	944172	1·14	733558	5·03	266442	34
27	677964	3·88	944104	1·14	733860	5·02	266140	33
28	678197	3·88	944036	1·14	734162	5·02	265838	32
29	678430	3·88	943967	1·14	734463	5·02	265537	31
30	678663	3·88	943899	1·14	734764	5·02	265236	30
31	9·678895	3·87	9·943830	1·14	9·735066	5·02	10·264934	29
32	679128	3·87	943761	1·14	735367	5·02	264633	28
33	679360	3·87	943693	1·15	735668	5·01	264332	27
34	679592	3·87	943624	1·15	735969	5·01	264031	26
35	679824	3·86	943555	1·15	736269	5·01	263731	25
36	680056	3·86	943486	1·15	736570	5·01	263430	24
37	680288	3·86	943417	1·15	736871	5·01	263129	23
38	680519	3·85	943348	1·15	737171	5·00	262829	22
39	680750	3·85	943279	1·15	737471	5·00	262529	21
40	680982	3·85	943210	1·15	737771	5·00	262229	20
41	9·681213	3·85	9·943141	1·15	9·738071	5·00	10·261929	19
42	681443	3·84	943072	1·15	738371	5·00	261629	18
43	681674	3·84	943003	1·15	738671	4·99	261329	17
44	681905	3·84	942934	1·15	738971	4·99	261029	16
45	682135	3·84	942864	1·15	739271	4·99	260729	15
46	682365	3·83	942795	1·16	739570	4·99	260430	14
47	682595	3·83	942726	1·16	739870	4·99	260130	13
48	682825	3·83	942656	1·16	740169	4·99	259831	12
49	683055	3·83	942587	1·16	740468	4·98	259532	11
50	683284	3·82	942517	1·16	740767	4·98	259233	10
51	9·683514	3·82	9·942448	1·16	9·741066	4·98	10·258934	9
52	683743	3·82	942378	1·16	741365	4·98	258635	8
53	683972	3·82	942308	1·16	741664	4·98	258336	7
54	684201	3·81	942239	1·16	741962	4·97	258038	6
55	684430	3·81	942169	1·16	742261	4·97	257739	5
56	684658	3·81	942099	1·16	742559	4·97	257441	4
57	684887	3·80	942029	1·16	742858	4·97	257142	3
58	685115	3·80	941959	1·16	743156	4·97	256844	2
59	685343	3·80	941889	1·17	743454	4·97	256546	1
60	685571	3·80	941819	1·17	743752	4·96	256248	0
	Cosine	D.	Sine	61°	Cotang.	D.	Tang.	M.

M.	Sine	D.	Cosine	D.	Tang.	D.	Cotang.	
0	9·685571	3·80	9·941819	1·17	9·743752	4·96	10·256248	60
1	685799	3·79	941749	1·17	744050	4·96	255950	59
2	686027	3·79	941679	1·17	744348	4·96	255652	58
3	686254	3·79	941609	1·17	744645	4·96	255355	57
4	686482	3·79	941539	1·17	744943	4·96	255057	56
5	686709	3·78	941469	1·17	745240	4·96	254760	55
6	686936	3·78	941398	1·17	745538	4·95	254462	54
7	687163	3·78	941328	1·17	745835	4·95	254165	53
8	687389	3·78	941258	1·17	746132	4·95	253868	52
9	687616	3·77	941187	1·17	746429	4·95	253571	51
10	687843	3·77	941117	1·17	746726	4·95	253274	50
11	9·688069	3·77	9·941046	1·18	9·747023	4·94	10·252977	49
12	688295	3·77	940975	1·18	747319	4·94	252681	48
13	688521	3·76	940905	1·18	747616	4·94	252384	47
14	688747	3·76	940834	1·18	747913	4·94	252087	46
15	688972	3·76	940763	1·18	748209	4·94	251791	45
16	689198	3·76	940693	1·18	748505	4·93	251495	44
17	689423	3·75	940622	1·18	748801	4·93	251199	43
18	689648	3·75	940551	1·18	749097	4·93	250903	42
19	689873	3·75	940480	1·18	749393	4·93	250607	41
20	690098	3·75	940409	1·18	749689	4·93	250311	40
21	9·690323	3·74	9·940338	1·18	9·749985	4·93	10·250015	39
22	690548	3·74	940267	1·18	750281	4·92	249719	38
23	690772	3·74	940196	1·18	750576	4·92	249424	37
24	690996	3·74	940125	1·19	750872	4·92	249128	36
25	691220	3·73	940054	1·19	751167	4·92	248833	35
26	691444	3·73	939982	1·19	751462	4·92	248538	34
27	691668	3·73	939911	1·19	751757	4·92	248243	33
28	691892	3·73	939840	1·19	752052	4·91	247948	32
29	692115	3·72	939768	1·19	752347	4·91	247653	31
30	692339	3·72	939697	1·19	752642	4·91	247358	30
31	9·692562	3·72	9·939625	1·19	9·752937	4·91	10·247063	29
32	692785	3·71	939554	1·19	753231	4·91	246769	28
33	693008	3·71	939482	1·19	753526	4·91	246474	27
34	693231	3·71	939410	1·19	753820	4·90	246180	26
35	693453	3·71	939339	1·19	754115	4·90	245885	25
36	693676	3·70	939267	1·20	754409	4·90	245591	24
37	693898	3·70	939195	1·20	754703	4·90	245297	23
38	694120	3·70	939123	1·20	754997	4·90	245003	22
39	694342	3·70	939052	1·20	755291	4·90	244709	21
40	694564	3·69	938980	1·20	755585	4·89	244415	20
41	9·694786	3·69	9·938908	1·20	9·755878	4·89	10·244122	19
42	695007	3·69	938836	1·20	756172	4·89	243828	18
43	695229	3·69	938763	1·20	756465	4·89	243535	17
44	695450	3·68	938691	1·20	756759	4·89	243241	16
45	695671	3·68	938619	1·20	757052	4·89	242948	15
46	695892	3·68	938547	1·20	757345	4·88	242655	14
47	696113	3·68	938475	1·20	757638	4·88	242362	13
48	696334	3·67	938402	1·21	757931	4·88	242069	12
49	696554	3·67	938330	1·21	758224	4·88	241776	11
50	696775	3·67	938258	1·21	758517	4·88	241483	10
51	9·696995	3·67	9·938185	1·21	9·758810	4·88	10·241190	9
52	697215	3·66	938113	1·21	759102	4·87	240898	8
53	697435	3·66	938040	1·21	759395	4·87	240605	7
54	697654	3·66	937967	1·21	759687	4·87	240313	6
55	697874	3·66	937895	1·21	759979	4·87	240021	5
56	698094	3·65	937822	1·21	760272	4·87	239728	4
57	698313	3·65	937749	1·21	760564	4·87	239436	3
58	698532	3·65	937676	1·21	760856	4·86	239144	2
59	698751	3·65	937604	1·21	761148	4·86	238852	1
60	698970	3·64	937531	1·21	761439	4·86	238561	0
	Cosine	D.	Sine	60°	Cotang.	D.	Tang.	M.

M.	Sine	D.	Cosine	D.	Tang.	D.	Cotang.	
0	9·	3·64	9·937531	1·21	9·761439	4·86	10·238561	60
1		3·64	937458	1·22	761731	4·86	238269	59
2	699407	3·64	937385	1·22	762023	4·86	237977	58
3	699626	3·64	937312	1·22	762314	4·86	237686	57
4	699844	3·63	937238	1·22	762606	4·85	237394	56
5		3·63	937165	1·22	762897	4·85	237103	55
6	700280	3·63	937092	1·22	763188	4·85	236812	54
7	700498	3·63	937019	1·22	763479	4·85	236521	53
8	700716	3·63	936946	1·22	763770	4·85	236230	52
9	700933	3·62	936872	1·22	764061	4·85	235939	51
10	701151	3·62	936799	1·22	764352	4·84	235648	50
11	9·701368	3·62	9·936725	1·22	9·764643	4·84	10·235357	49
12	701585	3·62	936652	1·23	764933	4·84	235067	48
13	701802	3·61	936578	1·23	765224	4·84	234776	47
14	702019	3·61	936505	1·23	765514	4·84	234486	46
15	702236	3·61	936431	1·23	765805	4·84	234195	45
16	702452	3·61	936357	1·23	766095	4·84	233905	44
17	702669	3·60	936284	1·23	766385	4·83	233615	43
18	702885	3·60	936210	1·23	766675	4·83	233325	42
19	703101	3·60	936136	1·23	766965	4·83	233035	41
20	703317	3·60	936062	1·23	767255	4·83	232745	40
21	9·703533	3·59	9·935988	1·23	9·767545	4·83	10·232455	39
22	703749	3·59	935914	1·23	767834	4·83	232166	38
23	703964	3·59	935840	1·23	768124	4·82	231876	37
24	704179	3·59	935766	1·24	768413	4·82	231587	36
25	704395	3·59	935692	1·24	768703	4·82	231297	35
26	704610	3·58	935618	1·24	768992	4·82	231008	34
27	704825	3·58	935543	1·24	769281	4·82	230719	33
28	705040	3·58	935469	1·24	769570	4·82	230430	32
29	705254	3·58	935395	1·24	769860	4·81	230140	31
30	705469	3·57	935320	1·24	770148	4·81	229852	30
31	9·705683	3·57	9·935246	1·24	9·770437	4·81	10·229563	29
32	705898	3·57	935171	1·24	770726	4·81	229274	28
33	706112	3·57	935097	1·24	771015	4·81	228985	27
34	706326	3·56	935022	1·24	771303	4·81	228697	26
35	706539	3·56	934948	1·24	771592	4·81	228408	25
36	706753	3·56	934873	1·24	771880	4·80	228120	24
37	706967	3·56	934798	1·25	772168	4·80	227832	23
38	707180	3·55	934723	1·25	772457	4·80	227543	22
39	707393	3·55	934649	1·25	772745	4·80	227255	21
40	707606	3·55	934574	1·25	773033	4·80	226967	20
41	9·707819	3·55	9·934499	1·25	9·773321	4·80	10·226679	19
42	708032	3·54	934424	1·25	773608	4·79	226392	18
43	708245	3·54	934349	1·25	773896	4·79	226104	17
44	708458	3·54	934274	1·25	774184	4·79	225816	16
45	708670	3·54	934199	1·25	774471	4·79	225529	15
46	708882	3·53	934123	1·25	774759	4·79	225241	14
47	709094	3·53	934048	1·25	775046	4·79	224954	13
48	709306	3·53	933973	1·25	775333	4·79	224667	12
49	709518	3·53	933898	1·26	775621	4·78	224379	11
50	709730	3·53	933822	1·26	775908	4·78	224092	10
51	9·709941	3·52	9·933747	1·26	9·776195	4·78	10·223805	9
52	710153	3·52	933671	1·26	776482	4·78	223518	8
53	710364	3·52	933596	1·26	776769	4·78	223231	7
54	710575	3·52	933520	1·26	777055	4·78	222945	6
55	710786	3·51	933445	1·26	777342	4·78	222658	5
56	710997	3·51	933369	1·26	777628	4·77	222372	4
57	711208	3·51	933293	1·26	777915	4·77	222085	3
58	711419	3·51	933217	1·26	778201	4·77	221799	2
59	711629	3·50	933141	1·26	778487	4·77	221512	1
60	711839	3·50	933066	1·26	778774	4·77	221226	0
	Cosine	D.	Sine	59°	Cotang.	D.	Tang.	M.

M.	Sine	D.	Cosine	D.	Tang.	D.	Cotang.	
0	9·711839	3·50	9·933066	1·26	9·778774	4·77	10·221226	60
1	712050	3·50	932990	1·27	779060	4·77	220940	59
2	712260	3·50	932914	1·27	779346	4·76	220654	58
3	712469	3·49	932838	1·27	779632	4·76	220368	57
4	712679	3·49	932762	1·27	779918	4·76	220082	56
5	712889	3·49	932685	1·27	780203	4·76	219797	55
6	713098	3·49	932609	1·27	780489	4·76	219511	54
7	713308	3·49	932533	1·27	780775	4·76	219225	53
8	713517	3·48	932457	1·27	781060	4·76	218940	52
9	713726	3·48	932380	1·27	781346	4·75	218654	51
10	713935	3·48	932304	1·27	781631	4·75	218369	50
11	9·714144	3·48	9·932228	1·27	9·781916	4·75	10·218084	49
12	714352	3·47	932151	1·27	782201	4·75	217799	48
13	714561	3·47	932075	1·28	782486	4·75	217514	47
14	714769	3·47	931998	1·28	782771	4·75	217229	46
15	714978	3·47	931921	1·28	783056	4·75	216944	45
16	715186	3·47	931845	1·28	783341	4·75	216659	44
17	715394	3·46	931768	1·28	783626	4·74	216374	43
18	715602	3·46	931691	1·28	783910	4·74	216090	42
19	715809	3·46	931614	1·28	784195	4·74	215805	41
20	716017	3·46	931537	1·28	784479	4·74	215521	40
21	9·716224	3·45	9·931460	1·28	9·784764	4·74	10·215236	39
22	716432	3·45	931383	1·28	785048	4·74	214952	38
23	716639	3·45	931306	1·28	785332	4·73	214668	37
24	716846	3·45	931229	1·29	785616	4·73	214384	36
25	717053	3·45	931152	1·29	785900	4·73	214100	35
26	717259	3·44	931075	1·29	786184	4·73	213816	34
27	717466	3·44	930998	1·29	786468	4·73	213532	33
28	717673	3·44	930921	1·29	786752	4·73	213248	32
29	717879	3·44	930843	1·29	787036	4·73	212964	31
30	718085	3·43	930766	1·29	787319	4·72	212681	30
31	9·718291	3·43	9·930688	1·29	9·787603	4·72	10·212397	29
32	718497	3·43	930611	1·29	787886	4·72	212114	28
33	718703	3·43	930533	1·29	788170	4·72	211830	27
34	718909	3·43	930456	1·29	788453	4·72	211547	26
35	719114	3·42	930378	1·29	788736	4·72	211264	25
36	719320	3·42	930300	1·30	789019	4·72	210981	24
37	719525	3·42	930223	1·30	789302	4·71	210698	23
38	719730	3·42	930145	1·30	789585	4·71	210415	22
39	719935	3·41	930067	1·30	789868	4·71	210132	21
40	720140	3·41	929989	1·30	790151	4·71	209849	20
41	9·720345	3·41	9·929911	1·30	9·790433	4·71	10·209567	19
42	720549	3·41	929833	1·30	790716	4·71	209284	18
43	720754	3·40	929755	1·30	790999	4·71	209001	17
44	720958	3·40	929677	1·30	791281	4·71	208719	16
45	721162	3·40	929599	1·30	791563	4·70	208437	15
46	721366	3·40	929521	1·30	791846	4·70	208154	14
47	721570	3·40	929442	1·30	792128	4·70	207872	13
48	721774	3·39	929364	1·31	792410	4·70	207590	12
49	721978	3·39	929286	1·31	792692	4·70	207308	11
50	722181	3·39	929207	1·31	792974	4·70	207026	10
51	9·722385	3·39	9·929129	1·31	9·793256	4·70	10·206744	9
52	722588	3·39	929050	1·31	793538	4·69	206462	8
53	722791	3·38	928972	1·31	793819	4·69	206181	7
54	722994	3·38	928893	1·31	794101	4·69	205899	6
55	723197	3·38	928815	1·31	794383	4·69	205617	5
56	723400	3·38	928736	1·31	794664	4·69	205336	4
57	723603	3·37	928657	1·31	794945	4·69	205055	3
58	723805	3·37	928578	1·31	795227	4·69	204773	2
59	724007	3·37	928499	1·31	795508	4·68	204492	1
60	724210	3·37	928420	1·31	795789	4·68	204211	0
	Cosine	D.	Sine	58°	Cotang.	D.	Tang.	M.

M.	Sine	D.	Cosine	D.	Tang.	D.	Cotang.	
0	9·724210	3·37	9·928420	1·32	9·795789	4·68	10·204211	60
1	724412	3·37	928342	1·32	796070	4·68	203930	59
2	724614	3·36	928263	1·32	796351	4·68	203649	58
3	724816	3·36	928183	1·32	796632	4·68	203368	57
4	725017	3·36	928104	1·32	796913	4·68	203087	56
5	725219	3·36	928025	1·32	797194	4·68	202806	55
6	725420	3·35	927946	1·32	797475	4·68	202525	54
7	725622	3·35	927867	1·32	797755	4·68	202245	53
8	725823	3·35	927787	1·32	798036	4·67	201964	52
9	726024	3·35	927708	1·32	798316	4·67	201684	51
10	726225	3·35	927629	1·32	798596	4·67	201404	50
11	9·726426	3·34	9·927549	1·32	9·798877	4·67	10·201123	49
12	726626	3·34	927470	1·33	799157	4·67	200843	48
13	726827	3·34	927390	1·33	799437	4·67	200563	47
14	727027	3·34	927310	1·33	799717	4·67	200283	46
15	727228	3·34	927231	1·33	799997	4·66	200003	45
16	727428	3·33	927151	1·33	800277	4·66	199723	44
17	727628	3·33	927071	1·33	800557	4·66	199443	43
18	727828	3·33	926991	1·33	800836	4·66	199164	42
19	728027	3·33	926911	1·33	801116	4·66	198884	41
20	728227	3·33	926831	1·33	801396	4·66	198604	40
21	9·728427	3·32	9·926751	1·33	9·801675	4·66	10·198325	39
22	728626	3·32	926671	1·33	801955	4·66	198045	38
23	728825	3·32	926591	1·33	802234	4·65	197766	37
24	729024	3·32	926511	1·34	802513	4·65	197487	36
25	729223	3·31	926431	1·34	802792	4·65	197208	35
26	729422	3·31	926351	1·34	803072	4·65	196928	34
27	729621	3·31	926270	1·34	803351	4·65	196649	33
28	729820	3·31	926190	1·34	803630	4·65	196370	32
29	730018	3·30	926110	1·34	803908	4·65	196092	31
30	730216	3·30	926029	1·34	804187	4·65	195813	30
31	9·730415	3·30	9·925949	1·34	9·804466	4·64	10·195534	29
32	730613	3·30	925868	1·34	804745	4·64	195255	28
33	730811	3·30	925788	1·34	805023	4·64	194977	27
34	731009	3·29	925707	1·34	805302	4·64	194698	26
35	731206	3·29	925626	1·34	805580	4·64	194420	25
36	731404	3·29	925545	1·35	805859	4·64	194141	24
37	731602	3·29	925465	1·35	806137	4·64	193863	23
38	731799	3·29	925384	1·35	806415	4·63	193585	22
39	731996	3·28	925303	1·35	806693	4·63	193307	21
40	732193	3·28	925222	1·35	806971	4·63	193029	20
41	9·732390	3·28	9·925141	1·35	9·807249	4·63	10·192751	19
42	732587	3·28	925060	1·35	807527	4·63	192473	18
43	732784	3·28	924979	1·35	807805	4·63	192195	17
44	732980	3·27	924897	1·35	808083	4·63	191917	16
45	733177	3·27	924816	1·35	808361	4·63	191639	15
46	733373	3·27	924735	1·36	808638	4·62	191362	14
47	733569	3·27	924654	1·36	808916	4·62	191084	13
48	733765	3·27	924572	1·36	809193	4·62	190807	12
49	733961	3·26	924491	1·36	809471	4·62	190529	11
50	734157	3·26	924409	1·36	809748	4·62	190252	10
51	9·734353	3·26	9·924328	1·36	9·810025	4·62	10·189975	9
52	734549	3·26	924246	1·36	810302	4·62	189698	8
53	734744	3·25	924164	1·36	810580	4·62	189420	7
54	734939	3·25	924083	1·36	810857	4·62	189143	6
55	735135	3·25	924001	1·36	811134	4·61	188866	5
56	735330	3·25	923919	1·36	811410	4·61	188590	4
57	735525	3·25	923837	1·36	811687	4·61	188313	3
58	735719	3·24	923755	1·37	811964	4·61	188036	2
59	735914	3·24	923673	1·37	812241	4·61	187759	1
60	736109	3·24	923591	1·37	812517	4·61	187483	0
	Cosine	D.	Sine	57°	Cotang.	D.	Tang.	M.

M.	Sine	D.	Cosine	D.	Tang.	D.	Cotang.	
0	9·736109	3·24	9·923591	1·37	9·812517	4·61	10·187482	60
1	736303	3·24	923509	1·37	812794	4·61	187206	59
2	736498	3·24	923427	1·37	813070	4·61	186930	58
3	736692	3·23	923345	1·37	813347	4·60	186653	57
4	736886	3·23	923263	1·37	813623	4·60	186377	56
5	737080	3·23	923181	1·37	813899	4·60	186101	55
6	737274	3·23	923098	1·37	814175	4·60	185825	54
7	737467	3·23	923016	1·37	814452	4·60	185548	53
8	737661	3·22	922933	1·37	814728	4·60	185272	52
9	737855	3·22	922851	1·37	815004	4·60	184996	51
10	738048	3·22	922768	1·38	815279	4·60	184721	50
11	9·738241	3·22	9·922686	1·38	9·815555	4·59	10·184445	49
12	738434	3·22	922603	1·38	815831	4·59	184169	48
13	738627	8·21	922520	1·38	816107	4·59	183893	47
14	738820	3·21	922438	1·38	816382	4·59	183618	46
15	739013	3·21	922355	1·38	816658	4·59	183342	45
16	739206	3·21	922272	1·38	816933	4·59	183067	44
17	739398	3·21	922189	1·38	817209	4·59	182791	43
18	739590	3·20	922106	1·38	817484	4·59	182516	42
19	739783	3·20	922023	1·38	817759	4·59	182241	41
20	739975	3·20	921940	1·38	818035	4·58	181965	40
21	9·740167	3·20	9·921857	1·39	9·818310	4·58	10·181690	39
22	740359	3·20	921774	1·39	818585	4·58	181415	38
23	740550	3·19	921691	1·39	818860	4·58	181140	37
24	740742	3·19	921607	1·39	819135	4·58	180865	36
25	740934	3·19	921524	1·39	819410	4·58	180590	35
26	741125	3·19	921441	1·39	819684	4·58	180316	34
27	741316	3·19	921357	1·39	819959	4·58	180041	33
28	741508	3·18	921274	1·39	820234	4·58	179766	32
29	741699	3·18	921190	1·39	820508	4·57	179492	31
30	741889	3·18	921107	1·39	820783	4·57	179217	30
31	9·742080	3·18	9·921023	1·39	9·821057	4·57	10·178943	29
32	742271	3·18	920939	1·40	821332	4·57	178668	28
33	742462	3·17	920856	1·40	821606	4·57	178394	27
34	742652	3·17	920772	1·40	821880	4·57	178120	26
35	742842	3·17	920688	1·40	822154	4·57	177846	25
36	743033	3·17	920604	1·40	822429	4·57	177571	24
37	743223	3·17	920520	1·40	822703	4·57	177297	23
38	743413	3·16	920436	1·40	822977	4·56	177023	22
39	743602	3·16	920352	1·40	823250	4·56	176750	21
40	743792	3·16	920268	1·40	823524	4·56	176476	20
41	9·743982	3·16	9·920184	1·40	9·823798	4·56	10·176202	19
42	744171	3·16	920099	1·40	824072	4·56	175928	18
43	744361	3·15	920015	1·40	824345	4·56	175655	17
44	744550	3·15	919931	1·41	824619	4·56	175381	16
45	744739	3·15	919846	1·41	824893	4·56	175107	15
46	744928	3·15	919762	1·41	825166	4·56	174834	14
47	745117	3·15	919677	1·41	825439	4·5	174561	13
48	745306	3·14	919593	1·41	825713	4·5	174287	12
49	745494	3·14	919508	1·41	825986	4·5	174014	11
50	745683	3·14	919424	1·41	826259	4·5	173741	10
51	9·745871	3·14	9·919339	1·41	9·826532	4·5	10·173468	9
52	746059	3·14	919254	1·41	826805	4·5	173195	8
53	746248	3·13	919169	1·41	827078	4·5	172922	7
54	746436	4·13	919085	1·41	827351	4·55	172649	6
55	746624	3·13	919000	1·41	827624	4·55	172376	5
56	746812	3·13	918915	1·42	827897	4·54	172103	4
57	746999	3·13	918830	1·42	828170	4·54	171830	3
58	747187	3·12	918745	1·42	828442	4·54	171558	2
59	747374	3·12	918659	1·42	828715	4·54	171285	1
60	747562	3·12	918574	1·42	828987	4·54	171013	0
	Cosine	D.	Sine	56°	Cotang.	D.	Tang.	M.

M.	Sine	D.	Cosine	D.	Tang.	D.	Cotang.	
0	9·747562	3·12	9·918574	1·42	9·828987	4·54	10·171013	60
1	747749	3·12	918489	1·42	829260	4·54	170740	59
2	747936	3·12	918404	1·42	829532	4·54	170468	58
3	748123	3·11	918318	1·42	829805	4·54	170195	57
4	748310	3·11	918233	1·42	830077	4·54	169923	56
5	748497	3·11	918147	1·42	830349	4·53	169651	55
6	748683	3·11	918062	1·42	830621	4·53	169379	54
7	748870	3·11	917976	1·43	830893	4·53	169107	53
8	749056	3·10	917891	1·43	831165	4·53	168835	52
9	749243	3·10	917805	1·43	831437	4·53	168563	51
10	749429	3·10	917719	1·43	831709	4·53	168291	50
11	9·749615	3·10	9·917634	1·43	9·831981	4·53	10·168019	49
12	749801	3·10	917548	1·43	832253	4·53	167747	48
13	749987	3·09	917462	1·43	832525	4·53	167475	47
14	750172	3·09	917376	1·43	832796	4·53	167204	46
15	750358	3·09	917290	1·43	833068	4·52	166932	45
16	750543	3·09	917204	1·43	833339	4·52	166661	44
17	750729	3·09	917118	1·44	833611	4·52	166389	43
18	750914	3·08	917032	1·44	833882	4·52	166118	42
19	751099	3·08	916946	1·44	834154	4·52	165846	41
20	751284	3·08	916859	1·44	834425	4·52	165575	40
21	9·751469	3·08	9·916773	1·44	9·834696	4·52	10·165304	39
22	751654	3·08	916687	1·44	834967	4·52	165033	38
23	751839	3·08	916600	1·44	835238	4·52	164762	37
24	752023	3·07	916514	1·44	835509	4·52	164491	36
25	752208	3·07	916427	1·44	835780	4·51	164220	35
26	752392	3·07	916341	1·44	836051	4·51	163949	34
27	752576	3·07	916254	1·44	836322	4·51	163678	33
28	752760	3·07	916167	1·45	836593	4·51	163407	32
29	752944	3·06	916081	1·45	836864	4·51	163136	31
30	753128	3·06	915994	1·45	837134	4·51	162866	30
31	9·753312	3·06	9·915907	1·45	9·837405	4·51	10·162595	29
32	753495	3·06	915820	1·45	837675	4·51	162325	28
33	753679	3·06	915733	1·45	837946	4·51	162054	27
34	753862	3·05	915646	1·45	838216	4·51	161784	26
35	754046	3·05	915559	1·45	838487	4·50	161513	25
36	754229	3·05	915472	1·45	838757	4·50	161243	24
37	754412	3·05	915385	1·45	839027	4·50	160973	23
38	754595	3·05	915297	1·45	839297	4·50	160703	22
39	754778	3·04	915210	1·45	839568	4·50	160432	21
40	754960	3·04	915123	1·46	839838	4·50	160162	20
41	9·755143	3·04	9·915035	1·46	9·840108	4·50	10·159892	19
42	755326	3·04	914948	1·46	840378	4·50	159622	18
43	755508	3·04	914860	1·46	840647	4·50	159353	17
44	755690	3·04	914773	1·46	840917	4·49	159083	16
45	755872	3·03	914685	1·46	841187	4·49	158813	15
46	756054	3·03	914598	1·46	841457	4·49	158543	14
47	756236	3·03	914510	1·46	841726	4·49	158274	13
48	756418	3·03	914422	1·46	841996	4·49	158004	12
49	756600	3·03	914334	1·46	842266	4·49	157734	11
50	756782	3·02	914246	1·47	842535	4·49	157465	10
51	9·756963	3·02	9·914158	1·47	9·842805	4·49	10·157195	9
52	757144	3·02	914070	1·47	843074	4·49	156926	8
53	757326	3·02	913982	1·47	843343	4·49	156657	7
54	757507	3·02	913894	1·47	843612	4·49	156388	6
55	757688	3·01	913806	1·47	843882	4·48	156118	5
56	757869	3·01	913718	1·47	844151	4·48	155849	4
57	758050	3·01	913630	1·47	844420	4 48	155580	3
58	758230	3·01	913541	1·47	844689	4·48	155311	2
59	758411	3·01	913453	1·47	844958	4·48	155042	1
60	758591	3·01	913365	1·47	845227	4·48	154773	0
	Cosine	D.	Sine	55°	Cotang.	D.	Tang.	M.

M.	Sine	D.	Cosine	D.	Tang.	D.	Cotang.	
0	9·758591	3·01	9·913365	1·47	9·845227	4·48	10·154773	60
1	758772	3·00	913276	1·47	845496	4·48	154504	59
2	758952	3·00	913187	1·48	845764	4·48	154236	58
3	759132	3·00	913099	1·48	846033	4·48	153967	57
4	759312	3·00	913010	1·48	846302	4·48	153698	56
5	759492	3·00	912922	1·48	846570	4·47	153430	55
6	759672	2·99	912833	1·48	846839	4·47	153161	54
7	759852	2·99	912744	1·48	847107	4·47	152893	53
8	760031	2·99	912655	1·48	847376	4·47	152624	52
9	760211	2·99	912566	1·48	847644	4·47	152356	51
10	760390	2·99	912477	1·48	847913	4·47	152087	50
11	9·760569	2·98	9·912388	1·48	9·848181	4·47	10·151819	49
12	760748	2·98	912299	1·49	848449	4·47	151551	48
13	760927	2·98	912210	1·49	848717	4·47	151283	47
14	761106	2·98	912121	1·49	848986	4·47	151014	46
15	761285	2·98	912031	1·49	849254	4·47	150746	45
16	761464	2·98	911942	1·49	849522	4·47	150478	44
17	761642	2·97	911853	1·49	849790	4·46	150210	43
18	761821	2·97	911763	1·49	850058	4·46	149942	42
19	761999	2·97	911674	1·49	850325	4·46	149675	41
20	762177	2·97	911584	1·49	850593	4·46	149407	40
21	9·762356	2·97	9·911495	1·49	9·850861	4·46	10·149139	39
22	762534	2·96	911405	1·49	851129	4·46	148871	38
23	762712	2·96	911315	1·50	851396	4·46	148604	37
24	762889	2·96	911226	1·50	851664	4·46	148336	36
25	763067	2·96	911136	1·50	851931	4·46	148069	35
26	763245	2·96	911046	1·50	852199	4·46	147801	34
27	763422	2·96	910956	1·50	852466	4·46	147534	33
28	763600	2·95	910866	1·50	852733	4·45	147267	32
29	763777	2·95	910776	1·50	853001	4·45	146999	31
30	763954	2·95	910686	1·50	853268	4·45	146732	30
31	9·764131	2·95	9·910596	1·50	9·853535	4·45	10·146465	29
32	764308	2·95	910506	1·50	853802	4·45	146198	28
33	764485	2·94	910415	1·50	854069	4·45	145931	27
34	764662	2·94	910325	1·51	854336	4·45	145664	26
35	764838	2·94	910235	1·51	854603	4·45	145397	25
36	765015	2·94	910144	1·51	854870	4·45	145130	24
37	765191	2·94	910054	1·51	855137	4·45	144863	23
38	765367	2·94	909963	1·51	855404	4·45	144596	22
39	765544	2·93	909873	1·51	855671	4·44	144329	21
40	765720	2·93	909782	1·51	855938	4·44	144062	20
41	9·765896	2·93	9·909691	1·51	9·856204	4·44	10·143796	19
42	766072	2·93	909601	1·51	856471	4·44	143529	18
43	766247	2·93	909510	1·51	856737	4·44	143263	17
44	766423	2·93	909419	1·51	857004	4·44	142996	16
45	766598	2·92	909328	1·52	857270	4·44	142730	15
46	766774	2·92	909237	1·52	857537	4·44	142463	14
47	766949	2·92	909146	1·52	857803	4·44	142197	13
48	767124	2·92	909055	1·52	858069	4·44	141931	12
49	767300	2·92	908964	1·52	858336	4·44	141664	11
50	767475	2·91	908873	1·52	858602	4·43	141398	10
51	9·767649	2·91	9·908781	1·52	9·858868	4·43	10·141132	9
52	767824	2·91	908690	1·52	859134	4·43	140866	8
53	767999	2·91	908599	1·52	859400	4·43	140600	7
54	768173	2·91	908507	1·52	859666	4·43	140334	6
55	768348	2·90	908416	1·53	859932	4·43	140068	5
56	768522	2·90	908324	1·53	860198	4·43	139802	4
57	768697	2·90	908233	1·53	860464	4·43	139536	3
58	768871	2·90	908141	1·53	860730	4·43	139270	2
59	769045	2·90	908049	1·53	860995	4·43	139005	1
60	769219	2·90	907958	1·53	861261	4·43	138739	0
	Cosine	D.	Sine	54°	Cotang.	D.	Tang.	M.

M.	Sine	D.	Cosine	D.	Tang.	D.	Cotang.	
0	9·769219	2·90	9·907958	1·53	9·861261	4·43	10·138739	60
1	769393	2·89	907866	1·53	861527	4·43	138473	59
2	769566	2·89	907774	1·53	861792	4·42	138208	58
3	769740	2·89	907682	1·53	862058	4·42	137942	57
4	769913	2·89	907590	1·53	862323	4·42	137677	56
5	770087	2·89	907498	1·53	862589	4·42	137411	55
6	770260	2·88	907406	1·53	862854	4·42	137146	54
7	770433	2·88	907314	1·54	863119	4·42	136881	53
8	770606	2·88	907222	1·54	863385	4·42	136615	52
9	770779	2·88	907129	1·54	863650	4·42	136350	51
10	770952	2·88	907037	1·54	863915	4·42	136085	50
11	9·771125	2·88	9·906945	1·54	9·864180	4·42	10·135820	49
12	771298	2·87	906852	1·54	864445	4·42	135555	48
13	771470	2·87	906760	1·54	864710	4·42	135290	47
14	771643	2·87	906667	1·54	864975	4·41	135025	46
15	771815	2·87	906575	1·54	865240	4·41	134760	45
16	771987	2·87	906482	1·54	865505	4·41	134495	44
17	772159	2·87	906389	1·55	865770	4·41	134230	43
18	772331	2·86	906296	1·55	866035	4·41	133965	42
19	772503	2·86	906204	1·55	866300	4·41	133700	41
20	772675	2·86	906111	1·55	866564	4·41	133436	40
21	9·772847	2·86	9·906018	1·55	9·866829	4·41	10·133171	39
22	773018	2·86	905925	1·55	867094	4·41	132906	38
23	773190	2·86	905832	1·55	867358	4·41	132642	37
24	773361	2·85	905739	1·55	867623	4·41	132377	36
25	773533	2·85	905645	1·55	867887	4·41	132113	35
26	773704	2·85	905552	1·55	868152	4·40	131848	34
27	773875	2·85	905459	1·55	868416	4·40	131584	33
28	774046	2·85	905366	1·56	868680	4·40	131320	32
29	774217	2·85	905272	1·56	868945	4·40	131055	31
30	774388	2·84	905179	1·56	869209	4·40	130791	30
31	9·774558	2·84	9·905085	1·56	9·869473	4·40	10·130527	29
32	774729	2·84	904992	1·56	869737	4·40	130263	28
33	774899	2·84	904898	1·56	870001	4·40	129999	27
34	775070	2·84	904804	1·56	870265	4·40	129735	26
35	775240	2·84	904711	1·56	870529	4·40	129471	25
36	775410	2·83	904617	1·56	870793	4·40	129207	24
37	775580	2·83	904523	1·56	871057	4·40	128943	23
38	775750	2·83	904429	1·57	871321	4·40	128679	22
39	775920	2·83	904335	1·57	871585	4·40	128415	21
40	776090	2·83	904241	1·57	871849	4·39	128151	20
41	9·776259	2·83	9·904147	1·57	9·872112	4·39	10·127888	19
42	776429	2·82	904053	1·57	872376	4·39	127624	18
43	776598	2·82	903959	1·57	872640	4·39	127360	17
44	776768	2·82	903864	1·57	872903	4·39	127097	16
45	776937	2·82	903770	1·57	873167	4·39	126833	15
46	777106	2·82	903676	1·57	873430	4·39	126570	14
47	777275	2·81	903581	1·57	873694	4·39	126306	13
48	777444	2·81	903487	1·57	873957	4·39	126043	12
49	777613	2·81	903392	1·58	874220	4·39	125780	11
50	777781	2·81	903298	1·58	874484	4·39	125516	10
51	9·777950	2·81	9·903203	1·58	9·874747	4·39	10·125253	9
52	778119	2·81	903108	1·58	875010	4·39	124990	8
53	778287	2·80	903014	1·58	875273	4·38	124727	7
54	778455	2·80	902919	1·58	875536	4·38	124464	6
55	778624	2·80	902824	1·58	875800	4·38	124200	5
56	778792	2·80	902729	1·58	876063	4·38	123937	4
57	778960	2·80	902634	1·58	876326	4·38	123674	3
58	779128	2·80	902539	1·59	876589	4·38	123411	2
59	779295	2·79	902444	1·59	876851	4·38	123149	1
60	779463	2·79	902349	1·59	877114	4·38	122886	0
	Cosine	D.	Sine	53°	Cotang.	D.	Tang.	M.

M.	Sine	D.	Cosine	D.	Tang.	D.	Cotang.	
0	9.779463	2.79	9.902349	1.59	9.877114	4.38	10.122886	60
1	779631	2.79	902253	1.59	877377	4.38	122623	59
2	779798	2.79	902158	1.59	877640	4.38	122360	58
3	779966	2.79	902063	1.59	877903	4.38	122097	57
4	780133	2.79	901967	1.59	878165	4.38	121835	56
5	780300	2.78	901872	1.59	878428	4.38	121572	55
6	780467	2.78	901776	1.59	878691	4.38	121309	54
7	780634	2.78	901681	1.59	878953	4.37	121047	53
8	780801	2.78	901585	1.59	879216	4.37	120784	52
9	780968	2.78	901490	1.59	879478	4.37	120522	51
10	781134	2.78	901394	1.60	879741	4.37	120259	50
11	9.781301	2.77	9.901298	1.60	9.880003	4.37	10.119997	49
12	781468	2.77	901202	1.60	880265	4.37	119735	48
13	781634	2.77	901106	1.60	880528	4.37	119472	47
14	781800	2.77	901010	1.60	880790	4.37	119210	46
15	781966	2.77	900914	1.60	881052	4.37	118948	45
16	782132	2.77	900818	1.60	881314	4.37	118686	44
17	782298	2.76	900722	1.60	881576	4.37	118424	43
18	782464	2.76	900626	1.60	881839	4.37	118161	42
19	782630	2.76	900529	1.60	882101	4.37	117899	41
20	782796	2.76	900433	1.61	882363	4.36	117637	40
21	9.782961	2.76	9.900337	1.61	9.882625	4.36	10.117375	39
22	783127	2.76	900240	1.61	882887	4.36	117113	38
23	783292	2.75	900144	1.61	883148	4.36	116852	37
24	783458	2.75	900047	1.61	883410	4.36	116590	36
25	783623	2.75	899951	1.61	883672	4.36	116328	35
26	783788	2.75	899854	1.61	883934	4.36	116066	34
27	783953	2.75	899757	1.61	884196	4.36	115804	33
28	784118	2.75	899660	1.61	884457	4.36	115543	32
29	784282	2.74	899564	1.61	884719	4.36	115281	31
30	784447	2.74	899467	1.62	884980	4.36	115020	30
31	9.784612	2.74	9.899370	1.62	9.885242	4.36	10.114758	29
32	784776	2.74	899273	1.62	885503	4.36	114497	28
33	784941	2.74	899176	1.62	885765	4.36	114235	27
34	785105	2.74	899078	1.62	886026	4.36	113974	26
35	785269	2.73	898981	1.62	886288	4.36	113712	25
36	785433	2.73	898884	1.62	886549	4.35	113451	24
37	785597	2.73	898787	1.62	886810	4.35	113190	23
38	785761	2.73	898689	1.62	887072	4.35	112928	22
39	785925	2.73	898592	1.62	887333	4.35	112667	21
40	786089	2.73	898494	1.63	887594	4.35	112406	20
41	9.786252	2.72	9.898397	1.63	9.887855	4.35	10.112145	19
42	786416	2.72	898299	1.63	888116	4.35	111884	18
43	786579	2.72	898202	1.63	888377	4.35	111623	17
44	786742	2.72	898104	1.63	888639	4.35	111361	16
45	786906	2.72	898006	1.63	888900	4.35	111100	15
46	787069	2.72	897908	1.63	889160	4.35	110840	14
47	787232	2.71	897810	1.63	889421	4.35	110579	13
48	787395	2.71	897712	1.63	889682	4.35	110318	12
49	787557	2.71	897614	1.63	889943	4.35	110057	11
50	787720	2.71	897516	1.63	890204	4.34	109796	10
51	9.787883	2.71	9.897418	1.64	9.890465	4.34	10.109535	9
52	788045	2.71	897320	1.64	890725	4.34	109275	8
53	788208	2.71	897222	1.64	890986	4.34	109014	7
54	788370	2.70	897123	1.64	891247	4.34	108753	6
55	788532	2.70	897025	1.64	891507	4.34	108493	5
56	788694	2.70	896926	1.64	891768	4.34	108232	4
57	788856	2.70	896828	1.64	892028	4.34	107972	3
58	789018	2.70	896729	1.64	892289	4.34	107711	2
59	789180	2.70	896631	1.64	892549	4.34	107451	1
60	789342	2.69	896532	1.64	892810	4.34	107190	0
	Cosine	D.	Sine	52°	Cotang.	D.	Tang.	M.

M.	Sine	D.	Cosine	D.	Tang.	D.	Cotang.	
0	9·789342	2·69	9·896532	1·64	9·892810	4·34	10·107190	60
1	789504	2·69	896433	1·65	893070	4·34	106930	59
2	789665	2·69	896335	1·65	893331	4·34	106669	58
3	789827	2·69	896236	1·65	893591	4·34	106409	57
4	789988	2·69	896137	1·65	893851	4·34	106149	56
5	790149	2·69	896038	1·65	894111	4·34	105889	55
6	790310	2·68	895939	1·65	894371	4·34	105629	54
7	790471	2·68	895840	1·65	894632	4·33	105368	53
8	790632	2·68	895741	1·65	894892	4·33	105108	52
9	790793	2·68	895641	1·65	895152	4·33	104848	51
10	790954	2·68	895542	1·65	895412	4·33	104588	50
11	9·791115	2·68	9·895443	1·66	9·895672	4·33	10·104328	49
12	791275	2·67	895343	1·66	895932	4·33	104068	48
13	791436	2·67	895244	1·66	896192	4·33	103808	47
14	791596	2·67	895145	1·66	896452	4·33	103548	46
15	791757	2·67	895045	1·66	896712	4·33	103288	45
16	791917	2·67	894945	1·66	896971	4·33	103029	44
17	792077	2·67	894846	1·66	897231	4·33	102769	43
18	792237	2·66	894746	1·66	897491	4·33	102509	42
19	792397	2·66	894646	1·66	897751	4·33	102249	41
20	792557	2·66	894546	1·66	898010	4·33	101990	40
21	9·792716	2·66	9·894446	1·67	9·898270	4·33	10·101730	39
22	792876	2·66	894346	1·67	898530	4·33	101470	38
23	793035	2·66	894246	1·67	898789	4·33	101211	37
24	793195	2·65	894146	1·67	899049	4·32	100951	36
25	793354	2·65	894046	1·67	899308	4·32	100692	35
26	793514	2·65	893946	1·67	899568	4·32	100432	34
27	793673	2·65	893846	1·67	899827	4·32	100173	33
28	793832	2·65	893745	1·67	900086	4·32	099914	32
29	793991	2·65	893645	1·67	900346	4·32	099654	31
30	794150	2·64	893544	1·67	900605	4·32	099395	30
31	9·794308	2·64	9·893444	1·68	9·900864	4·32	10·099136	29
32	794467	2·64	893343	1·68	901124	4·32	098876	28
33	794626	2·64	893243	1·68	901383	4·32	098617	27
34	794784	2·64	893142	1·68	901642	4·32	098358	26
35	794942	2·64	893041	1·68	901901	4·32	098099	25
36	795101	2·64	892940	1·68	902160	4·32	097840	24
37	795259	2·63	892839	1·68	902419	4·32	097581	23
38	795417	2·63	892739	1·68	902679	4·32	097321	22
39	795575	2·63	892638	1·68	902938	4·32	097062	21
40	795733	2·63	892536	1·68	903197	4·31	096803	20
41	9·795891	2·63	9·892435	1·69	9·903455	4·31	10·096545	19
42	796049	2·63	892334	1·69	903714	4·31	096286	18
43	796206	2·63	892233	1·69	903973	4·31	096027	17
44	796364	2·62	892132	1·69	904232	4·31	095768	16
45	796521	2·62	892030	1·69	904491	4·31	095509	15
46	796679	2·62	891929	1·69	904750	4·31	095250	14
47	796836	2·62	891827	1·69	905008	4·31	094992	13
48	796993	2·62	891726	1·69	905267	4·31	094733	12
49	797150	2·61	891624	1·69	905526	4·31	094474	11
50	797307	2·61	891523	1·70	905784	4·31	094216	10
51	9·797464	2·61	9·891421	1·70	9·906043	4·31	10·093957	9
52	797621	2·61	891319	1·70	906302	4·31	093698	8
53	797777	2·61	891217	1·70	906560	4·31	093440	7
54	797934	2·61	891115	1·70	906819	4·31	093181	6
55	798091	2·61	891013	1·70	907077	4·31	092923	5
56	798247	2·61	890911	1·70	907336	4·31	092664	4
57	798403	2·60	890809	1·70	907594	4·31	092406	3
58	798560	2·60	890707	1·70	907852	4·31	092148	2
59	798716	2·60	890605	1·70	908111	4·30	091889	1
60	798872	2·60	890503	1·70	908369	4·30	091631	0
	Cosine	D.	Sine	51°	Cotang.	D.	Tang.	M.

M.	Sine	D.	Cosine	D.	Tang.	D.	Cotang.	
0	9·798872	2·60	9·890503	1·70	9·908369	4·30	10·091631	60
1	799028	2·60	890400	1·71	908628	4·30	091372	59
2	799184	2·60	890298	1·71	908886	4·30	091114	58
3	799339	2·59	890195	1·71	909144	4·30	090856	57
4	799495	2·59	890093	1·71	909402	4·30	090598	56
5	799651	2·59	889990	1·71	909660	4·30	090340	55
6	799806	2·59	889888	1·71	909918	4·30	090082	54
7	799962	2·59	889785	1·71	910177	4·30	089823	53
8	800117	2·59	889682	1·71	910435	4·30	089565	52
9	800272	2·58	889579	1·71	910693	4·30	089307	51
10	800427	2·58	889477	1·71	910951	4·30	089049	50
11	9·800582	2·58	9·889374	1·72	9·911209	4·30	10·088791	49
12	800737	2·58	889271	1·72	911467	4·30	088533	48
13	800892	2·58	889168	1·72	911724	4·30	088276	47
14	801047	2·58	889064	1·72	911982	4·30	088018	46
15	801201	2·58	888961	1·72	912240	4·30	087760	45
16	801356	2·57	888858	1·72	912498	4·30	087502	44
17	801511	2·57	888755	1·72	912756	4·30	087244	43
18	801665	2·57	888651	1·72	913014	4·29	086986	42
19	801819	2·57	888548	1·72	913271	4·29	086729	41
20	801973	2·57	888444	1·73	913529	4·29	086471	40
21	9·802128	2·57	9·888341	1·73	9·913787	4·29	10·086213	39
22	802282	2·56	888237	1·73	914044	4·29	085956	38
23	802436	2·56	888134	1·73	914302	4·29	085698	37
24	802589	2·56	888030	1·73	914560	4·29	085440	36
25	802743	2·56	887926	1·73	914817	4·29	085183	35
26	802897	2·56	887822	1·73	915075	4·29	084925	34
27	803050	2·56	887718	1·73	915332	4·29	084668	33
28	803204	2·56	887614	1·73	915590	4·29	084410	32
29	803357	2·55	887510	1·73	915847	4·29	084153	31
30	803511	2·55	887406	1·74	916104	4·29	083896	30
31	9·803664	2·55	9·887302	1·74	9·916362	4·29	10·083638	29
32	803817	2·55	887198	1·74	916619	4·29	083381	28
33	803970	2·55	887093	1·74	916877	4·29	083123	27
34	804123	2·55	886989	1·74	917134	4·29	082866	26
35	804276	2·54	886885	1·74	917391	4·29	082609	25
36	804428	2·54	886780	1·74	917648	4·29	082352	24
37	804581	2·54	886676	1·74	917905	4·29	082095	23
38	804734	2·54	886571	1·74	918163	4·28	081837	22
39	804886	2·54	886466	1·74	918420	4·28	081580	21
40	805039	2·54	886362	1·75	918677	4·28	081323	20
41	9·805191	2·54	9·886257	1·75	9·918934	4·28	10·081066	19
42	805343	2·53	886152	1·75	919191	4·28	080809	18
43	805495	2·53	886047	1·75	919448	4·28	080552	17
44	805647	2·53	885942	1·75	919705	4·28	080295	16
45	805799	2·53	885837	1·75	919962	4·28	080038	15
46	805951	2·53	885732	1·75	920219	4·28	079781	14
47	806103	2·53	885627	1·75	920476	4·28	079524	13
48	806254	2·53	885522	1·75	920733	4·28	079267	12
49	806406	2·52	885416	1·75	920990	4·28	079010	11
50	806557	2·52	885311	1·76	921247	4·28	078753	10
51	9·806709	2·52	9·885205	1·76	9·921503	4·28	10·078497	9
52	806860	2·52	885100	1·76	921760	4·28	078240	8
53	807011	2·52	884994	1·76	922017	4·28	077983	7
54	807163	2·52	884889	1·76	922274	4·28	077726	6
55	807314	2·52	884783	1·76	922530	4·28	077470	5
56	807465	2·51	884677	1·76	922787	4·28	077213	4
57	807615	2·51	884572	1·76	923044	4·28	076956	3
58	807766	2·51	884466	1·76	923300	4·28	076700	2
59	807917	2·51	884360	1·76	923557	4·27	076443	1
60	808067	2·51	884254	1·77	923813	4·27	076187	0
	Cosine	D.	Sine	50°	Cotang.	D.	Tang.	M.

M.	Sine	D.	Cosine	D.	Tang.	D.	Cotang.	
0	9·808067	2·51	9·884254	1·77	9·923813	4·27	10·076187	60
1	808218	2·51	884148	1·77	924070	4·27	075930	59
2	808368	2·51	884042	1·77	924327	4·27	075673	58
3	808519	2·50	883936	1·77	924583	4·27	075417	57
4	808669	2·50	883829	1·77	924840	4·27	075160	56
5	808819	2·50	883723	1·77	925096	4·27	074904	55
6	808969	2·50	883617	1·77	925352	4·27	074648	54
7	809119	2·50	883510	1·77	925609	4·27	074391	53
8	809269	2·50	883404	1·77	925865	4·27	074135	52
9	809419	2·49	883297	1·78	926122	4·27	073878	51
10	809569	2·49	883191	1·78	926378	4·27	073622	50
11	9·809718	2·49	9·883084	1·78	9·926634	4·27	10·073366	49
12	809868	2·49	882977	1·78	926890	4·27	073110	48
13	810017	2·49	882871	1·78	927147	4·27	072853	47
14	810167	2·49	882764	1·78	927403	4·27	072597	46
15	810316	2·48	882657	1·78	927659	4·27	072341	45
16	810465	2·48	882550	1·78	927915	4·27	072085	44
17	810614	2·48	882443	1·78	928171	4·27	071829	43
18	810763	2·48	882336	1·79	928427	4·27	071573	42
19	810912	2·48	882229	1·79	928683	4·27	071317	41
20	811061	2·48	882121	1·79	928940	4·27	071060	40
21	9·811210	2·48	9·882014	1·79	9·929196	4·27	10·070804	39
22	811358	2·47	881907	1·79	929452	4·27	070548	38
23	811507	2·47	881799	1·79	929708	4·27	070292	37
24	811655	2·47	881692	1·79	929964	4·26	070036	36
25	811804	2·47	881584	1·79	930220	4·26	069780	35
26	811952	2·47	881477	1·79	930475	4·26	069525	34
27	812100	2·47	881369	1·79	930731	4·26	069269	33
28	812248	2·47	881261	1·80	930987	4·26	069013	32
29	812396	2·46	881153	1·80	931243	4·26	068757	31
30	812544	2·46	881046	1·80	931499	4·26	068501	30
31	9·812692	2·46	9·880938	1·80	9·931755	4·26	10·068245	29
32	812840	2·46	880830	1·80	932010	4·26	067990	28
33	812988	2·46	880722	1·80	932266	4·26	067734	27
34	813135	2·46	880613	1·80	932522	4·26	067478	26
35	813283	2·46	880505	1·80	932778	4·26	067222	25
36	813430	2·45	880397	1·80	933033	4·26	066967	24
37	813578	2·45	880289	1·81	933289	4·26	066711	23
38	813725	2·45	880180	1·81	933545	4·26	066455	22
39	813872	2·45	880072	1·81	933800	4·26	066200	21
40	814019	2·45	879963	1·81	934056	4·26	065944	20
41	9·814166	2·45	9·879855	1·81	9·934311	4·26	10·065689	19
42	814313	2·45	879746	1·81	934567	4·26	065433	18
43	814460	2·44	879637	1·81	934823	4·26	065177	17
44	814607	2·44	879529	1·81	935078	4·26	064922	16
45	814753	2·44	879420	1·81	935333	4·26	064667	15
46	814900	2·44	879311	1·81	935589	4·26	064411	14
47	815046	2·44	879202	1·82	935844	4·26	064156	13
48	815193	2·44	879093	1·82	936100	4·26	063900	12
49	815339	2·44	878984	1·82	936355	4·26	063645	11
50	815485	2·43	878875	1·82	936610	4·26	063390	10
51	9·815631	2·43	9·878766	1·82	9·936866	4·25	10·063134	9
52	815778	2·43	878656	1·82	937121	4·25	062879	8
53	815924	2·43	878547	1·82	937376	4·25	062624	7
54	816069	2·43	878438	1·82	937632	4·25	062368	6
55	816215	2·43	878328	1·82	937887	4·25	062113	5
56	816361	2·43	878219	1·83	938142	4·25	061858	4
57	816507	2·42	878109	1·83	938398	4·25	061602	3
58	816652	2·42	877999	1·83	938653	4·25	061347	2
59	816798	2·42	877890	1·83	938908	4·25	061092	1
60	816943	2·42	877780	1·83	939163	4·25	060837	0
	Cosine	D.	Sine	49°	Cotang.	D.	Tang.	M.

M.	Sine	D.	Cosine	D.	Tang.	D.	Cotang.	
0	9·816943	2·42	9·877780	1·83	9·939163	4·25	10·060837	60
1	817088	2·42	877670	1·83	939418	4·25	060582	59
2	817233	2·42	877560	1·83	939673	4·25	060327	58
3	817379	2·42	877450	1·83	939928	4·25	060072	57
4	817524	2·41	877340	1·83	940183	4·25	059817	56
5	817668	2·41	877230	1·84	940438	4·25	059562	55
6	817813	2·41	877120	1·84	940694	4·25	059306	54
7	817958	2·41	877010	1·84	940949	4·25	059051	53
8	818103	2·41	876899	1·84	941204	4·25	058796	52
9	818247	2·41	876789	1·84	941458	4·25	058542	51
10	818392	2·41	876678	1·84	941714	4·25	058286	50
11	9·818536	2·40	9·876568	1·84	9·941968	4·25	10·058032	49
12	818681	2·40	876457	1·84	942223	4·25	057777	48
13	818825	2·40	876347	1·84	942478	4·25	057522	47
14	818969	2·40	876236	1·85	942733	4·25	057267	46
15	819113	2·40	876125	1·85	942988	4·25	057012	45
16	819257	2·40	876014	1·85	943243	4·25	056757	44
17	819401	2·40	875904	1·85	943498	4·25	056502	43
18	819545	2·39	875793	1·85	943752	4·25	056248	42
19	819689	2·39	875682	1·85	944007	4·25	055993	41
20	819832	2·39	875571	1·85	944262	4·25	055738	40
21	9·819976	2·39	9·875459	1·85	9 944517	4·25	10·055483	39
22	820120	2·39	875348	1·85	944771	4·24	055229	38
23	820263	2·39	875237	1·85	945026	4·24	054974	37
24	820406	2·39	875126	1·86	945281	4·24	054719	36
25	820550	2·38	875014	1·86	945535	4·24	054465	35
26	820693	2·38	874903	1·86	945790	4·24	054210	34
27	820836	2·38	874791	1·86	946045	4·24	053955	33
28	820979	2·38	874680	1·86	946299	4·24	053701	32
29	821122	2·38	874568	1·86	946554	4·24	053446	31
30	821265	2·38	874456	1·86	946808	4·24	053192	30
31	9·821407	2·38	9·874344	1·86	9 947063	4·24	10·052937	29
32	821550	2·38	874232	1·87	947318	4·24	052682	28
33	821693	2·37	874121	1·87	947572	4·24	052428	27
34	821835	2·37	874009	1·87	947826	4·24	052174	26
35	821977	2·37	873896	1·87	948081	4·24	051919	25
36	822120	2·37	873784	1·87	948336	4·24	051664	24
37	822262	2·37	873672	1·87	948590	4·24	051410	23
38	822404	2·37	873560	1·87	948844	4·24	051156	22
39	822546	2·37	873448	1·87	949099	4·24	050901	21
40	822688	2·36	873335	1·87	949353	4·24	050647	20
41	9·822830	2·36	9·873223	1·87	9·949607	4·24	10·050393	19
42	822972	2·36	873110	1·88	949862	4·24	050138	18
43	823114	2·36	872998	1·88	950116	4·24	049884	17
44	823255	2·36	872885	1·88	950370	4·24	049630	16
45	823397	2·36	872772	1·88	950625	4·24	049375	15
46	823539	2·36	872659	1·88	950879	4·24	049121	14
47	823680	2·35	872547	1·88	951133	4·24	048867	13
48	823821	2·35	872434	1·88	951388	4·24	048612	12
49	823963	2·35	872321	1·88	951642	4·24	048358	11
50	824104	2·35	872208	1·88	951896	4·24	048104	10
51	9·824245	2·35	9·872095	1·89	9·952150	4·24	10·047850	9
52	824386	2·35	871981	1·89	952405	4·24	047595	8
53	824527	2·35	871868	1·89	952659	4·24	047341	7
54	824668	2·34	871755	1·89	952913	4·24	047087	6
55	824808	2·34	871641	1·89	953167	4·23	046833	5
56	824949	2·34	871528	1·89	953421	4·23	046579	4
57	825090	2·34	871414	1·89	953675	4·23	046325	3
58	825230	2·34	871301	1·89	953929	4·23	046071	2
59	825371	2·34	871187	1·89	954183	4·23	045817	1
60	825511	2·34	871073	1·90	954437	4·23	045563	0
	Cosine	D.	Sine	48°	Cotang.	D.	Tang.	M.

M.	Sine	D.	Cosine	D.	Tang.	D.	Cotang.	
0	9·825511	2·34	9·871073	1·90	9·954437	4·23	10·045563	60
1	825651	2·33	870960	1·90	954691	4·23	045309	59
2	825791	2·33	870846	1·90	954945	4·23	045055	58
3	825931	2·33	870732	1·90	955200	4·23	044800	57
4	826071	2·33	870618	1·90	955454	4·23	044546	56
5	826211	2·33	870504	1·90	955707	4·23	044293	55
6	826351	2·33	870390	1·90	955961	4·23	044039	54
7	826491	2·33	870276	1·90	956215	4·23	043785	53
8	826631	2·33	870161	1·90	956469	4·23	043531	52
9	826770	2·32	870047	1·91	956723	4·23	043277	51
10	826910	2·32	869933	1·91	956977	4·23	043023	50
11	9·827049	2·32	9·869818	1·91	9·957231	4·23	10·042769	49
12	827189	2·32	869704	1·91	957485	4·23	042515	48
13	827328	2·32	869589	1·91	957739	4·23	042261	47
14	827467	2·32	869474	1·91	957993	4·23	042007	46
15	827606	2·32	869360	1·91	958246	4·23	041754	45
16	827745	2·32	869245	1·91	958500	4·23	041500	44
17	827884	2·31	869130	1·91	958754	4·23	041246	43
18	828023	2·31	869015	1·92	959008	4·23	040992	42
19	828162	2·31	868900	1·92	959262	4·23	040738	41
20	828301	2·31	868785	1·92	959516	4·23	040484	40
21	9·828439	2·31	9·868670	1·92	9·959769	4·23	10·040231	39
22	828578	2·31	868555	1·92	960023	4·23	039977	38
23	828716	2·31	868440	1·92	960277	4·23	039723	37
24	828855	2·30	868324	1·92	960531	4·23	039469	36
25	828993	2·30	868209	1·92	960784	4·23	039216	35
26	829131	2·30	868093	1·92	961038	4·23	038962	34
27	829269	2·30	867978	1·93	961291	4·23	038709	33
28	829407	2·30	867862	1·93	961545	4·23	038455	32
29	829545	2·30	867747	1·93	961799	4·23	038201	31
30	829683	2·30	867631	1·93	962052	4·23	037948	30
31	9·829821	2·29	9·867515	1·93	9·962306	4·23	10·037694	29
32	829959	2·29	867399	1·93	962560	4·23	037440	28
33	830097	2·29	867283	1·93	962813	4·23	037187	27
34	830234	2·29	867167	1·93	963067	4·23	036933	26
35	830372	2·29	867051	1·93	963320	4·23	036680	25
36	830509	2·29	866935	1·94	963574	4·23	036426	24
37	830646	2·29	866819	1·94	963827	4·23	036173	23
38	830784	2·29	866703	1·94	964081	4·23	035919	22
39	830921	2·28	866586	1·94	964335	4·23	035665	21
40	831058	2·28	866470	1·94	964588	4·22	035412	20
41	9·831195	2·28	9·866353	1·94	9·964842	4·22	10·035158	19
42	831332	2·28	866237	1·94	965095	4·22	034905	18
43	831469	2·28	866120	1·94	965349	4·22	034651	17
44	831606	2·28	866004	1·95	965602	4·22	034398	16
45	831742	2·28	865887	1·95	965855	4·22	034145	15
46	831879	2·28	865770	1·95	966105	4·22	033891	14
47	832015	2·27	865653	1·95	966362	4·22	033638	13
48	832152	2·27	865536	1·95	966616	4·22	033384	12
49	832288	2·27	865419	1·95	966869	4·22	033131	11
50	832425	2·27	865302	1·95	967123	4·22	032877	10
51	9·832561	2·27	9·865185	1·95	9·967376	4·22	10·032624	9
52	832697	2·27	865068	1·95	967629	4·22	032371	8
53	832833	2·27	864950	1·95	967883	4·22	032117	7
54	832969	2·26	864833	1·96	968136	4·22	031864	6
55	833105	2·26	864716	1·96	968389	4·22	031611	5
56	833241	2·26	864598	1·96	968643	4·22	031357	4
57	833377	2·26	864481	1·96	968896	4·22	031104	3
58	833512	2·26	864363	1·96	969149	4·22	030851	2
59	833648	2·26	864245	1·96	969403	4·22	030597	1
60	833783	2·26	864127	1·96	969656	4·22	030344	0
	Cosine	D.	Sine	47°	Cotang.	D.	Tang.	M.

M.	Sine	D.	Cosine	D.	Tang.	D.	Cotang.	
0	9·833783	2·26	9·864127	1·96	9·969656	4·22	10·030344	60
1	833919	2·25	864010	1·96	969909	4·22	030091	59
2	834054	2·25	863892	1·97	970162	4·22	029838	58
3	834189	2·25	863774	1·97	970416	4·22	029584	57
4	834325	2·25	863656	1·97	970669	4·22	029331	56
5	834460	2·25	863538	1·97	970922	4·22	029078	55
6	834595	2·25	863419	1·97	971175	4·22	028825	54
7	834730	2·25	863301	1·97	971429	4·22	028571	53
8	834865	2·25	863183	1·97	971682	4·22	028318	52
9	834999	2·24	863064	1·97	971935	4·22	028065	51
10	835134	2·24	862946	1·98	972188	4·22	027812	50
11	9·835269	2·24	9·862827	1·98	9·972441	4·22	10·027559	49
12	835403	2·24	862709	1·98	972694	4·22	027306	48
13	835538	2·24	862590	1·98	972948	4·22	027052	47
14	835672	2·24	862471	1·98	973201	4·22	026799	46
15	835807	2·24	862353	1·98	973454	4·22	026546	45
16	835941	2·24	862234	1·98	973707	4·22	026293	44
17	836075	2·23	862115	1·98	973960	4·22	026040	43
18	836209	2·23	861996	1·98	974213	4·22	025787	42
19	836343	2·23	861877	1·98	974466	4·22	025534	41
20	836477	2·23	861758	1·99	974719	4·22	025281	40
21	9·836611	2·23	9·861638	1·99	9·974973	4·22	10·025027	39
22	836745	2·23	861519	1·99	975226	4·22	024774	38
23	836878	2·23	861400	1·99	975479	4·22	024521	37
24	837012	2·22	861280	1·99	975732	4·22	024268	36
25	837146	2·22	861161	1·99	975985	4·22	024015	35
26	837279	2·22	861041	1·99	976238	4·22	023762	34
27	837412	2·22	860922	1·99	976491	4·22	023509	33
28	837546	2·22	860802	1·99	976744	4·22	023256	32
29	837679	2·22	860682	2·00	976997	4·22	023003	31
30	837812	2·22	860562	2·00	977250	4·22	022750	30
31	9·837945	2·22	9·860442	2·00	9·977503	4·22	10·022497	29
32	838078	2·21	860322	2·00	977756	4·22	022244	28
33	838211	2·21	860202	2·00	978009	4·22	021991	27
34	838344	2·21	860082	2·00	978262	4·22	021738	26
35	838477	2·21	859962	2·00	978515	4·22	021485	25
36	838610	2·21	859842	2·00	978768	4·22	021232	24
37	838742	2·21	859721	2·01	979021	4·22	020979	23
38	838875	2·21	859601	2·01	979274	4·22	020726	22
39	839007	2·21	859480	2·01	979527	4·22	020473	21
40	839140	2·20	859360	2·01	979780	4·22	020220	20
41	9·839272	2·20	9·859239	2·01	9·980033	4·22	10·019967	19
42	839404	2·20	859119	2·01	980286	4·22	019714	18
43	839536	2·20	858998	2·01	980538	4·22	019462	17
44	839668	2·20	858877	2·01	980791	4·21	019209	16
45	839800	2·20	858756	2·02	981044	4·21	018956	15
46	839932	2·20	858635	2·02	981297	4·21	018703	14
47	840064	2·19	858514	2·02	981550	4·21	018450	13
48	840196	2·19	858393	2·02	981803	4·21	018197	12
49	840328	2·19	858272	2·02	982056	4·21	017944	11
50	840459	2·19	858151	2·02	982309	4·21	017691	10
51	9·840591	2·19	9·858029	2·02	9·982562	4·21	10·017438	9
52	840722	2·19	857908	2·02	982814	4·21	017186	8
53	840854	2·19	857786	2·02	983067	4·21	016933	7
54	840985	2·19	857665	2·03	983320	4·21	016680	6
55	841116	2·18	857543	2·03	983573	4·21	016427	5
56	841247	2·18	857422	2·03	983826	4·21	016174	4
57	841378	2·18	857300	2·03	984079	4·21	015921	3
58	841509	2·18	857178	2·03	984331	4·21	015669	2
59	841640	2·18	857056	2·03	984584	4·21	015416	1
60	841771	2·18	856934	2·03	984837	4·21	015163	0
	Cosine	D.	Sine	46°	Cotang.	D.	Tang.	M.

M.	Sine	D.	Cosine	D.	Tang.	D.	Cotang.	
0	9·841771	2·18	9·856934	2·03	9·984837	4·21	10·015163	60
1	841902	2·18	856812	2·03	985090	4·21	014910	59
2	842033	2·18	856690	2·04	985343	4·21	014657	58
3	842163	2·17	856568	2·04	985596	4·21	014404	57
4	842294	2·17	856446	2·04	985848	4·21	014152	56
5	842424	2·17	856323	2·04	986101	4·21	013899	55
6	842555	2·17	856201	2·04	986354	4·21	013646	54
7	842685	2·17	856078	2·04	986607	4·21	013393	53
8	842815	2·17	855956	2·04	986860	4·21	013140	52
9	842946	2·17	855833	2·04	987112	4·21	012888	51
10	843076	2·17	855711	2·05	987365	4·21	012635	50
11	9·843206	2·16	9·855588	2·05	9·987618	4·21	10·012382	49
12	843336	2·16	855465	2·05	987871	4·21	012129	48
13	843466	2·16	855342	2·05	988123	4·21	011877	47
14	843595	2·16	855219	2·05	988376	4·21	011624	46
15	843725	2·16	855096	2·05	988629	4·21	011371	45
16	843855	2·16	854973	2·05	988882	4·21	011118	44
17	843984	2·16	854850	2·05	989134	4·21	010866	43
18	844114	2·15	854727	2·06	989387	4·21	010613	42
19	844243	2·15	854603	2·06	989640	4·21	010360	41
20	844372	2·15	854480	2·06	989893	4·21	010107	40
21	9·844502	2·15	9·854356	2·06	9·990145	4·21	10·009855	39
22	844631	2·15	854233	2·06	990398	4·21	009602	38
23	844760	2·15	854109	2·06	990651	4·21	009349	37
24	844889	2·15	853986	2·06	990903	4·21	009097	36
25	845018	2·15	853862	2·06	991156	4·21	008844	35
26	845147	2·15	853738	2·06	991409	4·21	008591	34
27	845276	2·14	853614	2·07	991662	4·21	008338	33
28	845405	2·14	853490	2·07	991914	4·21	008086	32
29	845533	2·14	853366	2·07	992167	4·21	007833	31
30	845662	2·14	853242	2·07	992420	4·21	007580	30
31	9·845790	2·14	9·853118	2·07	9·992672	4·21	10·007328	29
32	845919	2·14	852994	2·07	992925	4·21	007075	28
33	846047	2·14	852869	2·07	993178	4·21	006822	27
34	846175	2·14	852745	2·07	993430	4·21	006570	26
35	846304	2·14	852620	2·07	993683	4·21	006317	25
36	846432	2·13	852496	2·08	993936	4·21	006064	24
37	846560	2·13	852371	2·08	994189	4·21	005811	23
38	846688	2·13	852247	2·08	994441	4·21	005559	22
39	846816	2·13	852122	2·08	994694	4·21	005306	21
40	846944	2·13	851997	2·08	994947	4·21	005053	20
41	9·847071	2·13	9·851872	2·08	9·995199	4·21	10·004801	19
42	847199	2·13	851747	2·08	995452	4·21	004548	18
43	847327	2·13	851622	2·08	995705	4·21	004295	17
44	847454	2·12	851497	2·09	995957	4·21	004043	16
45	847582	2·12	851372	2·09	996210	4·21	003790	15
46	847709	2·12	851246	2·09	996463	4·21	003537	14
47	847836	2·12	851121	2·09	996715	4·21	003285	13
48	847964	2·12	850996	2·09	996968	4·21	003032	12
49	848091	2·12	850870	2·09	997221	4·21	002779	11
50	848218	2·12	850745	2·09	997473	4·21	002527	10
51	9·848345	2·12	9·850619	2·09	9·997726	4·21	10·002274	9
52	848472	2·11	850493	2·10	997979	4·21	002021	8
53	848599	2·11	850368	2·10	998231	4·21	001769	7
54	848726	2·11	850242	2·10	998484	4·21	001516	6
55	848852	2·11	850116	2·10	998737	4·21	001263	5
56	848979	2·11	849990	2·10	998989	4·21	001011	4
57	849106	2·11	849864	2·10	999242	4·21	000758	3
58	849232	2·11	849738	2·10	999495	4·21	000505	2
59	849359	2·11	849611	2·10	999748	4·21	000253	1
60	849485	2·11	849485	2·10	10·000000	4·21	10·000000	0
	Cosine	D.	Sine	45°	Cotang.	D.	Tang.	M.

MATHEMATICS.

DAVIES'S COMPLETE SERIES.

ARITHMETIC.

Davies' Primary Arithmetic.
Davies' Intellectual Arithmetic.
Davies' Elements of Written Arithmetic.
Davies' Practical Arithmetic.
Davies' University Arithmetic.

TWO-BOOK SERIES.

First Book in Arithmetic, Primary and Mental.
Complete Arithmetic.

ALGEBRA.

Davies' New Elementary Algebra.
Davies' University Algebra.
Davies' New Bourdon's Algebra.

GEOMETRY.

Davies' Elementary Geometry and Trigonometry.
Davies' Legendre's Geometry.
Davies' Analytical Geometry and Calculus.
Davies' Descriptive Geometry.
Davies' New Calculus.

MENSURATION.

Davies' Practical Mathematics and Mensuration.
Davies' Elements of Surveying.
Davies' Shades, Shadows, and Perspective.

MATHEMATICAL SCIENCE.

Davies' Grammar of Arithmetic.
Davies' Outlines of Mathematical Science.
Davies' Nature and Utility of Mathematics.
Davies' Metric System.
Davies & Peck's Dictionary of Mathematics.

DAVIES'S NATIONAL COURSE OF MATHEMATICS.

ITS RECORD.

In claiming for this series the first place among American text-books, of whatever class, the publishers appeal to the magnificent record which its volumes have earned during the *thirty-five years* of Dr. Charles Davies's mathematical labors. The unremitting exertions of a life-time have placed *the modern series* on the same proud eminence among competitors that each of its predecessors had successively enjoyed in a course of constantly improved editions, now rounded to their perfect fruition, — for it seems almost that this science is susceptible of no further demonstration.

During the period alluded to, many authors and editors in this department have started into public notice, and, by borrowing ideas and processes original with Dr. Davies, have enjoyed a brief popularity, but are now almost unknown. Many of the series of to-day, built upon a similar basis, and described as "modern books," are destined to a similar fate; while the most far-seeing eye will find it difficult to fix the time, on the basis of any data afforded by their past history, when these books will cease to increase and prosper, and fix a still firmer hold on the affection of every educated American.

One cause of this unparalleled popularity is found in the fact that the enterprise of the author did not cease with the original completion of his books. Always a practical teacher, he has incorporated in his text-books from time to time the advantages of every improvement in methods of teaching, and every advance in science. During all the years in which he has been laboring he constantly submitted his own theories and those of others to the practical test of the class-room, approving, rejecting, or modifying them as the experience thus obtained might suggest. In this way he has been able to produce an almost perfect series of class-books, in which every department of mathematics has received minute and exhaustive attention.

Upon the death of Dr. Davies, which took place in 1876, his work was immediately taken up by his former pupil and mathematical associate of many years, Prof. W. G. Peck, LL.D., of Columbia College. By him, with Prof. J. H. Van Amringe, of Columbia College, the original series is kept carefully revised and up to the times.

DAVIES'S SYSTEM IS THE ACKNOWLEDGED NATIONAL STANDARD FOR THE UNITED STATES, for the following reasons: —

1st. It is the basis of instruction in the great national schools at West Point and Annapolis.

2d. It has received the *quasi* indorsement of the National Congress.

3d. It is exclusively used in the public schools of the National Capital.

4th. The officials of the Government use it as authority in all cases involving mathematical questions.

5th. Our great soldiers and sailors commanding the national armies and navies were educated in this system. So have been a majority of eminent scientists in this country. All these refer to "Davies" as authority.

6th. A larger number of American citizens have received their education from this than from any other series.

7th. The series has a larger circulation throughout the whole country than any other, being *extensively used in every State in the Union.*

DAVIES AND PECK'S ARITHMETICS.

OPTIONAL OR CONSECUTIVE.

The best thoughts of these two illustrious mathematicians are combined in the following beautiful works, which are the natural successors of Davies's Arithmetics, sumptuously printed, and bound in crimson, green, and gold:—

Davies and Peck's Brief Arithmetic.

Also called the "Elementary Arithmetic." It is the shortest presentation of the subject, and is *adequate* for all grades in common schools, being a thorough introduction to practical life, except for the specialist.

At first the authors play with the little learner for a few lessons, by object-teaching and kindred allurements; but he soon begins to realize that study is earnest, as he becomes familiar with the simpler operations, and is delighted to find himself master of important results.

The second part reviews the Fundamental Operations on a scale proportioned to the enlarged intelligence of the learner. It establishes the General Principles and Properties of Numbers, and then proceeds to Fractions. Currency and the Metric System are fully treated in connection with Decimals. Compound Numbers and Reduction follow, and finally Percentage with all its varied applications.

An Index of words and principles concludes the book, for which every scholar and most teachers will be grateful. How much time has been spent in searching for a half-forgotten definition or principle in a former lesson!

Davies and Peck's Complete Arithmetic.

This work certainly deserves its name in the best sense. Though complete, it is not, like most others which bear the same title, *cumbersome*. These authors excel in clear, lucid demonstrations, teaching the science pure and simple, yet not ignoring convenient methods and practical applications.

For turning out a thorough business man no other work is so well adapted. He will have a clear comprehension of the science as a whole, and a working acquaintance with details which must serve him well in all emergencies. Distinguishing features of the book are the logical progression of the subjects and the great variety of practical problems, not *puzzles*, which are beneath the dignity of educational science. A clear-minded critic has said of Dr. Peck's work that it is free from that juggling with numbers which some authors falsely call "Analysis" A series of Tables for converting ordinary weights and measures into the Metric System appear in the later editions.

PECK'S ARITHMETICS.

Peck's First Lessons in Numbers.

This book begins with pictorial illustrations, and unfolds gradually the science of numbers. It noticeably simplifies the subject by developing the principles of addition and subtraction simultaneously; as it does, also, those of multiplication and division.

Peck's Manual of Arithmetic.

This book is designed especially for those who seek sufficient instruction to carry them successfully through practical life, but have not time for extended study.

Peck's Complete Arithmetic.

This completes the series but is a much briefer book than most of the complete arithmetics, and is recommended not only for what it contains, but also for what is omitted.

It may be said of Dr. Peck's books more truly than of any other series published, that they are clear and simple in definition and rule, and that superfluous matter of every kind has been faithfully eliminated, thus magnifying the working value of the book and saving unnecessary expense of time and labor.

Algebra. The student's progress in Algebra depends very largely upon the proper treatment of the four *Fundamental Operations*. The terms *Addition*, *Subtraction*, *Multiplication*, and *Divison* in Algebra have a wider meaning than in Arithmetic, and these operations have been so defined as to *include* their arithmetical meaning; so that the beginner is simply called upon to *enlarge* his views of those fundamental operations. Much attention has been given to the explanation of the negative sign, in order to remove the well-known difficulties in the use and interpretation of that sign. Special attention is here called to "A Short Method of Removing Symbols of Aggregation," Art. 76. On account of their importance, the subjects of *Factoring*, *Greatest Common Divisor*, and *Least Common Multiple* have been treated at greater length than is usual in elementary works. In the treatment of *Fractions*, a method is used which is quite simple, and, at the same time, more general than that usually employed. In connection with *Radical Quantities* the roots are expressed by fractional exponents, for the principles and rules applicable to integral exponents may then be used without modification. The *Equation* is made the chief subject of thought in this work. It is defined near the beginning, and used extensively in every chapter. In addition to this, four chapters are devoted exclusively to the subject of *Equations*. All *Proportions* are equations, and in their treatment as such all the difficulty commonly connected with the subject of Proportion disappears. The chapter on Logarithms will doubtless be acceptable to many teachers who do not require the student to master Higher Algebra before entering upon the study of Trigonometry.

HIGHER MATHEMATICS.

Peck's Manual of Algebra.

Bringing the methods of Bourdon within the range of the Academic Course.

Peck's Manual of Geometry.

By a method purely practical, and unembarrassed by the details which rather confuse than simplify science.

Peck's Practical Calculus.

Peck's Analytical Geometry.

Peck's Elementary Mechanics.

Peck's Mechanics, with Calculus.

The briefest treatises on these subjects now published. Adopted by the great Universities: Yale, Harvard, Columbia, Princeton, Cornell, &c.

Macnie's Algebraical Equations.

Serving as a complement to the more advanced treatises on Algebra, giving special attention to the analysis and solution of equations with numerical coefficients.

Church's Elements of Calculus.

Church's Analytical Geometry.

Church's Descriptive Geometry. With plates. 2 vols.

These volumes constitute the "West Point Course" in their several departments. Prof. Church was long the eminent professor of mathematics at West Point Military Academy, and his works are standard in all the leading colleges.

Courtenay's Elements of Calculus.

A standard work of the very highest grade, presenting the most elaborate attainable survey of the subject.

Hackley's Trigonometry.

With applications to Navigation and Surveying, Nautical and Practical Geometry, and Geodesy.

GENERAL HISTORY.

Monteith's Youth's History of the United States.

A History of the United States for beginners. It is arranged upon the catechetical plan, with illustrative maps and engravings, review questions, dates in parentheses (that their study may be optional with the younger class of learners), and interesting biographical sketches of all persons who have been prominently identified with the history of our country.

Willard's United States. School and University Editions.

The plan of this standard work is chronologically exhibited in front of the titlepage. The maps and sketches are found useful assistants to the memory; and dates, usually so difficult to remember, are so systematically arranged as in a great degree to obviate the difficulty. Candor, impartiality, and accuracy are the distinguishing features of the narrative portion.

Willard's Universal History. New Edition.

The most valuable features of the "United States" are reproduced in this. The peculiarities of the work are its great conciseness and the prominence given to the chronological order of events. The margin marks each successive era with great distinctness, so that the pupil retains not only the event but its time, and thus fixes the order of history firmly and usefully in his mind. Mrs. Willard's books are constantly revised, and at all times written up to embrace important historical events of recent date. Professor Arthur Gilman has edited the last twenty-five years to 1882.

Lancaster's English History.

By the Master of the Stoughton Grammar School, Boston. The most practical of the "brief books." Though short, it is not a bare and uninteresting outline, but contains enough of explanation and detail to make intelligible the *cause and effect* of events. Their relations to the history and development of the American people is made specially prominent.

Willis's Historical Reader.

Being Collier's Great Events of History adapted to American schools. This rare epitome of general history, remarkable for its charming style and judicious selection of events on which the destinies of nations have turned, has been skilfully manipulated by Professor Willis, with as few changes as would bring the United States into its proper position in the historical perspective. As reader or text-book it has few equals and no superior.

Berard's History of England.

By an authoress well known for the success of her History of the United States. The social life of the English people is felicitously interwoven, as in fact, with the civil and military transactions of the realm.

Ricord's History of Rome.

Possesses the charm of an attractive romance. The fables with which this history abounds are introduced in such a way as not to deceive the inexperienced, while adding materially to the value of the work as a reliable index to the character and institutions, as well as the history of the Roman people.

A Brief History of Ancient Peoples.

With an account of their monuments, **literature, and manners.** 340 pages. 12mo. Profusely illustrated.

In this work the political history, which occupies nearly, if not all, the ordinary school text, is condensed to the salient and essential facts, in order to give room for a clear outline of the literature, religion, architecture, character, habits, &c., of each nation. Surely it is as important to know *something* about Plato as *all* about Cæsar, and to learn how the ancients wrote their books as how they fought their battles.

The chapters on Manners and Customs **and the Scenes** in Real Life represent the people of history as men and **women subject to the same wants, hopes and fears as ourselves, and so bring the distant past near to us. The Scenes, which are intended *only for reading*, are the result of a careful study of the unequalled collections of monuments in the London and Berlin Museums, of the ruins in Rome and Pompeii, and of the latest authorities on the domestic life of ancient peoples. Though intentionally written in a semi-romantic style, they are accurate pictures of what *might* have occurred, and some of them are simple transcriptions of the details sculptured in Assyrian alabaster or painted on Egyptian walls.**

HISTORY — *Continued.*

The extracts made from the sacred books of the East are not specimens of their style and teachings, but only gems selected often from a mass of matter, much of which would be absurd, meaningless, and even revolting. It has not seemed best to cumber a book like this with selections conveying no moral lesson.

The numerous cross-references, the abundant dates in parenthesis, the pronunciation of the names in the Index, the choice reading references at the close of each general subject, and the novel Historical Recreations in the Appendix, will be of service to teacher and pupil alike.

Though designed primarily for a text-book, a large class of persons — general readers, who desire to know something about the progress of historic criticism and the recent discoveries made among the resurrected monuments of the East, but have no leisure to read the ponderous volumes of Brugsch, Layard, Grote, Mommsen, and Ihne — will find this volume just what they need.

From **Homer B. Sprague,** ***Head Master Girls' High School, West Newton St., Boston, Mass.***

"I beg to recommend in strong terms the adoption of Barnes's 'History of Ancient Peoples' as a text-book. It is about as nearly perfect as could be hoped for. The adoption would give great relish to the study of Ancient History."

HE Brief History of France.

By the author of the "Brief United States," with all the attractive features of that popular work (which see) and new ones of its own.

It is believed that the History of France has never before been presented in such brief compass, and this is effected without sacrificing one particle of interest. The book reads like a romance, and, while drawing the student by an irresistible fascination to his task, impresses the great outlines indelibly upon the memory.

DRAWING.

BARNES'S POPULAR DRAWING SERIES.

Based upon the experience of the most successful teachers of drawing in the United States.

The Primary Course, consisting of a manual, ten cards, and three primary drawing books, A, B, and C.

Intermediate Course. Four numbers and a manual.

Advanced Course. Four numbers and a manual.

Instrumental Course. Four numbers and a manual.

The Intermediate, Advanced, and Instrumental Courses are furnished either in book or card form at the same prices. The books contain the usual blanks, with the unusual advantage of opening from the pupil, — placing the copy directly in front and above the blank, thus occupying but little desk-room. The cards are in the end more economical than the books, if used in connection with the patent blank folios that accompany this series.

The cards are arranged to be bound (or tied) in the folios and removed at pleasure. The pupil at the end of each number has a complete book, containing only his own work, while the copies are preserved and inserted in another folio ready for use in the next class.

Patent Blank Folios. No. 1. Adapted to Intermediate Course. No. 2. Adapted to Advanced and Instrumental Courses.

ADVANTAGES OF THIS SERIES.

The Plan and Arrangement. — The examples are so arranged that teachers and pupils can see, at a glance, how they are to be treated and where they are to be copied. In this system, copying and designing do not receive all the attention. The plan is broader in its aims, dealing with drawing as a branch of common-school instruction, and giving it a wide educational value.

Correct Methods. — In this system the pupil is led to rely upon himself, and not upon delusive mechanical aids, as printed guide-marks, &c.

One of the principal objects of any good course in freehand drawing is to educate the eye to estimate location, form, and size. A system which weakens the motive or removes the necessity of *thinking* is false in theory and ruinous in practice. The object should be to educate, not cram; to develop the intelligence, not teach tricks.

Artistic Effect. — The beauty of the examples is not destroyed by crowding the pages with useless and badly printed text. The Manuals contain all necessary instruction.

Stages of Development. — Many of the examples are accompanied by diagrams, showing the different stages of development.

Lithographed Examples. — The examples are printed in imitation of pencil drawing (not in hard, black lines) that the pupil's work may resemble them.

One Term's Work. — Each book contains what can be accomplished in an average term, and no more. Thus a pupil *finishes* one book before beginning another.

Quality — not Quantity. — Success in drawing depends upon the amount of *thought* exercised by the pupil, and *not* upon the large number of examples drawn.

Designing. — Elementary design is more skilfully taught in this system than by any other. In addition to the instruction given in the books, the pupil will find printed on the insides of the covers a variety of beautiful patterns.

Enlargement and Reduction. — The practice of enlarging and reducing from copies is not commenced until the pupil is well advanced in the course and therefore better able to cope with this difficult feature in drawing.

Natural Forms. — This is the only course that gives at convenient intervals easy and progressive exercises in the drawing of natural forms.

Economy. — By the patent binding described above, the copies need not be thrown aside when a book is filled out, but are preserved in perfect condition for future use. The blank books, only, will have to be purchased after the first introduction, thus effecting a saving of more than half in the usual cost of drawing-books.

Manuals for Teachers. — The Manuals accompanying this series contain practical instructions for conducting drawing in the class-room, with *definite* directions for drawing *each* of the examples in the books, instructions for designing, model and object drawing, drawing from natural forms, &c.

DRAWING — *Continued.*

Chapman's American Drawing-Book.

The standard American text-book and authority in all branches of art. A compilation of art principles. A manual for the amateur, and basis of study for the professional artist. Adapted for schools and private instruction.

CONTENTS. — "Any one who can Learn to Write can Learn to Draw." — Primary Instruction in Drawing. — Rudiments of Drawing the Human' Head. — Rudiments in Drawing the Human Figure. — Rudiments of Drawing. — The Elements of Geometry. — Perspective. — Of Studying and Sketching from Nature. — Of Painting. — Etching and Engraving — Of Modelling. — Of Composition. — Advice to the American Art-Student.

The work is of course magnificently illustrated with all the original designs.

Chapman's Elementary Drawing-Book.

A progressive course of practical exercises, or a text-book for the training of the eye and hand. It contains the elements from the larger work, and a copy should be in the hands of every pupil; while a copy of the "American Drawing-Book," named above, should be at hand for reference by the class.

Clark's Elements of Drawing.

A complete course in this graceful art, from the first rudiments of outline to the finished sketches of landscape and scenery.

Allen's Map-Drawing and Scale.

This method introduces a new era in map-drawing, for the following reasons: 1. It is a system. This is its greatest merit. — 2. It is easily understood and taught. — 3. The eye is trained to exact measurement by the use of a scale. — 4. By no special effort of the memory, distance and comparative size are fixed in the mind. — 5. It discards useless construction of lines. — 6. It can be taught by any teacher, even though there may have been no previous practice in map-drawing. — 7. Any pupil old enough to study geography can learn by this system, in a short time, to draw accurate maps. — 8. The system is not the result of theory, but comes directly from the school-room. It has been thoroughly and successfully tested there, with all grades of pupils. — 9. It is economical, as it requires no mapping plates. It gives the pupil the ability of rapidly drawing accurate maps.

FINE ARTS.

Hamerton's Art Essays (Atlas Series): —

No. 1. The Practical Work of Painting.

With portrait of Rubens. 8vo. Paper covers.

No. 2. Modern Schools of Art..

Including American, English, and Continental Painting. 8vo. Paper covers.

Huntington's Manual of the Fine Arts.

A careful manual of instruction in the history of art, up to the present time.

Boyd's Kames' Elements of Criticism.

The best edition of the best work on art and literary criticism ever produced in English.

Benedict's Tour Through Europe.

A valuable companion for any one wishing to visit the galleries and sights of the continent of Europe, as well as a charming book of travels.

Dwight's Mythology.

A knowledge of mythology is necessary to an appreciation of ancient art.

Walker's World's Fair.

The industrial and artistic display at the Centennial Exhibition.

DR. STEELE'S ONE-TERM SERIES, IN ALL THE SCIENCES.

Steele's 14-Weeks Course in Chemistry.
Steele's 14-Weeks Course in Astronomy.
Steele's 14-Weeks Course in Physics.
Steele's 14-Weeks Course in Geology.
Steele's 14-Weeks Course in Physiology.
Steele's 14-Weeks Course in Zoölogy.
Steele's 14-Weeks Course in Botany.

Our text-books in these studies are, as a general thing, dull and uninteresting. They contain from 400 to 600 pages of dry facts and unconnected details. They abound in that which the student cannot learn, much less remember. The pupil commences the study, is confused by the fine print and coarse print, and neither knowing exactly what to learn nor what to hasten over, is crowded through the single term generally assigned to each branch, and frequently comes to the close without a definite and exact idea of a single scientific principle.

Steele's "Fourteen-Weeks Courses" contain only that which every well-informed person should know, while all that which concerns only the professional scientist is omitted. The language is clear, simple, and interesting, and the illustrations bring the subject within the range of home life and daily experience. They give such of the general principles and the prominent facts as a pupil can make familiar as household words within a single term. The type is large and open; there is no fine print to annoy; the cuts are copies of genuine experiments or natural phenomena, and are of fine execution.

In fine, by a system of condensation peculiarly his own, the author reduces each branch to the limits of a single term of study, while sacrificing nothing that is essential, and nothing that is usually retained from the study of the larger manuals in common use. Thus the student has rare opportunity to *economize his time*, or rather to employ that which he has to the best advantage.

A notable feature is the author's charming "style," fortified by an enthusiasm over his subject in which the student will not fail to partake. Believing that Natural Science is full of fascination, he has moulded it into a form that attracts the attention and kindles the enthusiasm of the pupil.

The recent editions contain the author's "Practical Questions" on a plan never before attempted in scientific text-books. These are questions as to the nature and cause of common phenomena, and are not directly answered in the text, the design being to test and promote an intelligent use of the student's knowledge of the foregoing principles.

Steele's Key to all His Works.

This work is mainly composed of answers to the Practical Questions, and solutions of the problems, in the author's celebrated "Fourteen-Weeks Courses" in the several sciences, with many hints to teachers, minor tables, &c. Should be on every teacher's desk.

Prof. J. Dorman Steele is an indefatigable student, as well as author, and his books have reached a fabulous circulation. It is safe to say of his books that they have accomplished more tangible and better results in the class-room than any other ever offered to American schools, and have been translated into more languages for foreign schools. They are even produced in raised type for the blind.

NATURAL SCIENCE—*Continued.*

BOTANY.

Wood's Object-Lessons in Botany.

Wood's American Botanist and Florist.

Wood's New Class-Book of Botany.

The standard text-books of the United States in this department. In style they are simple, popular, and lively; in arrangement, easy and natural; in description, graphic and scientific. The Tables for Analysis are reduced to a perfect system. They include the flora of the whole United States east of the Rocky Mountains, and are well adapted to the regions west.

Wood's Descriptive Botany.

A complete flora of all plants growing east of the Mississippi River.

Wood's Illustrated Plant Record.

A simple form of blanks for recording observations in the field.

Wood's Botanical Apparatus.

A portable trunk, containing drying press, knife, trowel, microscope, and tweezers, and a copy of Wood's "Plant Record," — the collector's complete outfit.

Willis's Flora of New Jersey.

The most useful book of reference ever published for collectors in all parts of the country. It contains also a Botanical Directory, with addresses of living American botanists.

Young's Familiar Lessons in Botany.

Combining simplicity of diction with some degree of technical and scientific knowledge, for intermediate classes. Specially adapted for the Southwest.

Wood & Steele's Botany.

See page 33.

AGRICULTURE.

Pendleton's Scientific Agriculture.

A text-book for colleges and schools; treats of the following topics: Anatomy and Physiology of Plants; Agricultural Meteorology; Soils as related to Physics; Chemistry of the Atmosphere; of Plants; of Soils; Fertilizers and Natural Manures; Animal Nutrition, &c. By E. M. Pendleton, M. D., Professor of Agriculture in the University of Georgia.

From PRESIDENT A. D. WHITE, *Cornell University.*

"*Dear Sir:* I have examined your 'Text-book of Agricultural Science,' and it seems to me excellent in view of the purpose it is intended to serve. Many of your chapters interested me especially, and all parts of the work seem to combine scientific instruction with practical information in proportions dictated by sound common sense."

From PRESIDENT ROBINSON, *of Brown University.*

"It is scientific in method as well as in matter, comprehensive in plan, natural and logical in order, compact and lucid in its statements, and must be useful both as a text-book in agricultural colleges, and as a hand-book for intelligent planters and farmers."

PHYSIOLOGY.

Jarvis's Elements of Physiology.

Jarvis's Physiology and Laws of Health.

The only books extant which approach this subject with a proper view of the true object of teaching Physiology in schools, viz., that scholars may know how to take care of their own health. In bold contrast with the abstract *Anatomies*, which children learn as they would Greek or Latin (and forget as soon), to *discipline the mind*, are these text-books, using the *science* as a secondary consideration, and only so far as is necessary for the comprehension of the *laws of health*.

Steele's Physiology.

See page 33.

ASTRONOMY.

Willard's School Astronomy.

By means of clear and attractive illustrations, addressing the eye in many cases by analogies, careful definitions of all necessary technical terms, a careful avoidance of verbiage and unimportant matter, particular attention to analysis, and a general adoption of the simplest methods, Mrs. Willard has made the best and most attractive *elementary* Astronomy extant.

McIntyre's Astronomy and the Globes.

A complete treatise for intermediate classes. Highly approved.

Bartlett's Spherical Astronomy.

The West Point Course, for advanced classes, with applications to the current wants of Navigation, Geography, and Chronology.

Steele's Astronomy.

See page 33.

NATURAL HISTORY.

Carll's Child's Book of Natural History.

Illustrating the animal, vegetable, and mineral kingdoms, with application to the arts. For beginners. Beautifully and copiously illustrated.

Anatomical Technology. Wilder & Gage.

As applied to the domestic cat. For the use of students of medicine.

ZOÖLOGY.

Chambers's Elements of Zoölogy.

A complete and comprehensive system of Zoölogy, adapted for academic instruction, presenting a systematic view of the animal kingdom as a portion of external nature.

Steele's Zoölogy.

See page 33.

LITERATURE.

Gilman's First Steps in English Literature.

The character and plan of this exquisite little text-book may be best understood from an analysis of its contents: Introduction. Historical Period of Immature English, with Chart; Definition of Terms; Languages of Europe, with Chart; Period of Mature English, with Chart; a Chart of Bible Translations, a Bibliography or Guide to General Reading, and other aids to the student.

Cleveland's Compendiums. 3 vols. 12mo.

English Literature. American Literature.
English Literature of the XIXth Century.

In these volumes are gathered the cream of the literature of the English-speaking people for the school-room and the general reader. Their reputation is national. More than 125,000 copies have been sold.

Boyd's English Classics. 6 vols. Cloth. 12mo.

Milton's Paradise Lost. Thomson's Seasons.
Young's Night Thoughts. Pollok's Course of Time.
Cowper's Task, Table Talk, &c. Lord Bacon's Essays.

This series of annotated editions of great English writers in prose and poetry is designed for critical reading and parsing in schools. Prof. J. R. Boyd proves himself an editor of high capacity, and the works themselves need no encomium. As auxiliary to the study of belles-lettres, &c., these works have no equal.

Pope's Essay on Man. 16mo. Paper.

Pope's Homer's Iliad. 32mo. Roan.

The metrical translation of the great poet of antiquity, and the matchless "Essay on the Nature and State of Man," by Alexander Pope, afford superior exercise in literature and parsing.

POLITICAL ECONOMY.

Champlin's Lessons on Political Economy.

An improvement on previous treatises, being shorter, yet containing everything essential, with a view of recent questions in finance, &c., which is not elsewhere found.

ÆSTHETICS.

Huntington's Manual of the Fine Arts.

A view of the rise and progress of art in different countries, a brief account of the most eminent masters of art, and an analysis of the principles of art. It is complete in itself, or may precede to advantage the critical work of Lord Kames.

Boyd's Kames's Elements of Criticism.

The best edition of this standard work; without the study of which none may be considered proficient in the science of the perceptions. No other study can be pursued with so marked an effect upon the taste and refinement of the pupil.

ELOCUTION.

Watson's Practical Elocution.

A scientific presentment of accepted principles of elocutionary drill, with blackboard diagrams and full collection of examples for class drill. Cloth. 90 pages, 12mo.

Taverner Graham's Reasonable Elocution.

Based upon the belief that true elocution is the right interpretation of thought, and guiding the student to an intelligent appreciation, instead of a merely mechanical knowledge, of its rules.

Zachos's Analytic Elocution.

All departments of elocution — such as the analysis of the voice and the sentence, phonology, rhythm, expression, gesture, &c. — are here arranged for instruction in classes, illustrated by copious examples.

SPEAKERS.

Northend's Little Orator.

Northend's Child's Speaker.

Two little works of the same grade but different selections, containing simple and attractive pieces for children under twelve years of age.

Northend's Young Declaimer.

Northend's National Orator.

Two volumes of prose, poetry, and dialogue, adapted to intermediate and grammar classes respectively.

Northend's Entertaining Dialogues.

Extracts eminently adapted to cultivate the dramatic faculties, as well as entertain.

Oakey's Dialogues and Conversations.

For school exercises and exhibitions, combining useful instruction.

James's Southern Selections, for Reading and Oratory.

Embracing exclusively Southern literature.

Swett's Common School Speaker.

Raymond's Patriotic Speaker.

A superb compilation of modern eloquence and poetry, with original dramatic exercises. Nearly every eminent modern orator is represented.

MODERN LANGUAGES.

A COMPLETE COURSE IN THE GERMAN.

By James H. Worman, A.M., Professor of Modern Languages in the Adelphi Academy, Brooklyn, L. I.

Worman's First German Book.

Worman's Second German Book.

Worman's Elementary German Grammar.

Worman's Complete German Grammar.

These volumes are designed for intermediate and advanced classes respectively.

Though following the same general method with "Otto" (that of "Gaspey"), our author differs essentially in its application. He is more practical, more systematic more accurate, and besides introduces a number of invaluable features which have never before been combined in a German grammar.

Among other things, it may be claimed for Professor Worman that he has been *the first* to introduce, in an American text-book for learning German, a system of analogy and comparison with other languages. Our best teachers are also enthusiastic about his methods of inculcating the art of speaking, of understanding the spoken language, of correct pronunciation; the sensible and convenient original classification of nouns (in four declensions), and of irregular verbs, also deserves much praise. We also note the use of heavy type to indicate etymological changes in the paradigms and, in the exercises, the parts which specially illustrate preceding rules.

Worman's Elementary German Reader.

Worman's Collegiate German Reader.

The finest and most judicious compilation of classical and standard German literature. These works embrace, progressively arranged, selections from the masterpieces of Goethe, Schiller, Korner, Seume, Uhland, Freiligrath, Heine, Schlegel, Holty, Lenau, Wieland, Herder, Lessing, Kant, Fichte, Schelling, Winkelmann, Humboldt, Ranke, Raumer, Menzel, Gervinus, &c., and contain complete Goethe's "Iphigenie," Schiller's "Jungfrau;" also, for instruction in modern conversational German, Benedix's "Eigensinn."

There are, besides, biographical sketches of each author contributing, notes, explanatory and philological (after the text), grammatical references to all leading grammars, as well as the editor's own, and an adequate Vocabulary.

Worman's German Echo.

Worman's German Copy-Books, 3 Numbers.

On the same plan as the most approved systems for English penmanship, with progressive copies.

CHAUTAUQUA SERIES.

First and Second Books in German.

By the natural or Pestalozzian System, for teaching the language without the help of the Learner's Vernacular. By James H. Worman, A. M.

These books belong to the new Chautauqua German Language Series, and are intended for beginners learning to *speak* German. The peculiar features of its method are:—

1. **It teaches the language by direct appeal to illustrations of the objects referred to,** and does not allow the student to guess what is said. He speaks from the first hour *understandingly* and *accurately*. Therefore,

2. **Grammar is taught both analytically and synthetically** throughout the course. The beginning is made with the auxiliaries of tense and mood, because their kinship with the English makes them easily intelligible; then follow the declensions of nouns, articles, and other parts of speech, always systematically arranged. It is easy to confuse the pupil by giving him one person or one case at a time. This pernicious practice is discarded. Books that beget unsystematic habits of thought are worse than worthless.

FRENCH.

Worman's First Book in French.

The first book in the companion series to the successful German Series by the same author, and intended for those wishing to *speak* French. The peculiar features of Professor Worman's new method are:—

1. The French language is taught without the help of English.
2. It appeals to *pictorial illustrations for the names of objects.*
3. The learner speaks from the first hour *understandingly.*
4. Grammar is taught to prevent missteps in composition.
5. The laws of the language are taught *analytically* to make them the learner's own inferences (= deductions).
6. Rapidity of progress by dependence upon *association* and *contrasts.*
7. Strictly *graded lessons* and conversations on *familiar, interesting,* and *instructive topics,* providing the words and idioms of every-day life.
8. *Paradigms* to give a systematic treatment to variable inflections.
9. *Heavy type* for inflections, to make the eye a help to the mind.
10. *Hair line type* for the silent letters, and links for words to be connected, in order to teach an accurate pronunciation.

Worman's French Echo.

This is not a mass of meaningless and parrot-like phrases thrown together for a tourist's use, to bewilder him when in the presence of a Frenchman.

The "Echo de Paris" is a *strictly progressive conversational book,* beginning with simple phrases and leading by frequent repetition to a mastery of the *idioms* and of *the every-day language* used in business, on travel, at a hotel, in the chit-chat of society.

It presupposes an elementary knowledge of the language, such as may be acquired from the First French Book by Professor Worman, and furnishes *a running French text,* allowing the learner of course to find the meaning of the words (in the appended Vocabulary), and forcing him, by the absence of English in the text, to *think in French.*

CHER MONSIEUR WORMAN,—Vous me demandez mon opinion sur votre "Echo de Paris" et quel usage j'en fais. Je ne saurais mieux vous répondre qu'en reproduisant une lettre que j'écrivais dernièrement à un collègue qui était, me disait-il, "bien fatigué de ces insipides livres de dialogues."

"Vous ne connaissez donc pas," lui disais-je, "'l'Echo de Paris,' édité par le Professor Worman? C'est un véritable trésor, merveilleusement adapté au développement de la conversation familière et pratique, telle qu'on la veut aujourd'hui. Cet excellent livre met successivement en scène, d'une manière vive et intéressante, *toutes* les circonstances possibles de la vie ordinaire. Voyez l'immense avantage il vous transporte en France; du premier mot, je m'imagine, et mes élèves avec moi, que nous sommes à Paris, dans la rue, sur une place, dans une gare, dans un salon, dans une chambre, voire même à la cuisine; je parle comme avec des Français; les élèves ne songent pas à traduire de l'anglais pour me répondre; ils pensent en français; ils sont Français pour le moment par les yeux, par l'oreille, par la pensée. Quel autre livre pourrait produire cette illusion? . . ."

Votre tout dévoué,

A. DE ROUGEMONT.

Illustrated Language Primers.

FRENCH AND ENGLISH. GERMAN AND ENGLISH.

SPANISH AND ENGLISH.

The names of common objects properly illustrated and arranged in easy lessons.

Pujol's Complete French Class-Book.

Offers in one volume, methodically arranged, a complete French course—usually embraced in series of from five to twelve books, including the bulky and expensive lexicon. Here are grammar, conversation, and choice literature, selected from the best French authors. Each branch is thoroughly handled; and the student, having diligently completed the course as prescribed, may consider himself, without further application, *au fait* in the most polite and elegant language of modern times.

TEACHERS' AIDS AND SCHOOL REQUISITES.

CHARTS AND MAPS.

Baade's Reading Case.

This remarkable piece of school-room furniture is a receptacle containing a number of primary cards. By an arrangement of slides on the front, one sentence at a time is shown to the class. Twenty-eight thousand transpositions may be made, affording a variety of progressive exercises which no other piece of apparatus offers. One of its best features is, that it is so exceedingly simple as not to get out of order, while it may be operated with one finger.

Clark's Grammatical Chart.

Exhibits the whole science of language in one comprehensive diagram.

Davies's Mathematical Chart.

Elementary mathematics clearly taught to a full class at a glance.

De Rupert's Philological and Historical Chart.

This very comprehensive chart shows the birth, development, and progress of the literatures of the world; their importance, their influence on each other, and the century in which such influence was experienced; with a list for each country of standard authors and their best works. Illustrating also the division of languages into classes, families, and groups. Giving date of settlement, discovery, or conquest of all countries, with their government, religion, area, population, and the percentage of enrolment for 1872, in the primary schools of Europe and America.

Eastman's Chirographic Chart. Family Record.

Giffins's Number Chart.

Teaches addition, subtraction, multiplication, and division. Size, 23 x 31 inches.

Marcy's Eureka Tablet.

A new system for the alphabet, by which it may be taught without fail in nine lessons.

McKenzie's Elocutionary Chart.

Monteith's Pictorial Chart of Geography.

A crayon picture illustrating all the divisions of the earth's surface commonly taught in geography.

WM. L. DICKINSON, *Superintendent of Schools, Jersey City, says.*

"It is an admirable amplification of the system of pictorial illustration adopted in all good geographies. I think the chart would be a great help in any primary department."

Monteith's Reference Maps. School and Grand Series.

Names all laid down in small type so that to the pupil at a short distance they are outline maps, while they serve as *their own key* to the teacher.

Page's Normal Chart.

The whole science of elementary sounds tabulated.

Scofield's School Tablets.

On five cards, exhibiting ten surfaces. These tablets teach orthography, reading, object-lessons, color, form, &c.

Watson's Phonetic Tablets.

Four cards and eight surfaces; teaching pronunciation and elocution phonetically. For class exercises.

Whitcomb's Historical Chart.

A student's topical historical chart, from the creation to the present time, including results of the latest chronological research. Arranged with spaces for summary, that pupils may prepare and review their own chart in connection with any text-book.

Willard's Chronographers.

Historical. Four numbers: Ancient chronographer, English chronographer, American chronographer, temple of time (general). Dates and events represented to the eye.

APPARATUS.

Bock's Physiological Apparatus.

A collection of twenty-seven anatomical models.

Harrington's Fractional Blocks.

Harrington's Geometrical Blocks.

These patent blocks are *hinged*, so that each form can be dissected.

Kendall's Lunar Telluric Globe.

Moon, globe, and tellurian combined.

Steele's Chemical Apparatus.

Steele's Geological Cabinet.

Steele's Philosophical Apparatus.

Wood's Botanical Apparatus.

RECORDS.

Cole's Self-Reporting Class Book.

For saving the teacher's labor in averaging. At each opening are a full set of tables showing any scholar's standing at a glance, and entirely obviating the necessity of computation.

Tracy's School Record. {Desk edition. / Pocket edition.}

For keeping a simple but exact record of attendance, deportment, and scholarship. The larger edition contains also a calendar, an extensive list of topics for compositions and colloquies, themes for short lectures, suggestions to young teachers, &c.

Benet's Individual Records.

Brooks's Teacher's Register.

Presents at one view a record of attendance, recitations, and deportment for the whole term.

Carter's Record and Roll-Book.

This is the most complete and convenient record offered to the public. Besides the usual spaces for general scholarship, deportment, attendance, &c., for each name and day, there is a space in red lines enclosing six minor spaces in blue for recording recitations.

National School Diary.

A little book of blank forms for weekly report of the standing of each scholar, from teacher to parent. A great convenience.

REWARDS.

National School Currency.

A little box containing certificates in the form of money. The most entertaining and stimulating system of school rewards. The scholar is paid for his merits and fined for his short-comings. Of course the most faithful are the most successful in business. In this way the use and value of money and the method of keeping accounts are also taught. One box of currency will supply a school of fifty pupils.